PENGUIN BOOKS
EINSTEIN IN LOVE

Dennis Overbye is a reporter for *The New York Times* and a critically acclaimed science writer who has contributed to *Time*, *The New York Times Magazine*, and many other publications. His first book, *Lonely Hearts of the Cosmos*, won the American Institute of Physics Science Writing Award and was nominated for a National Book Critics Circle Award in Nonfiction and a *Los Angeles Times* Book Award in Science. He lives with his wife in New York City.

Praise for *Einstein in Love*

Nominated for the *Los Angeles Times* Book Award for Science
A Dual Main Selection of the QPB Book Club

"Lyrical and precise in its re-creation of time and setting, clear and engaging in its presentation of science matters, this is the story of an adventure that changed the world. Overbye is blessed with a novelist's grasp of character, transcending easy judgment—a principled respect for the humanity of Einstein and Maric, which rescues them from history, and restores to them at last, their flawed luminous selves."
—Thomas Pynchon, author of *Gravity's Rainbow*

"This is the best account of Einstein the human being ever written, and there have been many good ones. Dennis Overbye's ever-fascinating story of sexy Einstein is written with flawless prose and something like a jeweler's eye for the right detail. A great accomplishment in narrative and the history of science. You finish this book feeling that you know Einstein the person the way you know a friend."
—Richard Preston, author of *The Hot Zone* and *The Cobra Event*

"*Einstein in Love* is meticulously researched and beautifully written. Overbye has done a remarkable job of chronicling the history of physics and explaining the science behind it with artful clarity. . . . With *Einstein in Love*, Overbye accomplishes an important goal: he humanizes a legend and reminds readers that even icons have dark sides."
—*San Francisco Chronicle*

"*Einstein in Love* . . . [is] the most complete and accessible account of Einstein's life and work, from his birth in 1879 to the death of his mother in 1920. . . . The book provides what is arguably the best nontechnical account to date of the genesis of the idea of the special theory of relativity."
—*The Washington Post Book World*

"A superb biography. . . . Einstein here is a lusty, moody, baffling, brilliant lad whose early loves were physics and Mileva. . . . Their romance was too tempestuous not to be utterly doomed. . . . What makes this biography so fascinating is Overbye's novelistic detailing of the texture of Einstein's thoughts."
—*Esquire*

"Through objective research and excellent writing, Overbye succeeds in creating a captivating portrait of Einstein's marriage to Mileva Maric, and of their world. . . . Overbye's grasp of the science results in a masterpiece of writing."
—*Library Journal*

"An expertly crafted account of the life, loves, and sciences of Einstein during the first two decades of the twentieth century. . . . Overbye is able to cut through the haze of myth that has inevitably enveloped Einstein. . . . An ambitious project that works: Overbye has created an accurate, detailed, and fascinating portrait of Einstein as a flawed, complex young man."
—*Kirkus Reviews*

"A bemused biography seen through the prism of his tumultuous love life. . . . Overbye writes a clean, unresisting prose that guarantees this book to be easy sailing not just through the romance, but the science as well."
—*Elle*

"Of the many recent and imminent books on Einstein, Overbye's may have the most compelling title and the most fulfilling approach. Overbye's aim—which he accomplishes with the precision of a scientist and the ear of a musician—is to portray Einstein the man, not the myth. In the end the reader may come to like Einstein less but appreciate his achievements even more."
—*Publishers Weekly* (starred review)

"Overbye is one of the most seductive science writers working today; his story infuses Einstein's affairs with women with edgy desire and his affair with physics with a nearly comparable obsessiveness."
—Deborah Blum, *Minneapolis Star Tribune*

"It's hard to imagine a more skillful rendering of the flesh-and-blood package—Einstein as son, difficult student, musician, coffee-club maven, longing-filled lover, husband, science-establishment reject, working stiff and evolving giant of modern physics."
—*American Scientist*

"Dennis Overbye's engaging tale is partly a biography of the young Einstein, partly an exposition of early twentieth-century physics, and—best of all—an account of that mysterious place where heart and mind and flesh tangle in ways that no past, present, or future Einstein will ever unravel."
—Chet Raymo, *Commonweal*

"Overbye says that his 'goal has been to bring the youthful Einstein to life, to illuminate the young man who performed the deeds for which the old man, the icon, is revered.' He has done that admirably."
—*Scientific American*

"*Einstein in Love* is startlingly successful in turning an icon back into a man—one with destructively high self-regard. 'If everyone had a life like mine,' Einstein said to his sister Maja, 'there would be no need for novels.' As Dennis Overbye has admirably demonstrated, you couldn't make it up."
—*The Independent* (London)

"This book provides a lyrical and compelling introduction to the ideas and arguments that flowed out of two of the most important decades, not simply in Einstein's life, but also in the history of modern physics and, indeed, in the shaping of our perceptions of reality."
—*Sunday Telegraph* (London)

EINSTEIN IN LOVE

A SCIENTIFIC ROMANCE

Dennis Overbye

PENGUIN BOOKS

PENGUIN BOOKS

Published by the Penguin Group

Penguin Putnam Inc., 375 Hudson Street,
New York, New York 10014, U.S.A.

Penguin Books Ltd, 80 Strand, London WC2R 0RL, England

Penguin Books Australia Ltd, Ringwood, Victoria, Australia

Penguin Books Canada Ltd, 10 Alcorn Avenue,
Toronto, Ontario, Canada M4V 3B2

Penguin Books (N.Z.) Ltd, 182-190 Wairau Road,
Auckland 10, New Zealand

Penguin Books Ltd, Registered Offices:
Harmondsworth, Middlesex, England

First published in the United States of America by Viking Penguin,
a member of Penguin Putnam Inc., 2000
Published in Penguin Books 2001

1 3 5 7 9 10 8 6 4 2

Grateful acknowledgment is made for permission to reprint excerpts from the letters
of Albert Einstein from the Albert Einstein Archives of the Hebrew University of Jerusalem
and *The Collected Papers of Albert Einstein*. Copyright 1979–2000 by Hebrew University and
Princeton University Press. Reprinted by permission of Princeton University Press.

THE LIBRARY OF CONGRESS HAS CATALOGED THE HARDCOVER EDITION AS FOLLOWS:
Overbye, Dennis, 1944–
Einstein in love: a scientific romance / Dennise Overbye.
p.   cm.
ISBN 0-670-89430-3 (hc.)
ISBN 0 14 10.0221 2 (pbk.)
1. Einstein, Albert, 1879–1955—Relations with women—non-fiction.   2. Einstein, Albert,
1879–1955—non-fiction.   3. Zurich (Switzerland)—non-fiction.   4. Physicist—non-fiction.   I. Title.
PS3565. V415 E39 2000
813'.54—dc21   00–038192

Printed in the United States of America
Set in Bembo
Designed by Carla Bolte

FOR NANCY

CONTENTS

PROLOGUE: THE SACRED AND THE PROFANE

THE ROAD TO SPLÜGEN PASS CLIMBS WARILY FROM THE OLD ITALIAN city of Chiavenna in a series of hairpin switchbacks and one-lane tunnels carved out of rock, up along Alpine cliffs. It passes from green forest shadows to a stony, bright, windswept summit and the Swiss border, all in a few exquisitely acrophobic miles. The road was first built by the Romans, and it feels as if it has barely been improved since then, with only a few flimsy-looking wooden posts serving as the hint of a guardrail between a swerving traveler and the abyss.

In the spring of 1901 Albert Einstein left home for the last time at the age of twenty-two and set off up that rugged landscape with his lover and former physics classmate, twenty-six-year-old Mileva Maric. They must have been a handsome couple, he with his glowing dark eyes, a scrum of fuzz on his upper lip, and stocky, stubborn build, striding cockily along, and Mileva, all cheeks and eyes, limping beside him like a determined little bird, both jabbering away about atoms and heat conduction, electricity, light waves, and the invisible, all-pervading aether.

It was the kind of heady moment every pair of young self-styled bohemians should have: Albert and Mileva were striking out at last from the tyranny of their families' disapproval and the oppressiveness of narrow-minded teachers for the springlit slopes of new opportunity. On the other side of the mountains there lay, for Albert, his first postuniversity job, a modest teaching position that nevertheless offered the promise of a modicum of independence. For Mileva, studying for her second try at the university graduation exams that would certify her as a physicist, beyond the peaks was the prospect, after months of enforced separation, of a life with her sweetheart.

But there are no guardrails for the heart. The precariousness of the Splügen Pass road would prove an apt metaphor for the course that Albert and Mileva's life together would take. For both, the trip would mark a turning point that would lead them down unexpected paths, entwining their lives forever. This book is, in part, an attempt to tell their story—the story of Albert and Mileva, or "Johnnie" and "Dollie," as they affectionately nicknamed each other—a tale entwined in turn with the history of science and the twentieth century, and still not well known or understood, despite the glaring celebrity of the man Albert eventually became.

As of this writing Albert Einstein has been dead for forty-five years, but in his absence he seems more present than ever. He remains the scientist most likely to make front-page newspaper headlines, as modern science confirms yet another of his bizarre-sounding hypotheses, published long ago. Within the last two years, astronomers have discovered that a strange repulsive force, known as the cosmological constant, which Einstein dreamed up in desperation while trying to explain why gravity didn't cause the universe to collapse in on itself, seems to be shoving the galaxies farther and farther apart. The hottest thing in physics labs these days is Einstein-Bose condensate, an exotic new form of matter whose existence Einstein first predicted in 1932; the substance itself was first created only in 1995. Even Einstein's brain, preserved for four decades, made news in the summer of 1999, when neuroscientists at McMaster University in Ontario announced that his parietal lobe, a region associated with math and spatial relationships, was 15 percent larger than a normal person's. *Time* magazine capped the millennium by naming Einstein its Man of the Century.

From a distance, the trajectory of Einstein's life looks mythic. The one-time humble patent clerk, with his corona of white hair and the haunted eyes, who overturned the universe and gave us the formula for God's fire, who was chased by war and Promethean guilt to wander sockless like a holy fool through the streets of Princeton, making oracular pronouncements about God and nature, has become an icon not just of science but of humanity in the face of the unknown. His visage, peering benificently out at us from coffee mugs, posters, calendars, and T-shirts, is familiar in every corner of the world. Behind the iconic face, however, was a human being, one capable—as all human beings are—of behaving in distinctly un-iconic ways.

I first made the acquaintance of this lesser-known Einstein in 1990, in New Orleans, during a meeting of the American Association for the Advancement of Science. The AAAS meeting annually draws a few thousand scientists to debate and discuss issues ranging from better ways to explore Mars to the ethics of the Human Genome Project. At the time, in one of the stranger episodes in recent Einstein scholarship, a small clique of revisionist historians was advancing the notion that Einstein had cheated Mileva, who became his first wife, out of her proper share of credit for the theory of relativity. One slow afternoon at the meeting, I stumbled into a heated debate on this subject and was mesmerized. It wasn't that I necessarily thought the assertion was true, but it seemed amazing to me that such a debate was even occurring, as Einstein had died in 1955; he was arguably the most famous man in the world, the very author of our modernity. In my naiveté about history, I'd always presumed the key questions about Einstein—how he invented his theories, the nature of his relations with lovers and loved ones—had long since been answered.

In fact, I had never been particularly interested in Einstein the man. Like everyone else, I grew up with the image of the cosmic saint, whose only peer was God. It was hard to imagine that he had ever been young. But as I sat in the auditorium in New Orleans that afternoon it came as a curious relief to me—and I suspect to some of Einstein's modern colleagues, who've had to labor in the shadow of his enormity—to hear young Albert described from the stage as a philanderer, a draft dodger, a flirt, a lover, a hustler, an artist, an errant son, an egregious poet, and a scuffling physicist, whose girlfriend was a feminist and a mathematician. Before my eyes, in a kind of miracle of time-reversal, Einstein shed fifty years. *So the old boy had some juice in him after all,* I thought.

The claim that Mileva had a part in the authorship of relativity came largely from a selective reading of passages in letters, recently published for the first time, that Albert had written to her during their student years. In their correspondence, he had talked about the scientific issues that relativity would ultimately resolve, as well as about details of their courtship, Albert's fights with his mother, and, most spectacularly, the birth of the couple's illegitimate daughter, Lieserl, in 1902. (Mileva's half of the correspondence for the most part seems not to have survived.) When I began to look further into the alleged controversy over relativity's authorship, I found that the letters, fifty-one in all, were only the most sensational part of an avalanche of newly uncovered material about Einstein, material that had the potential to transform our whole understanding of the man, his life, and his science.

In the early 1980s, after years of legal wrangling, a collaboration between Princeton University Press and the Hebrew University in Jerusalem, to which Einstein willed his papers, had began to move quietly forward, with John Stachel, a feisty and stubborn physics professor at Boston University, leading the project to publish Einstein's papers. Stachel and his colleagues had the daunting job of making their way through a mountain of some 43,000 documents that Einstein had left behind in Princeton when he died. At the same time, they were scouring the world to find other Einstein documents that might be unknown, unreleased, or scattered piecemeal in dusty archives. In 1987 the first of some thirty projected volumes had appeared, covering Einstein's youth and university years, and including the love letters to Mileva. From its pages, a young Einstein strode forth, revealing himself for the first time, warts and all.

I like to think that if I had been around in the first or second decade of the twentieth century, I would have camped out on Einstein's doorstep or followed him and his friends to lectures and coffeehouses, so that I could write about him and the commotion he was causing in physics circles. But science journalism did not exist as an art form back then. The letters and other material being uncovered by the Einstein Papers Project and other independent scholars seemed to offer the next best thing to being there, a way to eavesdrop on Albert, Mileva, and their friends and to pose the questions their contemporaries were too polite to ask. The letters were a chance to catch the scents and sounds and sights of a century being created, to listen in and report on the first scratching on cosmic doors. I envisioned the letters as a base camp from which to do the book I realized I myself would like to read about Einstein—a portrait of the physicist as a young scuffler.

Strictly speaking, this is not a biography of Einstein—there is already an abundance of those on the bookstore shelves. Instead, my goal has been to bring the youthful Einstein to life, to illuminate the young man who performed the deeds for which the old man, the icon, is revered. Over the last seven years I've gone through five eyeglass prescriptions, reading hundreds upon hundreds of published and unpublished letters, squinting with my research assistant, Val Tekavec, at Einstein's cramped handwriting. I've followed him as he discusses everything from the details of space-time metrics to how his children should brush their teeth. I've tracked down every place Albert and Mileva lived, separately or together, and have walked the streets of their neighborhoods, eaten cheap wurst in student-quarter cafes, as they must have done. I've read Einstein's high school transcripts and his divorce papers. I've clambered on the razor-edged Säntis Mountain where Einstein almost lost his life as a teenager, walked in the Engadine Alps where he and Marie Curie took a famous hike in 1913, and snaked up perilous switchbacks retracing Albert and Mileva's trip up to Lake Como and over Splügen Pass (where their lost daughter, Lieserl, was conceived in 1901). I've sought out descendants of Einstein and his friends, in order to plague them with embarrassing questions about their ancestors' behavior.

No history, especially a narrative one, can escape the charge that it is, on some level, a fiction, an inexact blending of a writer's subjective choices, interests, and prejudices with the data of the documentary record. While the Albert Einstein portrayed in this book is of necessity partly my own creation, thoughts or feelings that I attribute to Albert or Mileva are drawn from letters or other of their writings. When I speculate about what someone's thought processes might have been, I've taken pains to clearly signal in the text that I'm out there alone on the thin ice of history.

Doubtless there will be among the readers of this book physicists who are uncomfortable with the detailed treatment of Einstein's romantic and family affairs, while there will be nonscientific readers put off by discussions of Einstein's physics. But no exposition of Einstein could pretend to completion if it did not explore both the sacred and the profane aspects of his existence. Physics was Einstein's music; it was the tune he tried to play in the beginning, with Mileva, and we can no more penetrate Einstein's life without it than we could comprehend Mozart without listening to the great operas. So I have tried to play a little of Einstein's music as best I can convey, in the hope that the reader can make out at least some of the haunting melody of the gently curving cosmos.

As much as he may sometimes have wished it were otherwise, however, physics was not all of Einstein's life. He lived on the Earth, with a belly and a heart. As he once wrote in a poem to his friend the young Peter Bucky, "The upper half thinks and plans, but the lower half determines our fate." History would show that, among other things, Albert was an astute and even poetic observer of the human condition, and of his own.

UPON HIS DEATH, Albert Einstein left all his papers and his copyright to the Hebrew University in Jerusalem. His papers and letters reside there, while copies of most of them are held in archives at Princeton University and Boston University. I would like to thank Ze'ev Rosenkranz, curator of the Albert Einstein Archives at the Hebrew University in Jerusalem, and Walter Lippincott of Princeton University Press for access to these papers and for permission to quote from them. This book would not have been possible without the tremendous effort they and their colleagues put into finding, annotating, and publishing Einstein's papers. I am accordingly also indebted to the editors of the Einstein Papers Project, especially Robert Schulmann, who provided invaluable advice and encouragement and a friendly ear during the long years of this project, and Michel Janssen, who spent many hours patiently explaining the philosophical subtleties of relativity.

Numerous archivists helped light my way into the past. In particular I would like to thank Liz Bolton and Margaret Goostray of Boston University's Mugar Library, where a duplicate of the Einstein archive is kept, for their assistance and hospitality when I visited there for a summer. Brigitte Emmer of the Institute für Zeitgeschichte in Munich helped me find my way through the Nicolai archive to Ilse Einstein's letters. I am also indebted to Beat Glaus in the Wissenschaftshistorische Sammlung of the ETH Liary in Zurich, Huldrych Gastpar of the Schweizerisches Literaturarchiv in ern, and Traute Hirt of the Staatsarchiv des Kantons Zürich, who helped me find Albert and Mileva's divorce papers. Lori Olson of the American Heritage Center at the University of Wyoming sorted through the uncatalogued papers of Wolfgang Zuelzer to find the notes of his interview with Margot Einstein and other crucial correspondence with the Einstein estate. Zdenek Pousta gave me an impromptu tour of Charles University in Prague and pointed the way to Einstein's office. Jürgen Staudte spent a rainy afternoon showing me around the Einstein Tower in Potsdam.

Aude Einstein and Evelyn Einstein each shared Einstein family lore and gave me access to letters and other materials that would otherwise have been unavailable. Over cake in Baden, Charles Schärer generously shared his memories of Anneli Meyer-Schmid. Gerald Holton of Harvard University has shared his thoughts generously in many conversations. I have also benefited from advice and discussions with Abraham Pais, John Stachel, Heinrich Medicus, John Wheeler, Jozsef Illy, Jürgen Renn, Julian Barbour, Giuseppe Castagnetti, John Earmann, Peter Bucky, Paul Einstein, Peter Galison, Alice Calaprice, Peter Skiff, Katie Klemenich, Margit Schermer, Ruth Marton, and Milan Popovic.

I want to give special thanks to Valerie Tekavec, who was my German tutor and became my translator, research assistant, and lifeline when it seemed as if the whole world were disappearing into bottomless stacks of German correspondence. Val's astute readings of Einstein's letters to Mileva and to his children, and to his best friend Michele Besso, brought Einstein to life as a person and as a colorful and witty writer. Karin Evans also translated large batches of Einstein correspondence.

Trudi Guhl and Ernst Bucher provided a home away from home for me in Switzerland with enough jazz and good food to help me forget the vicissitudes of Einstein scholarship on the road. Beth O'Sullivan and Al Mailman lent me their house in Cambridge during my summer in the archives. The late Clair Rosenfield always had a room for me as well, and I still miss her. Mary Jo Vath and Fred Escher welcomed me as a member of the household on East 7 Street. Other friends whose encouragement and advice were crucial include Susan Brown, whose bibliography proved to be a gold mine; Lisa Starger, baker of the Einstein birthday cake; Tom Franzel; Jim Polk; Laura Kaplan; Valerie Wacks; Jane and Manuel Brombergs; Lee Smolin; Michael Turner; the late David Schramm; Allan Sandage; and Timothy Ferris. Catherine Arra was a stalwart companion and coach during the years on the Einstein trail. I have been blessed with the faith and support of my family: Gordon Overbye; Olive Overbye; my late father, Milan Overbye; and my stepfather, Jack Schneider.

My colleagues at the *New York Times* are a tremendous source of inspiration, setting an intimidating example of dedication and talent. I want to thank Cornelia Dean, Laura Chang, and John Wilson, in particular, for enduring my occasional absences with good cheer.

Michel Janssen, Alan Lightman, John Norton, Robert Schulmann, Lee Smolin, and Michael Turner generously gave their time to read all or part of the manuscript and offered many comments and corrections. The many errors that doubtless remain are my own.

Once again my editor, the peripatetic Rick Kot, and my agent, Kris Dahl, performed splendidly far beyond the call of duty. As ever, I am thankful for the friendship and support of Natalie Angier.

Finally, I am grateful to Nancy Wartik for her inspiration and advice, and for being the star at the center of my own cosmos.

PART ONE

Runaways

If everybody lived a life like mine
there would be no need for novels.

Albert Einstein to his sister, Maja, 1899

I

ON THE ROAD

ZURICH, 1897. ALBERT EINSTEIN, EIGHTEEN, SITS DEPRESSED AND SORRY for himself in his room. His short, compact body is wrapped in a threadbare bathrobe. Dark curls frame a sensitive face punctuated by large brown eyes, a fleshy nose, and a small soft mouth. A wispy moustache haunts his upper lip; stubble inhabits his teenage cheeks because he can't find the energy to venture out, even for a shave. June light blows through the windows looking out on Unionstrasse in the heart of Zurich's sprawling fabled student quarter. Downhill in the heart of the old city, students throng the alleys of Niederdorfstrasse. Clatter and tobacco smoke fill the cafes along the banks of the Limmat River. The sun is laughing down on everyone in the world except him.

The spring holiday known as Whitsuntide celebrating Pentecostal Sunday is approaching, and Albert will go to Selina Caprotti's house by the Zurichsee and play Mozart, flirt, smoke, and talk physics and philosophy with his new friend, the brilliant but indecisive Michele Besso. But the prospect seems hollow. The memory of Whitsuntide Past weighs on him; he carries a debt of bad behavior, and now it gives him a certain bittersweet satisfaction to pay the price.

He takes pen in hand. "Dear Mommy," he writes to Pauline Winteler, endearingly if inaccurately, firmly declining her invitation to spend Whitsuntide with her family in the country town of Aarau. "It would be more than unworthy of me to buy a few days of bliss at the cost of new pain, of which I have already caused too much to the dear child through my fault." The dear child, the one he must avoid at all costs, is Pauline's youngest and most beautiful daughter, Marie, whose delicate soul he has crushed with his manly thoughtlessness. From now on, he swears, he will mind his own business and avoid romantic adventures. "Strenuous intellectual work and looking at God's nature are the reconciling, fortifying, yet relentlessly strict angels that shall lead me through all of life's troubles. If only I were able to give some of this to the good child! and yet, what a peculiar way this is to weather the storms of life—in many a lucid moment I appear to myself as an ostrich who buries his head in the desert sand so as not to perceive the danger."[1]

ALL HIS LIFE, Albert Einstein has been trouble for women.

Pauline Koch Einstein was only three weeks past her twenty-first birthday when Albert's huge misshapen head squeezed out of her womb, scaring her half to death. He was her first child. She had been only eighteen when she married Hermann Einstein and traded the cushioned life of a grain merchant's daughter for the arms of a failed mathematician and peripatetic entrepreneur. Life hadn't been easy since then, and it wasn't ever going to get any better.

Hard times, of course, had long been the legacy of Jews in what had recently become southern Germany. In the town of Heilbronn, for example, for three hundred years Jews had not been allowed inside the city gates after dark. But in the second half of the nineteenth century, as the Prussian genius Otto von Bismarck was brandishing Francophobia like a welding torch to unite the states of the German Confederation into a new nation, the Jews of southern Germany began to climb from centuries of persecution and pogroms into the middle class, casting their traditions aside as they tried to assimilate with their Roman Catholic neighbors.

Julius Koch, Pauline's father, was a choleric and cunning man who had worked hard and risen from modest circumstances as a baker to amass a small fortune with his brother in the grain trade.[2] Success and comfort didn't bring graciousness, however, or diminish his appetite for a good deal. When he decided that a man of his means should become an art collector, Koch sought out the cheapest paintings and painters he could find, often winding up with copies instead of originals. He bargained a mediocre artist into painting family portraits in exchange for room and board. Pauline and her siblings grew up with the family of Julius's brother in a communal household in which the wives traded off cooking and other chores. In an unpublished memoir, Albert's younger sister, Maja, credited the success of this arrangement to the good humor and maternal disposition of Julius's wife, Jette Bernheim, calling her the soul of the household.

Hermann Einstein, the second man in Pauline's life, was everything Julius was not—or perhaps was not everything that Julius was. With his walrus moustache and pince-nez, Hermann looked formidably Prussian, but he was in fact a cheerful if indecisive man, prone to endlessly mulling over every possibility and point of view.[3] Inclined toward mathematics, he was one of seven children, and his family's limited means precluded higher education for all of them. At the time of his marriage he was an aspiring featherbed merchant, but as events were to prove over and over again, he had no head for business.

Albert was born in 1879 in Ulm, a small city famous today for its high cathedral and its ancient fishermen's quarter—a Venice on the Danube. Ulm lies in a region of rolling hills in the southwestern corner of Germany called Swabia where the Rhine and the Danube, lying only a few miles apart, flow in opposite directions. In the German scheme of things, Swabia was a kind of Middle Earth, an easygoing place inhabited by friendly peasants and minor burghers speaking a colorful dialect. Albert's arrival did not augur any distinction for the Swabian race. When he was born his grandmother Jette exclaimed that he was too fat. The back of his skull was so large that Pauline feared he was deformed, a fear that was reinforced when he was slow in learning to speak. He was well past two before he made any attempt at language. His most memorable utterance was at two and a half, when his sister, Maja, was born. Apparently expecting some kind of a toy, he demanded to know why she didn't have any wheels. Until the age of seven he had the curious habit of repeating softly to himself every sentence he said.

He was a pretty, dark-haired youngster. In the earliest surviving photograph he looks like a miniature adult dressed in a black frock, bow tie, and black shoes, leaning against a chair with a half-lidded bemused look on his plump face. But beneath this calm demeanor lurked a demon with his grandfather Koch's temper. When he was angry his whole face turned yellow, and the tip of his nose turned white. Once he threw a bowling ball at Maja. Another time he clobbered her in the head with a hoe.

A year after Albert's birth the family moved to Munich, to join the nascent electrical industry with Hermann's younger brother, Jakob, who had been fortunate enough to study electrical engineering at the Stuttgart Polytechnic Institute. The electrical industry had been revolutionized in 1867 when Werner von Siemens invented the dynamo, which used the energy of burning coal or water-driven turbines to generate powerful currents and high voltages. By the 1880s electricity was the wave of the future, and Jakob, who was running a gas-fitting and plumbing company in Munich, was eager to get into the game.[4] The Einstein brothers' company would marry Jakob's technical expertise to Hermann's in-laws' money. Hermann, despite his lack of a commanding disposition, took care of the administrative side of the business. In a curious reflection of the Kochs' living arrangement in Ulm, Uncle Jakob and his wife, Ida, lived with Hermann's family in a large suburban villa, sheltered from the street by a shaded yard.

Pauline Einstein was a streetwise, stern-backed, gray-eyed exemplar of tough love, and she raised her children to take care of themselves. After one tour of Munich's busy streets Albert was set loose to navigate his way home on his own while Pauline secretly monitored his progress. In a newspaper interview later in her life, she would ascribe the success of her household to discipline. Patient, persevering, warm but practical, she was given to elaborate needlework. One of her products was a tablecloth emblazoned with the slogan *Sich regen bringt Segen,* meaning roughly, "Keeping busy brings blessings," a more positive version of the old adage that the devil finds work for idle hands.[5] Music was Pauline's other indulgence. She played the piano and she endeavored to bequeath a love of it to her children. With typical brio, Albert began violin lessons when he was five. He threw a chair at his teacher and chased her from the house.

The Einstein villa became a frequent gathering place for the sprawling tribe of Einstein and Koch relatives scattered across Germany and northern Italy, including, significantly, Albert's cousins Elsa, Paula, and Hermine, the daughters of his aunt and Pauline's sister Fanny. (As an example of how complex and interrelated Jewish family structures were in that part of the world, Fanny was in turn married to another Einstein, Rudolf, a textile manufacturer in Hechingen and son of Hermann's uncle Rafael.)

Albert tended to keep to himself during these gatherings. The temperamental child was growing into a solemn and persistent youth, given to pursuits like building houses of cards to a height of fourteen stories. His more typical playmates were the chickens and pigeons, or the small boat he sailed in a pail of water.[6]

In accordance with the tenor of the times, the Einsteins kept a secular household, observing none of the traditional Jewish holidays or rites. The city of Munich, however, required that all public school students receive religious instruction. Albert was going to a Catholic *Volksschule* at the time, but the Einsteins' secularism didn't extend as far as bringing up their children Catholic. A distant relative was imported to tutor Albert on the Jewish faith.

Albert responded more enthusiastically than anyone could have dreamed. To the bemusement of his family, he began following on his own the traditional religious practices his clan had spurned in the quest for modernity. For several years he refused to eat pork, and composed little hymns in praise of God that he sang to himself on the way to school in the morning.

This "religious paradise of youth," as Albert later recalled it, came to a crashing end in a collision with science when he was about twelve.[7] The collision was inevitable, given that his parents were secular people striving to make a living on the most revolutionary development of the age. Albert was a child of technological optimism, swaddled since birth in the mysterious hum of the electric revolution. He had been nudged along a scientific path by Uncle Jakob, by Uncle Caesar, who visited often from Brussels, and by Max Talmey, a Polish student of medicine at the university in Munich who was introduced to the Einsteins by his older brother and subsequently became a weekly dinner guest.[8] Talmey was drawn to Albert, who was only ten when they met but was already intellectually sophisticated enough to hold his own in conversation with a university student. Talmey plied him with books, in particular popular works by Aaron Bernstein, the Carl Sagan of his time, whose writings emphasized the unity presumed to underlie natural phenomena. In one of them the author imagined traveling through a telegraph line with an electrical signal.[9]

All this reading culminated, when Albert was around twelve, in what he later described as an orgy of "fanatic freethinking, coupled with the impression that youth is intentionally being deceived by the State through lies; it was a crushing impression."[10]

In place of God came mathematics. Talmey gave Albert a geometry book that he referred to for the rest of his life as the "holy" book. It ignited a mathematical fire in his brain. Albert was already an adept at the subject and bragged to his sister that he had found an original proof of the Pythagorean theorem.[11] During the following summer vacation he worked his way through the entire gymnasium mathematics curriculum, including calculus, sitting by himself for days on end proving theorems and solving problems in textbooks that Hermann brought home for him.

That was enough for Talmey. Left in the dust mathematically, he switched to philosophy. Together they picked their way through Kant's *Critique of Pure Reason,* among other things. As difficult then as now, Kant was nonetheless hot stuff among college kids. And as it turned out, his philosophy had tremendous implications for science. Kant's theories could be read to imply a kind of outer-space–inner-space connection between the external world of the senses and the internal mental world, that is, mathematics. Was it possible that the key to understanding the universe was in the structure of our own minds?

Another of Albert's early favorites was Arthur Schopenhauer, whose exaltation of the individual's standing and going his own way against the unthinking herd might have been a comfort to a solitary youth.

ALBERT, as it turned out, had plenty of opportunities to practice Schopenhauerian virtue. His precocity displayed itself only fitfully to those charged with his formal education. He got good grades in math, but his Greek professor at Munich's Luitpold Gymnasium, where he was enrolled at the age of nine, is said to have informed him in front of the whole class that he would never amount to anything at all. At the *Volksschule,* according to

Maja, he had been considered slow; at the Luitpold Gymnasium he was thought impudent. His classmates called him "Biedermeier," which roughly translated meant "wonk" or "nerd." The antipathy was mutual.

Although the Luitpold was one of the most modern schools in Germany, with up-to-date lab equipment donated by an electrical company, and its pedagogy organized along lines recommended by the great physicist and philosopher Ernst Mach, a devotee of the empirical tabula rasa school of learning, Albert still complained that it was a factory of rote learning.

Honest Biedermeier soldiered on, bringing home the grades. He came back to the villa, read books and solved mathematical puzzles in the garden, played Mozart and Beethoven sonatas with his mother or his sister, even sat down at the piano himself and picked out arpeggios in a meditative way, and arm-wrestled intellectually with Uncle Jakob.

LATER, when he was an old man and fame and the atomic bomb and quantum carelessness hung around his neck, Einstein would claim that the mystery of the universe first presented itself to him when he was six and his father showed him a compass, its quivering needle unfailingly pointing to magnetic north. It's an iconic story, the young boy tumbling to the invisible order behind chaotic reality, a story told and retold, most recently in the movie *IQ,* in which Walter Matthau, as Einstein, wears the compass around his neck and gives it to Timothy Robbins, who is courting his (fictional) niece.

Whether literally true or not, the compass story is metaphorically apt. Magnetism, after all, was the force that pumped electricity through the coils of dynamos and out along wires to fire lights and jolt streetcars into motion. It was here on this invisible order of the aether, this secret mover, that the Einstein brothers had staked their claim to the future.

At first they flourished. Soon after moving to Munich the Einsteins were exhibiting lamps, dynamos, and a telephone system at the city's International Electrical Exhibition in 1882. Many of the devices were designed by Jakob himself, who along with various employees was issued six patents for dynamos and electric meters. Albert knew enough about his father's work to help out on occasion and to explain the workings of a telephone to his classmates.[12] The firm, rechristened the Elektro-Technische Fabrik J. Einstein & Co. in 1885, grew to two hundred employees and went from building devices to installing power and lighting networks. In 1885 it was awarded a contract to provide the first electric lighting for the famed Bavarian celebration, Oktoberfest. Three years later the Einsteins wired the entire town of Schwabing, then a Munich village of 10,000 souls and now the student quarter.

Eight of Jakob's dynamos were prominently displayed at the 1891 International Electrical Exhibition in Frankfurt. Humming and thundering away in the main machine hall, three of them were in service, pouring out as much as 100 horsepower—75,000 watts—to the other buildings, including the tavern. More than a million people, including the kaiser himself, attended the fair, an island of the future, glittering with artificial lights. The technological tourists marveled at waterfalls and trams galvanized into life by the mysterious flow of electricity, and watched a special electrical ballet with dancers representing Prometheus, Alessandro Volta, and Luigi Galvani.

At the time of the Frankfurt exposition the Munich factory made arc lamps, electrical meters, and dynamos to be sold and installed throughout Bavaria and northern Italy. Hermann and Jakob were posed to become an industrial and technological force in southern Germany. Between them and their wives and families they reasoned that they had the brains and resources to make it in the gold rush of the 1890s, but capitalism was about to quash their dreams.

Around 1890 they had received contracts to install power in the northern Italian towns of Varese and Susa. Increasingly, however, as the Einstein brothers sought to expand their business to the installation of complete power systems, they began to run up against burdensome capital requirements. It took upwards of a million marks in operating capital to be able to compete in the market for large-scale power plants run by water or gas and capable of generating alternating current that could be transmitted long distances. (The per capita income in Germany then was about 600 marks.) In bidding for a contract to light Munich itself, the Einsteins found themselves up against giant firms like the German Edison Company, Allgemeine Elektrizitäts-Gesellschaft, and Siemens, the giant company formed by the inventor of the dynamo himself, with thousands of employees. The Einsteins had to mortgage their house to raise the necessary funds, but it was to no avail. In 1893 Schuckert, another big firm, beat them out of the Munich contract. By the next year Einstein & Co. was kaput.

An Italian engineer with whom they had been working, Lorenzo Garrone, convinced the Einsteins, however, that there might be opportunities for a small firm like theirs in northern Italy, where they had already done some business. The brothers sold the villa (which was promptly razed, garden and all, for an apartment house) and, lured by the lucrative prospect of building a hydroelectric power system for the town of Pavia, outside Milan, left for Italy. The Einstein ensemble settled into a sumptuous house in Pavia that had belonged to a well-known poet, Ugo Foscolo.

All, that is, except for Albert, who was progressing satisfactorily, if grumpily, through the best education system in the world. Since he had three years to go to finish the gymnasium—which would ensure entrance to a good university—Hermann and Pauline decided to park him with a relative in Munich. The move had been Jakob's idea, and one can hardly imagine the rending of heart that must have accompanied Pauline's decision to leave not only the region in which she had been born but her son as well.

By Maja's account, Albert at first put up a brave front, sending laconic letters to Milan disguising the fact that he was in fact lonely and depressed. Without his family around him everything that was oppressive about the gymnasium and German society in general was only magnified. And one aspect of that culture suddenly stood out more dreadfully than all the rest, as the machinery of state descended to metal-stamp his delicate, unprotected soul. As a child, Albert had once been frightened by the sight of troops marching past his window, for to him, as he later recalled it, they looked like spooky automatons being jerked along with no will of their own.[13] He had made his parents promise him that he would never have to be a soldier.

Unfortunately, the German state had different ideas. By law, every male was required to perform military duty. The only way for a young man to avoid it was to leave the country before his seventeenth birthday and renounce his citizenship. Anyone who left later and then didn't report was declared a deserter.

What happened next has been told in several ways. One day that winter, Albert's Greek professor, one Herr Degenhart, summoned the boy and invited him to leave the gymnasium. When Albert protested that he hadn't done anything wrong, Degenhart replied that Albert's mere presence in the classroom was disruptive.[14] Some biographers, notably Philipp Frank, have maintained that Albert had already made a plan to escape the gymnasium before this incident. As Maja tells it, however, Herr Degenhart's sarcastic suggestion was the last straw. Albert was dying to leave. A good idea was a good idea.

In a way this was the payoff, the ultimate extension of those exercises of being left alone and finding his way home when he was young. Albert went to the old Einstein family doctor, whom he talked into writing a note saying he had nervous disorders requiring extended home rest. Luitpold's principal released Albert from his bondage on December 29, 1894. He packed his math books, his Kant, his compass, and his violin, and headed for the train station, a Biedermeier in flight, his beak cocked south toward Switzerland, where the Alps loomed like a cracked jail door, aglow from beyond with the light of Italy and of freedom.

Albert arrived on his startled parents' doorstep in Pavia vowing never to return to Munich or Germany. He announced that he would renounce his German citizenship. Presented with a fait accompli, the Einsteins had no choice but to go along. Their son had shown that he could navigate the byways of Europe as safely as he could the streets of Munich.

It was springtime for Albert, both literally and metaphorically. As he later recalled it, there was hardly anything that he didn't like about Italy, a blast of light and color after the gray heavy baroqueness of Munich. Legend and anecdote paint a picture of a footloose sixteen-year-old wandering about the musical landscape, seeking out Michelangelo, drunk on the light, the air, the music, the food, the operatic people, and the company of his own family. About the worst that could be said was that Pavia and the other towns were a little dirtier than Munich.

Albert spent the next few months working for his father and uncle—allegedly helping them solve a tricky design problem at one point—as well as reading and visiting friends and family around the Italian countryside. He hiked south over the Ligurian Alps to Genoa to stay with his uncle Jacob Koch—and spent the summer of 1895 in the alpine village of Airolo being befriended by a future prime minister, Luigi Lazzatti.[15] He was also penning science essays and little philosophical notes inspired by his reading of Leibniz: "It is wrong to infer from the imperfection of our thinking that objects are imperfect," read one tantalizing sample of his wisdom.[16]

Hermann, worried about where this lighter-than-air existence was going to lead, sat his son down and urged him to learn a sensible trade like electrical engineering so that he could earn a living and perhaps take over the family business someday.[17] Albert responded with a plan of his own: In the fall he would take the entrance exams for the Swiss Federal Polytechnical School in Zurich, which did not require a gymnasium degree. To show the world that he wasn't completely crazy, Albert had obtained a certificate from his gymnasium math teacher attesting to his brilliance and intellectual maturity.[18]

The Swiss Federal Polytechnical School, commonly called the Polytechnic, was a new kind of college, a so-called *technische Hochschule,* primarily devoted to producing teachers. As such, it stood a notch below the great universities such as Heidelberg, Berlin, or Göttingen that Albert would have been entitled to attend had he finished the Luitpold. (In fact, under German law, a degree from a *Hochschule* would not even qualify him to be an officer in the unlikely event he ever fulfilled his military duty. Hermann, as Albert's guardian, had filed a request asking that his son be relieved of his citizenship in the state of Württemberg, of which, by birth, Albert was a citizen, and it was granted on January 28, 1896, thereby eliminating the German army from Albert's future.)

Its secondary status notwithstanding, the Polytechnic was not eager to take in young Einstein. In general, entrants to the school were at least eighteen and had achieved the *Matura,* or high school diploma; Albert was two years younger and had few credentials other than his own cockiness and the math teacher's certificate. The polytechnic director, Albin Herzog, wrote to Gustav Maier,[19] an Einstein family friend in Zurich: "According to my experience it is not advisable to withdraw a student from the institution in which he had begun his studies even if he is a so-called child prodigy." Herzog's advice was that Albert complete his general studies, but if the Einsteins insisted, he would waive the age rule and let Albert take the entrance exam.

Which Albert did in September of 1896, doing well enough on the math and physics part of the exam that Heinrich Weber, the head of the physics department, invited him to attend lectures if he stayed in Zurich. But Albert did so poorly in languages and history that Herzog sent him back for another year in secondary school.

The place Herzog sent him was just thirty minutes outside of Zurich in the pretty town of Aarau, in the canton of Aargau. The Aargau Canton School there was one of the best-regarded in Switzerland and had a liberal reputation. It was just down the road from the site of an experimental school founded by the famous educational reformer Heinrich Pestalozzi.[20]

The fall semester was already starting, so Albert was shuffled quickly out to Aarau, nestled alongside the Aare River, which streamed from the Bernese Alps to the Rhine. Maier arranged for him to stay with the family of Jost Winteler, who taught history and philology at the school. This was not an unusual arrangement; that same year Albert's cousin Robert Koch—son of Julie and Caesar—also attended Aargau and boarded with the Wintelers.

Albert quickly grew to feel more at home in the Winteler house than he ever had in his own. Papa Winteler, as Albert called him, was a handsome, distinguished-looking man with a curly beard. A former journalist, ornithologist, and outdoorsman, Winteler was a liberal democrat and shared Albert's disapproval of German militarism. He and Mama Winteler—coincidentally also named Pauline—had seven children. Their boisterous household across the street from the school was a place of books, music, parties, and spirited discussion. Winteler was always organizing kite-flying expeditions or treks into the surrounding countryside or to nearby peaks. He was also in the habit of holding conversations with his birds.

Mama Winteler likewise indulged Albert as if he were one of her own. Albert, turning gregarious, found himself spending idyllic hours sitting around in his bathrobe pontificating and philosophizing in the company of women while the steam of coffee rose over his face and sunlight slanted over the pastoral landscape and in through the windows.

The Albert that held court in his bathrobe was no longer a Biedermeier, but a rude and self-possessed teenager, handsome, with a shock of curly hair, dark luminous eyes imbued with attitude, and a genius for insolence, a boy whose favorite story was about the teacher (in his Munich gymnasium) who complained that Albert's mere *presence* in his classroom was subversive. A man of the world, a flirt and a seducer, the "impudent Swabian" Albert was something of a terror to his less-traveled, less-experienced schoolmates, with only a violin to reign him in socially.

Hans Byland, one of Albert's classmates, remembered him thus to the biographer Carl Seelig:[21] "Sure of himself, his gray felt hat pushed back on his thick, black hair, he strode energetically up and down in a rapid, I might almost say, crazy, tempo of a restless spirit which carries a whole world in itself. Nothing escaped the sharp gaze of his bright brown eyes. Whoever approached him immediately came under the spell of his superior personality. A sarcastic curl of his rather full mouth with the protruding lower lip did not encourage philistines to fraternize with him. Unhampered by convention, his attitude towards the world was that of the laughing philosopher, and his witty mockery pitilessly lashed any conceit or pose."

Byland went on to recall that he heard Albert play a Mozart sonata on the violin one day at school and was shocked by the power and grace of his performance. He felt that he was hearing Mozart for the first time in all the composer's clear beauty. There was so much fire in Einstein's playing that Byland barely recognized him. "So this was the genius and the unregenerate mocker. He could not help himself! He was one of those split personalities who know how to protect, with a prickly exterior, the delicate realm of their intense emotional life." The mask never slipped for long, however. Often the last notes would barely have rung out before Albert would wreck the mood with some wisecrack, bringing the whole party back down to earth. "He loathed any display of sentimentality," Byland concluded, "and kept a cool head even in a slightly hysterical atmosphere." Even his choice of music betrayed his unease with sentimentality and emotion: Bach and Mozart he loved, but not Beethoven and certainly not Wagner.

In all Aarau nobody was more taken by Albert's charms than Marie Winteler, Jost and Pauline's eighteen-year-old daughter, who was attending the Aargau teachers' school. Marie was quite beautiful (the prettiest of the daughters, according to Robert Schulmann of the Einstein Papers Project, who once saw a photograph of her, but was unable to obtain it for publication) and played the piano, which made her a natural to accompany Einstein musically and evidently in other ways as well.

By Christmas Marie and Albert (who spent the holiday in Aarau) were an item. Marie was by now writing to Albert's mother, Pauline, who was pleased with the match and passed back greetings and approval. A steady stream of notes from the Einsteins to the Wintelers expressed how thrilled they were that Albert was in such a fine home, so happy, and so well cared for.

When Albert returned to Pavia on a spring vacation, his mother teased him about the fact that he was no longer attracted to the girls who had enchanted him in the past (including, apparently, his little cousin Paula, about whom he later said, "Whoever she has not lied to, he does not know the meaning of happiness").[22] Albert reported back to Marie that his mother was quite taken by her, even though they had never met. As for him, Marie's letters, he wrote, had made him understand the meaning of homesickness, and how much his "dear little sunshine" counted for his happiness.[23]

"You mean more to my soul than the whole world did before," he went on, the "insignificant little sweetheart that knows nothing and understands nothing."

Despite being older than Albert at an age when even two years could represent a chasm of experience, Marie knew that she wasn't on his level intellectually. In fact, she worried that she was too fluffy for him and fretted that he would lose interest in her. Faced with such trepidation, Albert was more than willing to shower her with epistolary scolds and kisses. "If you were here at the moment, I would defy all reason and would give you a kiss for punishment and would have a good laugh at you as you deserve, sweet little angel! And as to whether I will be patient? What other choice do I have with my beloved, naughty little angel?"

"We loved each other dearly, but it was a pure love," said Marie much later.[24]

Albert's year in Aarau marked the beginning of what would be a lifelong entanglement with the sprawling and boisterous Winteler family. His sister, Maja, following in her brother's footsteps through the Aarau scene, would marry Marie's brother Paul. Anna, their sister, would marry Albert's best friend, Michele Besso. (Anna, who had a brassy personality, reportedly claimed that she had spied on Albert and Marie while they were kissing.)[25] Both the Wintelers and the Einsteins were eager for a betrothal between Albert and Marie, but at the moment Albert had other priorities, namely a date in the fall with the *Matura* exam and the Polytechnic.

Albert was flourishing at the cantonal school, which was smaller and less authoritarian than the Luitpold. The professors seemed less perturbed when Albert was being Albert. Science teacher Friedrich Mühlberg on a geology field trip: "Now Einstein, how do the strata run here? From below upwards or vice versa?"

"It is pretty much the same to me whichever way they run, Professor."

By the end of his year in Aarau, despite his father's earlier career advice, Albert had decided to study theoretical physics. In fact, Albert was fascinated by electricity, but invariably, as he pointed out in an essay during his graduation exam at Aarau, his interests ran to the mathematical and the abstract—not the practical.[26] Allowed to roam during his Italian interlude, his imagination had focused not on how to further manipulate electromagnetic forces into doing society's bidding, but on the nature of electromagnetism itself. In a paper he sent to his uncle Caesar he had tried to explain how magnetism could be caused by a deformation of the aether that was alleged to pervade all space. And perhaps in imitation of the author Aaron Bernstein, who had imagined himself squeezing through a telegraph wire, Albert had tried to imagine what he could see if he were traveling along on a light wave, like some futuristic surfer. What he could see was that there was some kind of paradox there, but also that thought experiments like this were great fun.

"Besides," he added in his remarks on his future, "I am also attracted by a certain independence offered by the scientific profession."

The advantages of that vaunted independence were perhaps underscored when Albert's Aarau idyll was interrupted by yet another business crisis for his father. In 1896 the deal to build a hydroelectric system in Pavia fell through when the Einsteins bungled the negotiations for water rights. Believing that they had been deceived, the city fathers booted the brothers off the project and their company collapsed once again. At this point Jakob had had enough, and with Hermann's blessing he went off to work for another firm. Hermann moved to Milan and started yet another business to manufacture dynamos and motors. By now he had not only lost all his wife's money but also was in debt to the rest of the family, particularly Rudolf Einstein, the husband of Fanny Koch, Pauline's sister.

Convinced that this new venture was doomed to failure as well, Albert hopped on a train and went to Heilbronn, where Rudolf and Fanny lived, to dissuade Rudolf from loaning any more money to his father.[27] His arguments were to no avail, and so Hermann lifted off on uncertain wings into the electrical business on his own. As usual, things went well for about a year—Hermann won concessions for the lighting systems in the towns of

Canneto sull'Oglio and Isola della Scala—until Rudolf began pressuring him to pay back his loans. For a young man with no professed practical ability, Albert was to prove only too good a business prophet.

IN JUNE, Albert and some classmates went for a three-day hike, the Swiss version of a commencement party, on the Säntis, a toothlike ridge soaring 2,500 meters above the rolling landscape of the Appenzell region in eastern Switzerland, home of Heidi, cows, cheese, and ancient agrarian ritual. One day it began to rain, and Albert was wearing poor shoes. Clambering along the top of this razor edge less suited to people than to the small black birds that ride the drafts up and down the cliffs and perch on crags, Albert slipped and began sliding down the slope toward a sheer dropoff. His career as mountain man, physicist, and musician was about to go sailing into space over the rocks when a classmate, Adolf Fisch, stretched out his alpenstock.[28] Albert grabbed it and was hauled from the abyss.

Three months later he passed the *Matura* and returned to Zurich to attend the Polytechnic at last. The parting with Marie was sad but hopeful. She had accepted a temporary teaching job in Olsberg, a small mountain town an hour and a half from the nearest train station. But she could spend weekends in Aarau, which was only twenty minutes from Zurich. Albert assured her that they could still write. Marie offered to do his laundry.

STANDING, newly arrived, on the platform of Zurich's Hauptbahnhof, a violin case and a suitcase in his hands, peering down busy Bahnhofstrasse, Albert had only to lift his eyes slightly upwards and to the left, across the narrow Limmat, to see the pale-mustard neoclassical buildings of the Polytechnic and the University of Zurich looming on a hillside.

Cradled by steep green mountains whose slopes are dotted with villas, Old Zurich with its Roman ruins, cathedrals, banks, hotels, and restaurants runs south from the train station to the lake, the Zürichsee, straddling the Limmat for about a mile. To the east and the west of the old town, trolleys climb the steep hillsides of the Zürichberg and the Ütliberg, respectively, to the stars, or at least to the clouds. Thus bounded and protected, Zurich has been a strategic and favored location throughout history. By the time the city and its surrounding canton joined the Swiss Confederation in the fourteenth century, it was already a grand center of commerce. Its character was further forged by the fire and brimstone offered weekly from the pulpit of the Grossmünster—a twin-towered cathedral whose earliest construction dates from 1100—by the Swiss reformer Huldrych Zwingli in the early sixteenth century. Zurich grew up proud of its Calvinist heritage and stern discipline, ruthlessly spare, pure, democratic, independent.

In 1888 the city had quadrupled its population from 28,000 to 120,000 by expanding its boundaries outward from the central core to encompass eleven outer districts. Within its total area of twenty-five square miles, it was said, were over nine hundred inns and five hundred clubs.[29]

The technological revolution ruffling the rest of the world proceeded apace in Zurich. Electric lights had first graced the town in 1855 during the "Sechselaeuten" festival when, as a demonstration of the new invention, the entire Grossmünster and both sides of the Limmat were illuminated. The following year the town fathers began installing public gas lighting, replacing old oil lamps. In 1880 electric arc lamps lit the Federal Choral Festival. The first telephones were installed the same year. In 1884 electric trams began to replace the horse-drawn kind.

In 1896 fin de siècle Europe trembled on the brink of a dozen revolutions. In Vienna Sigmund Freud had begun to ruminate on dreams and sexual hysteria. France was convulsed by the Dreyfus affair. In Paris, where a young Pablo Picasso had begun to paint, Impressionism was still in its glory, and the poet Stéphane Mallarmé was experimenting with randomness and silence. X rays and radioactivity were discovered, and the science of physics, pronounced "finished" only a few years before, was exploding into wreaths of complexity and mystery. The Olympic Games were reborn in Athens. Karl Benz began building automobiles. The first automated telephone exchange and the first movie camera were constructed. Electric light bulbs, running on direct current, were slowly replacing gas lamps, making everybody but the Einsteins rich.

The Eiffel Tower, finished in 1889, dominated the Paris skyline, outraging traditionalists with its stark engineering aesthetic. Technological optimism was the rampant theme of the fading century. Everywhere the old forms were being discarded like so much clutter, dropping like faded blossoms. The Austrian physicist Ernst Mach argued for a new aesthetic in science, devoid of old ideas and abstractions, freed of metaphysical baggage and shorn of all content save what could be built logically from the evidence of the senses. "Physics," he had written in 1882, "is experience arranged in economical order."[30]

Zurich was one of those metaphysically clean towns, sheltered by political stability and freedom, fresh air, prosperity, and hallowed historical neutrality. It was rubbed clean of preconception by adherence to Zwingli's ethic, democracy (at least for white men), and ruthless commerce. Carl Jung, the psychoanalyst, who moved to Zurich from the old university town of Basel in 1900, found it to be a breath of fresh air.[31] "Zurich relates to the world not by the intellect," he wrote, "but by commerce. Yet here the air was free and I had always valued that. Here you were not weighed down by the brown fog of the centuries, even though one missed the rich background of culture." Rosa Luxemburg, the founder of the German Communist Party, and her cohorts were already living in the city when Albert arrived. In time, Lenin, Joyce, and the Dadaists would all seek shelter in Zurich's bracing air.

Along the east bank of the Limmat on the lower slopes of the Zurichberg, a community of outcasts, freethinkers, and students filled a warren of boardinghouses and cafes. The Federal Polytechnical School sat beside the university smack in the middle of this enclave, on Ramistrasse, a busy boulevard that crossed the hillside. Its main entrance, set back from the street by a small courtyard, is today a colonnaded dome reminiscent of the Roman Pantheon that towers over the other university buildings. Passing through the arched oak doors, a visitor or young student enters a vaulted echo chamber of arches, balconies, gleaming floors, and magisterial staircases lit only vaguely by skylights and tall distant windows.

The doors on the other side of the building open out onto a vast stone terrace at eye level with a forest of spired clock towers and weathered green domes. From the double-barreled Grossmünster nearly underfoot to the delicate Fraumünster to the blocky thirteenth-century St. Peter's with the largest clock face in Europe to the distant slender Predigerkirche in the west, right on downtown to the ancient domed observatory tower on Uraniastrasse, the view contains more history than the eye can bear. On the hour, if you listen carefully, you can hear the bells in all those towers begin to chime, adding their voices one after another to the wind according to the corroded rhythms and haphazard schedules of their designers, at first nearby and then from afar in succeeding waves of ever more distant and fainter notes—*bong bong, ding ding dong*—like an asynchronous chorus, until the entire spire-pricked sky seems engaged in a subliminal argument of tintinnabulation.

COFFEEHOUSE WARS

OCTOBER IS A CLOUDY MONTH IN ZURICH. FOG HANGS AND DRIFTS across the dark green hillsides, kissing the gardens on the flanks of Ütliberg, shrouding the mountaintops, painting the watery landscape of the Limmat and the Zürichsee as gray as the cobblestones of Bahnhofstrasse. At this time of year the cafes of the sprawling student quarter fill with radical talk, leather jackets, attitude, and tobacco smoke. The waiters gaze paternally over the hubbub, making change briskly from their black leather purses.

As a new student, Albert gratefully joined the melee. One of his favorite pastimes was to stroll the sidewalks puffing on a yard-long pipe that his uncle had given him, twirling it about so that his head was completely wreathed in pungent smoke.[1] With the 100 francs a month that Aunt Julie was sending him, he rented a room on Unionstrasse just off a pleasant quiet square called Baschigplatz. From there the daily walk to the Polytechnic took him up Plattenstrasse past houses full of foreign students.

Exactly five people arrived to comprise Section VI A, physics and mathematics, of the class of 1900 at the Polytechnic. Marcel Grossmann, the tall and aristocratic scion of an old agricultural manufacturing family, was the aspiring mathematician of the group. Besides being blessed with a handsome cleft jaw, Grossmann was a keen judge of character and talent, and quickly befriended Albert, taking him home to meet his family in nearby Thalwil. "This Einstein will one day be a very great man," he announced to his parents.[2] Another classmate was Jakob Ehrat, from the old river-castle town of Schaffhausen, who often sat next to Albert in lectures and classes. Ehrat, a worrier, lacked Grossmann's poise or Einstein's arrogance. His presence at the Poly was due more to hard work than to genius, and he required frequent pep talks from Albert, especially at exam time. Ehrat, too, often took Albert home to his mother. She grew very fond of him, and Albert would sometimes drop by and visit her if he felt lonely. Louis Kollros, from a watchmaking town on the French border, rounded out the male contingent. Finally, there was the Hungarian woman, Mileva Maric.

At twenty-one, Mileva was older than the rest of the group. Her journey to the gates of physics had been more arduous and single-minded than that of any of the men in the class, even Albert. The eldest daughter of a wealthy landowner and judge in far-off Serbia, a rough-and-ready land of dark peasant stock, Gypsies, and brigands that was presently part of the Hungarian half of the Austro-Hungarian empire, she was beautiful in an intense, brooding way. She had smooth cat cheeks and thick wavy black hair. In a photograph taken that year, she is wearing a dress with a big bow. Her dark eyes stare frankly and even a little saucily; her mouth is wide and generous with thick, sensuous lips drawn in the barest hint of a smile. She walked with a noticeable limp. She had a beautiful voice, but she rarely said much in class.

Albert was relatively well traveled for his years, but it is fair to assume that he had never come across anything as exotic as a woman physicist or even a woman would-be physicist. He had spent his life so far in the company of women like the twin Paulines, Marie, Maja, and his relatives, whose interests ranged from the merely bourgeois to the purely frivolous, as in the case of Aunt Julie. Mileva, however, was nothing if not dead serious. There was not an unserious bone in her awkward little body, just a heartbreaking modesty and singleness of purpose. She radiated mystery and determination. Mileva Maric was the real outsider that Albert only pretended to be.

Albert and Mileva spent their first year at the Poly stalking each other warily around lab tables, libraries, and music recitals. No documentary trace of their interactions during that year has been uncovered—no letters, no class notes, no concert ticket stubs—but it seems clear that their getting friendly was not a quick process. Mileva was the kind of woman whose light shone for a very small circle. She lived in a boardinghouse on Plattenstrasse, in Albert's neighborhood, and spent most of her time with a small circle of Eastern European women who were attending the University of Zurich. Her shyness and unease in a strange culture, and her steadfast studiousness—the very same qualities that attracted Albert—made her difficult to know.

Meanwhile, Albert was writing to Marie and sending his laundry to her in Olsberg, but their relationship was fraying. On at least one occasion Marie walked the hour and half to the station, and was thrilled just to see Albert's handwriting on the address of his laundry bundle, only to find that he hadn't even included a card. Nevertheless, she made the return trip in the rain to send back his freshly cleaned clothes.[3]

Most likely, part of the problem was that Albert was making the inevitable comparisons, Marie's worst fear. Perhaps, from the vantage point of intellectual Zurich and the presence of Mileva, Marie began to appear too provincial and flouncy—the best of the bourgeois, certainly, but to Albert, bourgeois nonetheless. But there was a deeper, more mysterious, current to his feelings. In later life Albert would refer again and again to his fear of going mad with desire if he saw Marie again and his need to stay in a "high fortress of calm."[4] Perhaps he felt that the strength of his crush on her had tugged him from his emotional moorings and he was in danger of losing his cool ironic detachment and being swept away—a feeling he didn't like or trust any more than he liked emotional music. The bourgeois that Albert feared most was not the bourgeois in Marie but the bourgeois in himself.

In any event, suddenly the fact that Marie had moved so far away seemed ominous—never mind that he himself had moved first. In November, just before they were to meet for a weekend in Aarau, he suggested that they should stop corresponding.

This was such an abrupt thought that Marie didn't seem to want to take notice of it, until midway through her next letter, when she suddenly addressed the puzzling comment in midsentence.[5] "My love, I do not quite understand a passage in your letter. You write that you do not want to correspond with me any longer, but why not, sweetheart?" she blurted.

"You scold me rudely that I don't want to write to you how and why I've come here. But you dear wicked one, don't you know that there exist a lot more beautiful and more clever things that one can chatter and tell about than something so stupid?" Making fun of Albert, her "darling curly head," for his rash conclusions, she promised to scold him in turn for his rudeness when next they met.

She sent him a teapot. After a delay so long that she was afraid he was ill, Albert responded by calling the gift stupid. His criticism didn't matter, she wrote back to him calmly and undeterred, as long he was going to brew some good tea in it. She then compounded the error by threatening to come and visit him along with her mother, wipe his worried brow, and rearrange the furniture in his room.

To a man gradually descending into the rigors of differential calculus and projective geometry, Marie's kittenish ways might have seemed as inconsequential and maddening as they had once been enchanting. At the end of November, two weeks before another rendezvous, she sounds like a woman looking for a commitment but not certain if she is ready to deal with the consequences: "That you do not want to give me an answer, just you wait, Albert, you'll get quite a wicked punishment for it, I have still 12 days left in which to devise one (and all the same I am so glad that you don't want to give me one and that you think the whole thing stupid, isn't it curious that I like this even much better than an answer, but then again, I would like one all the same, I don't know why)."[6]

No answer was forthcoming, but Marie was both smitten and persistent in her way. As it happened, one of her "little blockheads," a first-grader, was named Albert and bore a passing resemblance to her hero. "I want to help him with his lessons and, in general, love him so much. It's sometimes quite strange, something comes over me when he looks at me and I always believe that you are looking at your little sweetheart."

Classes claimed little of Albert's attention that first year. Out of six "obligatory" first-year courses, five were in mathematics. The lone physics offering, a course on mechanics, did not start until the second semester. Albert was disinclined to study mathematics—so much so that Hermann Minkowski, one of his math professors, later remarked that Einstein had been a "lazy dog"[7]—and even more disinclined to attend lectures. Luckily Albert had friends like Marcel, who never missed a lecture and kept meticulous notes, and Mileva, who had cultivated a habit of showing up for things.

Nevertheless, physics was in the air and in the coffeehouses, books, and salons, where Albert could easily imbibe it. At the end of the nineteenth century, science was poised on the verge of a great triumph. This triumph had two components—atoms and mechanics—and each offered an answer to one of the Great Questions philosophers and physicists had been wrestling with since there had been physicists and philosophers. Many of the brightest and the best minds of the 1890s believed that, together, these two concepts formed the basis for a theory that could reduce the splendid and seemingly inexhaustible variety of natural phenomena, from teapots to rainbows, to a few fundamental principles the same way that all the complications of geometry could be traced to a few axioms and propositions, the lucid elegant reality lurking behind apparent chaos.

Those principles—the laws of action and reaction, force and acceleration and gravitation, which described the motions of billiard balls and constrained even the movements of worlds—had first been stated two centuries before by Isaac Newton, arguably the greatest scientist who ever lived. But Newton's was only one loud voice in a 2,500-year debate about the nature of reality, space, time, and their contents.

One of those great continuing arguments was about whether the true nature of reality could be better grasped by understanding the diversity or the unity of the rich phenomena available to the senses. What was more important? That which separates a horse from a zebra and confers a distinct identity on each, or the basic biological properties common to both? In the latter camp there was Plato, for example, the ultimate idealist, who described the world as a pale miserable shadow of the real world of forms behind it. Attempting to resolve this dichotomy between the sacred and the profane, Leucippus and, more famously, his student Democritus, who lived in Thrace two and a half millennia ago, proposed to explain how variety could arise out of sameness by suggesting that the world was made of identical, little indivisible nuggets called atoms. The different ways of assembling these atoms was what gave rise to the diversity of the world. "By convention there is sweet," said Democritus, "by convention there is bitterness, by convention hot and cold, by convention color; but in reality there are only atoms and the void."[8] At the core of reality was simplicity and permanence, on the surface complexity and change.

Aristotle, Plato's successor, was a phenomenologist, and rejected both atoms and the void. Because God, after all, had to be present everywhere, Aristotle preferred to fill the void with a formless *aether*, literally "the breath of the gods." His views on almost everything, adapted to Catholic orthodoxy, prevailed for centuries. The aether was modernized in the seventeenth century by René Descartes, who pictured the planets as being dragged around the sun by aether vortices,[9] and in the nineteenth century by James Clerk Maxwell, whose equations revolutionized the understanding of electricity and magnetism, and who envisioned whole systems of aether vortices, interlocking like gears, to explain various electromagnetic effects.

The concept of atoms meanwhile had fallen out of favor for almost two millennia. Atomism was considered tantamount to atheism until Newton, a fan of laws and of atoms, managed to bridge the gap in his *Principia,* published in 1687: "God in the beginning," he wrote, "formed matter in solid massy, hard, impenetrable, movable particles, of such sizes and figures, and with such other properties, and in such proportion to space, as most conducive to the end for which he formed them."[10]

Newton's God was a master clockwork mechanic. Surprisingly, however, the greatest triumph of Newtonian thought came not from the consideration of clocks or levers or even the flight of cannonballs wherein the notions of force and inertia made immediate visible sense, but from the study of the billowing of hot gas. The invention of steam and combustion engines had made thermodynamics one of the key new sciences of the nineteenth century and had given rise to a set of far-reaching laws governing heat, energy, and matter in bulk.

By the latter part of the century men like Maxwell, Hermann von Helmholtz, a German physician turned physicist who had indeed invented the mathematical concept of energy, and, most of all, the Austrian Ludwig Boltzmann were claiming that they could account for the laws of thermodynamics by assuming that a gas was just a vast assemblage of atoms bouncing around off one another like tiny balls in accordance with Newton's familiar laws.

Because no one had ever seen atoms, though, their existence was controversial, and atom wars were raging throughout Europe. At a scientific meeting in Vienna in 1895 Boltzmann had engaged in a heated public debate with his friend Wilhelm Ostwald, a University of Leipzig chemist who was part of a romantic backlash against mechanism called *Naturphilosophie,* which espoused a holistic view of nature that traced its roots to Goethe, among others; rejecting atoms as a mathematical fiction, the *Naturphilosophen* believed that everything was energy.

Indeed, the thermodynamic argument was an abstract and mathematical kind of proof—if "proof" was indeed what it was—that left unanswered precisely *what* these so-called atoms were: Vortices in the all-pervading aether? Condensations? Lumps of something else, like rocks in a stream, that the aether did not permeate? The same scientists who were investigating gases had set their sights on the other great advance of nineteenth-century physics, the electromagnetic theory of light, hoping that they could likewise explain the transmission of light waves mechanically, as vibrations or compression of the aether, the way air transmits sound.

Just as scientists could almost taste their triumph, however, experiments from around Europe were churning out whole new categories of phenomena to be explained. In 1895 Wilhelm Roentgen had discovered mysterious "X rays" that could pass through matter and fog photographic film. A year later the Frenchman Henri Bequerel accidentally found that a hunk of pitchblende, a simple rock, *just sitting there,* was giving off invisible rays that could similarly expose otherwise shielded plates. Two years later, Marie and Pierre Curie, working in a poorly ventilated shed in Paris, would isolate radium and polonium as the radioactive elements in pitchblende. Meanwhile, J. J. Thomson in Cambridge, England, had managed to show that so-called cathode rays were composed of negatively charged little particles that he called "electrons," the stuff of electricity and perhaps the key to understanding matter.

From the sideline seats in the cafes of Bahnofstrasse, Albert and his new friends followed the new discoveries over coffee and plates of cheap wurst.

Albert's dark good looks, his wit, his old family connections, and his violin opened many a door and heart in Zurich. He quickly established a reputation as a fool for music who would dash out of the house half dressed with his instrument in hand when he heard a piano sonata being performed next door.[11] He played Mozart for his landladies, their daughters, and their friends, and stomped out when the upstairs ladies brought their knitting needles along. "I wouldn't dream of disturbing your knitting," he muttered. He played Handel at the homes of his history professor, Alfred Stern, where the music was often preceded by arguments about physics, and of the chairman of the mathematics department, Adolf Hurwitz, and his daughters Lisbeth and Eva. On visits to the Wintelers he played Schubert. Saturday nights in Zurich he often spent playing music with his old classmate Hans Byland at the lakeside home of Selina Caprotti, an older woman who often entertained the students.

It was at the Caprotti house one night just before school started that Albert met a young engineer with the symphonic name of Michele Angelo Besso.[12] Short and dark, with curly hair, a bushy black beard, and fierce eyes, Besso was gregarious, smart, and possessed of a nervous eclectic curiosity. Perhaps most important, he had a deep philosophical interest in physics. He and Albert hit it off immediately.

Besso belonged to another of those rambling Jewish families. His father was an insurance executive who had moved to Switzerland from Trieste, Italy, married a Swiss woman, and become a naturalized citizen.[13] Michele had been born on the outskirts of Zurich, the first of five children. When he was a few years old, the family moved back to Trieste. A precocious child with a gift for mathematics, Michele was nonetheless expelled from the local gymnasium for insubordination and sent to live with an uncle in Rome. After a year at the University of Rome, he transferred to the Polytechnic, graduated with an engineering degree in 1895, and got a job at an electrical factory in Winterthur, a suburb of Zurich.

Albert regarded Michele as a kind of cosmic schlemiel, brilliant but scattered and lacking in will. "He is an awful weakling without a spark of healthy humaneness, who cannot rouse himself to any action in life or scientific creation, but an extraordinarily fine mind, whose working, though disorderly, I watch with great delight," Albert wrote, "Yesterday evening I talked shop with him for almost four hours."[14]

It was at a Caprotti soiree that Albert introduced Michele to Anna Winteler, Marie's older sister. The two fell in love and soon married. Besso returned the favor by suggesting that Albert read up on an Austrian physicist and philosopher named Ernst Mach, of whom Besso was something of a distant disciple.

An incorrigible skeptic and something of a cult figure, Mach was the major spoilsport at the party of fin de siècle physics, arguing among other things that atoms were metaphysical nonsense, at best a convenient fiction. He was a follower of the French positivist philosopher August Comte, who had tried to orient philosophers toward the verifiable, or "positive," kinds of statements that science could make rather than the negativistic critiques then prevalent. Comte had organized the sciences into a hierarchy based on how closely they hewed to the principle that experience was the only basis for knowledge; mathematics and physics were at the top, sociology and psychology at the bottom.

Mach had gone Comte one better, enunciating a whole epistemology of the scientific process, a recipe for truth that was starkly empirical. Proper science began with gathering facts, he argued, and proceeded to categorize and then generalize them into laws. The laws, however, had no real standing beyond serving as a shorthand, a kind of mnemonic, for recalling the actual facts, the sensory details. Atoms were beyond the realm of the senses; therefore Mach refused to admit them to the realm of scientific discourse.

"Intelligible as it is," Mach wrote in 1894, "that the efforts of thinkers have always been bent upon the 'reduction of all physical processes into the motions of atoms,' it must yet be affirmed that this is a chimerical ideal. This ideal has often played an effective part in popular lectures, but in the workshop of the serious inquirer it has discharged scarcely the least function."[15]

Albert admired Mach's ruthless skepticism, but he himself was already inclined to accept the reality of atoms. One of the things he liked about playing with his uncle's pipe was that he could imagine he was seeing atoms in action as he watched the smoke drifting, clumping, and dispersing.

That first year in Zurich, Albert and his friends perfected a kind of cafe lifestyle. Here was the drill: Rise late and breakfast on a pastry cadged from his landlady or on Baschigplatz. Spend the morning either hanging around the physics lab, dozing through a couple of lectures, or sailing on the Zurichsee. In the afternoon, after or instead of class, camp out at the Café Metropole with a book. Einstein wasn't much for drinking alcohol; beer, he used to say, was a recipe for stupidity.[16] Coffee and tobacco were his main vices, and even at an early age his teeth were stained brown from smoking. At night, play the violin, wander home, and rattle the door—"It's me, Einstein. I've forgotten my key again." Or hop on the trolley and ride across town to Ütlibergstrasse, where a network of promenades and trails wound through meadows, woods, and garden plots up to the top of the mountain, which offered a spectacular view over the city. In nice weather he and his friends might rent a sailboat and go out on the Zurichsee, where they would argue about physics and scribble in notebooks.

Einstein's undeniable charm also percolated into the classroom, at least as far as his classmates. Margarete von Uexkull, a biology student from Kovno who shared a physics lab course with Albert as well as a rooming house with Mileva, recalled in a memoir that she had found herself in a bitter argument with her corpulent little dictator of a professor one day.[17] Suddenly, from behind her tormentor, she saw Albert's dark shining eyes beseeching her to hold her temper. After the professor stomped off in a rage, Albert offered his opinion that the man was crazy. He offered to take her notebook home and cook up some acceptable results. It turned out he already had eight other notebooks that he was doctoring for people.

The life of a student vagabond agreed with him. In March when he came home to Milan for spring vacation, Pauline wrote to Marie bragging about how good he looked and how healthy his appetite was. "There is so much laughing, joking, and music making going on that there is not enough time for anything else."[18]

His letters home had dwindled so drastically that Pauline was often reduced to writing to Marie asking for news. But the trips to Aarau were dwindling also, and by spring they had stopped completely. Marie was heartbroken, and she suspected another woman.

As the weather turned mild an uncharacteristic gloom descended on Albert, who was lonely and bored, and feeling a little guilty about Marie. "I seek solitude in order to complain quietly about it," he complained to Pauline when she wrote to invite him to Whitsuntide. Pauline sent him back roses, which plunged him into a reverie of his happy days in Aarau: Albert pacing in a tattered bathrobe in his room thinking philosophically portentous thoughts while sunlight streamed in the windows. Rosy-cheeked Marie on a red footstool launching sweet words that flew up about him like birds mocking his seriousness.[19] In retrospect it all began to seem like some kind of bourgeois heaven. "And afterwards I feel so silly and vacillating between laughter and tears—and finally the beloved piano resounds like my soul, calm or mad, depending on what just happens to be its mood, and if the latter is the case, then I also think of the lovely hours and the little red footstool and whatever else goes together with it."

His friendship with the very unbourgeois Mileva was growing nonetheless. Toward the end of the school year, in July, they went on a hike, a *Wanderung,* together—not a likely form of recreation for someone with a deformed hip. She later told him that he had given her great joy during their trip.

Maybe too much joy. In the fall, when Albert made his way back over to Plattenstrasse, he discovered that Mileva had not returned to Zurich. She had quit school.

3

THE ROSE OF HUNGARY

THE VILLAGE OF KAC LAY A THOUSAND KILOMETERS AND MANY WORLDS east of Zurich's musical green skyline, at the far end of what was then Hungary, sixty kilometers up the muddy Danube from Belgrade. More than anyplace else, home to Mileva was a two-hundred-acre country villa nicknamed The Spire, for the towers that crowned each gable end and the central bell tower of the three-story manor house.[1] The villa probably belonged to her grandparents. It was here, amid the corn and sunflowers, that Mileva had passed the happiest days of her childhood. According to an account of her life published in 1969 by her fellow Serb Desanka Trbuhovic-Gjuric, *In the Shadow of Albert Einstein: The Tragic Life of Mileva Einstein-Maric,* based on family tales and local anecdotes, Mileva spent countless hours curled in the tower attic next to the bell dreaming away summer and winter afternoons like some fairy-tale princess, reading, watching the shifting landscape, talking to her father, and waiting for the "magical bell" to ring, signaling to peasants scattered across the countryside to come in for dinner.

In the few pictures that date from that time she is the determined little child who is going to break your heart, all eyes and fat cheeks, her face a weathervane perpetually poised between sunshine and thunder. She looks both stubborn and vulnerable stepping forward in thick lace-up boots on stubby legs.

In her heart, her father never tired of telling her, beat the blood of bandits. Kac is part of the Vojvodina, a region of fertile plains stretching north from the Danube. Vojvodina was scar tissue from the long struggle between the Austro-Hungarian Empire to the north and west and the Ottoman Empire to the south and east. The boundary between these historical enemies was often marked by the Danube River, which cuts across the northeast corner of what would come to be called, for a period in the twentieth century, Yugoslavia. The push of the rival armies up and down the Balkan Peninsula had pumped refugees and nationalities around like water. As a tactic to shore up their frontier, the Austrians at one time had encouraged with land grants (and restrictions against emigration) anyone and everyone to settle and develop the marshy region north of the Danube. In 1690, 40,000 Serbian families fled north from the Turks under cover of the retreating Austrian army and settled across the river in that spot. Thereafter the ethnic stew in the Vojvodina had a strong Serbian flavor, and the region's capital, Novi Sad, became known as "the Athens of Serbia." The whole area became a hotbed of Serbian nationalism, somewhat to the nervousness and suspicion of the Hungarian overseers and the other ethnic groups in the area. The exiled Serbs harassed the occupiers of their homeland from gunboats on the Danube, by launching guerrilla strikes, and by shipping weapons south to their compatriots. They cultivated an image of themselves as brigands and cutthroat businessmen.

Milos Maric, born near Novi Sad in 1845, was a descendant of that great Serb migration. He was a rugged youth, the strong silent type, shrewd but fair, and with a self-made man's exaggerated respect for learning. His grandson once described him as someone you could trust but must fear.[2] When he was sixteen, Milos had enrolled in the army. His duties took him around the Austro-Hungarian Empire, during which time he learned German and became a crack administrator. In 1866 his regimental travels brought him to Titel, a Vojvodina town on the Tisza River. By then he was something of a heartthrob, tall with an elegant drooping moustache. When he tromped into a town with his regiment, dusty with duty and glowering with Serbian pride, women all over Vojvodina swooned.[3] Because he could have had any girl in town, it was a slight surprise to the neighbors when he chose to marry the shy and unassuming Marija Ruzic, one of three daughters born to one of the richest men in Titel. They seemed an unlikely couple; she was as quiet and modest as he was imposing and outgoing. Rumors would float for years that Milos never quite gave up playing the field. The couple was married in 1867 in a Greek Orthodox ceremony led by the same priest who had baptized Marija, and they settled temporarily with the bride's parents in Titel.

Their first child, Mileva, was born eight years later in 1875. She came into the world with a dislocated hip, apparently a sort of congenital abnormality not uncommon to the Montenegrin people from whom Mileva was partly descended. A sister, Zorka, would later bear the same burden. When Mileva began to walk, she limped, to the dismay of her parents, who despaired that nobody would want to marry a lame woman.

Miza, as she was known within her family, was a sickly child who seemed to have inherited her mother's shyness. But she was also bright and determined, possessed of a plucky but stormy temperament. She was a serious girl, capable of an elegant scowl, who experienced everything intensely. From an early age, apparently, music was one of her passions. When she was eight she started taking piano lessons. The descriptions of her attempts to dance in spite of her limp make for a heartbreaking tableau: Mileva turns around on her toes as if trying to leave the earth, hopping, in her father's words, like "a little wounded bird."[4]

When Mileva was eight years old, her sister, Zorka, was born, followed two years later by a brother, Milos. But Miza remained her father's favorite. Reveling in the role of patriarch, Milos liked to refer to his little band as *klipani,* or "hooligans." In return they dubbed him "Onkel Klipan."[5]

Milos was ambitious for his children and insisted that only German be spoken in his home, knowing that a command of that language was essential to success in the Austro-Hungarian world. Mileva was given German books of fairy tales to read, as well as Serbian folk poetry. Having fashioned himself a man of action rather than of words, Milos had never studied at university, but he placed a great deal of value on education. He liked to tell his children, "Learn as much as you want, I won't deny it you; but don't count on my getting to know your professors."

Soon after Mileva was born, Milos had resigned from the army and taken a civil service job as a court clerk in Ruma, on the other side of the Danube.[6] It was there, in 1882, that Mileva took her first steps on the long road to the Polytechnic. School was a bittersweet pleasure. The other children teased her and made fun of her deformity, and she was no good at games, leaving her feeling ugly and isolated. She liked learning, though, and her teachers were impressed with her quickness and her lively sense of fantasy and imagination. She seemed especially gifted at mathematics. "Pay attention to the child!" one of her teachers is reported to have told her father.[7] "It is an unusual phenomenon."

For the next twelve years Mileva zigged and zagged her way through the educational institutions of eastern Hungary, following her father in his various postings and her own academic star: the elementary *Volksschule* in Ruma; the Serbian Higher Girls' School in Novi Sad, where they nicknamed her "the saint"; the Kleine Real Schule in Sremska Mitrovica, near Ruma again; the Royal Serbian Gymnasium in Sabac; and finally the Royal Classical Gymnasium in Zagreb, the capital of Croatia, where Milos had been appointed to the High Court of Justice. At every school she was at the top of her class, making up for what she thought was a lack of feminine allure with cleverness and hard work. By now Mileva's interest in math was said to be insatiable. Judging by a reproduction of one of her drawings in the Serbian biography, she had also developed into a talented artist.

At every step of this journey Milos was his daughter's sturdiest and surest support, nudging her and advising her on schools, petitioning the Zagreb gymnasium to waive the Greek requirement from her admission, because she hadn't studied it before. (They made her take it anyway, and she breezed through the course.) Most significant, he used his pull in Zagreb to gain her the ultimate distinction for a woman in an Austro-Hungarian high school. As it happened, the Royal Classical had superb physics facilities—so good, in fact, that they later became the property of Zagreb University Medical School—but they were not available to girls. So once again Milos and Mileva applied to the authorities for an exception, and once again it was granted. Thus in 1894 Mileva Maric became one of the first women in the empire to sit in such a classroom next to boys. Although her final exams were delayed when she came down with a bad cold or tuberculosis, in the end she scored the highest math and physics grades in her class.

Her next stop was to be Switzerland. In Sabac, Mileva had become good friends with another ambitious Serbian girl, Ruzica Drazic, who suffered from tuberculosis. Ruzica wanted to go to Switzerland and study at the University of Zurich when she passed her *Matura,* partly because of the alleged beneficial effects of mountain air, and partly because the country was a liberal political oasis. Given her own recent bout with illness, Mileva decided to do the same.

At the end of the century, hypochondriacs, tuberculars, and melancholics were flocking by train and carriage from all over Europe up steep Swiss mountainsides to breathe the clean air, eat healthy food, bathe in hot springs, and gaze at unpolluted scenery in fashionable mountaintop spas. The whole country had taken on the atmosphere of a spa.

Even more important, moreover, Zurich in particular was known as an educational haven for women. For years European universities had allowed women to audit classes but not to enroll as regular students. The University of Zurich had begun admitting women in 1865, and in 1867 had been the first institution in all of Europe to grant a Ph.D. to a woman. After the university finally opened the door, women (and men) from less progressive countries, particularly those of Central and Eastern Europe, began to pour into Zurich. The student population, and the Russian contingent in particular, gained a certain amount of notoriety for being a hotbed of socialists and suffragettes, as well as a haven to the likes of Rosa Luxemburg, whose apartment on Universitätstrasse was a pole of the political underground. Switzerland was sometimes said to be the center of German Marxism.

Mixed in with these firebrands and freethinkers were a number of Balkan women who because they were apolitical went uncelebrated locally, but were heroines in their homelands. One of the earliest was Helene Druskowicz, born in 1856, who had written her doctoral dissertation on

Lord Byron's *Don Juan,* and was a rising academic star in literature and music. Another was Draga Ljocic, a Serb who had obtained her medical degree in Zurich and then performed heroically as a doctor in the ongoing Serbian wars.

Mileva moved to Zurich in November 1894, and after a couple of unremarkable years living and studying at the Higher Daughters' School, she graduated and passed the *Matura* exam, her ticket to the university, in the summer of 1896. The financial resources that would allow Mileva to pursue her schooling had never been in doubt. Milos had retired from the government in 1895, at the tender age of forty-nine, the same year he inherited The Spire, and entered the comfortable life of a country squire.

Mileva immediately enrolled in the University of Zurich medical school. She moved in with a family on Plattenstrasse, a tree-lined avenue of apartments and old houses near Baschigplatz. Her old friend Ruzica Drazic, from the Royal Serbian Gymnasium in Sabac, finally arrived in Zurich that same summer to study at the university. Ruzica had received a scholarship from the Serbian government, and moved into a boardinghouse up the street at 50 Plattenstrasse. Those lodgings, run by the Engelbrecht family, were a sort of mecca for Balkan and East European students, and Mileva spent much time there. Through Ruzica she got to know Milana Bota, a fellow Serb from Krusevac studying psychology, the sisters Helene and Adolfine Kaufler from Vienna, and the Croats Ruzica Saj and Ada Broch from Zagreb.

"You would never think that she had such a smart head," Milana wrote to her mother.[8] "She seems to be a very good girl, and very smart and serious; small, delicate, brunette, ugly." Among this group Mileva began stepping out from the role of the quiet, limping little ugly duckling that she always imagined herself to be. While to the world at large she was quiet, shy, or glum, reserved or snobbish, within the intimate company of her friends, her personality sometimes took a saucy turn. Once she kept them laughing all night with her imitations of a Bulgarian acquaintance. "She was very witty and knew how to make the whole group laugh irresistibly, while remaining serious herself," recalled Ida Kaufler.[9] "When she was pleased about something, a transient smile passed across her face. Her pleasure was joyful and infectious, and she was beautiful in those moments. Her eyes glowed with an inner fire. Her cheerfulness cast a shine on herself and everyone around her."

Mileva, Milana, and Ruzica Drazic became inseparable, sharing jokes and confidences as well as outsider status in the prudish Swiss society. In a picture from those times the three are identically coiffed, staring intensely off to the left, the taller and fairer Milana in back, Mileva and Ruzica, nearly twins, shoulder to shoulder in front. Their white dresses blend into the background while their faces glow like three blooms on the same plant. There were parties full of music and debate, and long walks into the nearby mountains. Frequently they were enlivened by Mileva singing and playing on her tamburitza. People often remarked on her lovely voice.

One summer of medical school turned out to be enough for Mileva, however, and she returned to her first loves, her enduring academic strengths in physics and mathematics. Across Ramistrasse from the hospital sat the pale-yellow hulk of the Federal Polytechnic, since 1855 the finest technical university in Central Europe. It was as if she had been following some invisible path of footprints from Ruma to Novi Sad to Sabac to Zagreb to Zurich to the left bank of the Limmat when she finally looked up to find herself at 101 Ramistrasse.

At the end of the summer she took a special entrance exam for the Polytechnic in mathematics and representational geometry and was accepted for the fall semester into the first-year course of section VIa, mathematics and physics. She was only the fifth woman to be admitted to study physics at the Polytechnic, the first having been a Norwegian, Marie Elisabeth Stephansen, in 1891. It went without saying that she was the only woman in her entering class of five, but for her there was nothing new in that.

But one of them was this irreverent little German. Romance was unknown to Mileva, and Albert's interest evidently caught her off guard. In the absence of letters or a diary, there is no way to know what Mileva was feeling as she rode back home across the Alps after her first year at the Polytechnic, but judging from what happened next, and according to the family tradition, she was confused; subterranean tides were tugging at the moorings of resolve that had served her so well.[10] For a young woman on the verge of a profession, love and romance could be a menace as much as a promise, a surefire way to derail years of purpose. Albert was not the only one worried about the charms of the bourgeoisie.

Back at The Spire, perhaps ensconced again in its little bell tower, she had much to tell her father, and to celebrate. She had survived the first-year mathematics gauntlet with fair to good grades. On the Polytechnic's scale of 1 to 6, Mileva scored 4.5 in most of her courses, her low point being a 3.5 in projective geometry. She also told Milos about Albert. Her father gave Mileva some tobacco to give to him.

In October Mileva headed west again, gifts in hand, but she didn't return to Zurich, and Albert was never to get his treat. Instead Mileva went to Heidelberg, the old university town on the Neckar River in southwestern Germany, known for its dueling societies and beer halls, where she checked into the Hotel Ritter. The Neckar Valley was famous for its beauty, but

Mileva found it enveloped in a pea-soup fog that limited visibility to the nearest tree trunk. No matter how much she strained her eyes as she wandered the riverbank and forests, all that she could see was as desolate and gray as infinity.

"I do not think that the structure of the human brain is to be blamed for man's inability to understand the concept of infinity," she wrote to Albert after he had tracked her down through her friends.[11] "He would certainly be able to understand it if, when young, and while developing his sense of perception, he were allowed to venture out into the universe rather than being cooped up on earth or, worse yet, confined within four walls in a provincial backwater. If someone can conceive of infinite happiness, he should be able to comprehend the infinity of space—I should think it much easier."

If Albert was disappointed by Mileva's sudden departure, he had tried not to let on. He had sent her a four-page letter with the admonition that she could write back to him if she grew bored. It was more than a month before she broke her silence (even though, as she pointed out, she still wasn't bored) to react to gossip from back in Zurich, to moon about the meaning of the foggy landscape, and to apologize for not bringing him the tobacco. The gift had been meant, she explained, to whet his appetite for their little land of brigands.

As for the gossip, she thought it fitting that a mutual friend who had just had his heart broken should go off to the woods and become a forester. "But it serves him right, why does he need to fall in love nowadays?"

Whatever Mileva was doing in those gloomy woods, it was not furthering her career. Though Heidelberg did have a distinguished faculty and venerable tradition, for Mileva it was in many respects a step down from the Polytechnic. Women weren't allowed to matriculate, but rather had to obtain permission from professors on a course-by-course basis in order to attend lectures.

As it happened, there was first-rate physics being done in the labs of the foggy Neckar Valley by Philipp Lenard, a brilliant experimentalist who had been the assistant and successor to Heinrich Hertz, the radio wave impresario, in Bonn. Lenard was an expert with one of the physicist's favorite toys, the cathode ray tube, which was simply a long glass tube, from which all or most of the air had been evacuated, with wires in each end. Experimenters had found that when they hooked the wires to a battery, the gas left in the tube could be made to glow. If the end of the tube opposite the negative terminal (the "cathode") was painted with a fluorescent compound, a small dot would glow as if some kind of invisible "cathode rays" were zinging down the tube. Lenard had a chip on his shoulder because he kept being beaten out—unfairly he thought—on the big discoveries. Roentgen, for example, had used a tube of Lenard's design to discover X rays, and had failed to credit him. Lenard was now involved in a race with several of his contemporaries to prove that the cathode rays in his tubes were actually particles of electricity, so-called electrons, implying that matter, in fact, was made of electricity. (His bitterness would increase the following year when J. J. Thomson, an English physicist at the Cavendish Laboratory in Cambridge, beat him to that discovery as well.)

In the meantime, Heidelberg's undergraduates had the privilege of hearing Lenard lecture straight from the cutting edge of science. In her letter, Mileva gave a charming report to Albert on Lenard's lecture about the kinetic theory of heat in gases, which ascribes heat to the motion of molecules. Lenard was an enthusiastic supporter of atoms. He began with a calculation demonstrating that oxygen molecules at room temperature move at speeds of 400 meters per second. "And after calculating and calculating, the good professor set up equations, differentiated, integrated, substituted, and finally showed that the molecules in question actually do move at such a velocity," she wrote, "but that they only move the distance of 1/100 of a hair's breadth."

A long silence now ensued between Albert and Mileva. As time went on, Albert became too embarrassed to write because he hadn't already written—or at least that was the excuse he eventually offered.[12] In the meantime he took his violin around to the usual parties, playing frequently for his new landlady, Frau Markwalder, her daughter, and her daughter's friends. In January Michele Besso married Anna Winteler; later that year they had a son, Vero.

Albert avoided Aarau. After her teaching stint, Marie Winteler had returned to the family household and was studying French and Italian with an eye toward getting a job as a governess.

Albert's own family was in sad shape after a brief burst of affluence. Although his father's factory had gone under yet again, Hermann was planning another go at the business of installing power plants. Watching from afar, Albert felt the pangs of his own failure two years earlier to dissuade his uncle Rudolf from financing this folly. Out of fear for their finances his poor parents had not had a happy moment for years, while Albert, an adult man, was powerless to help them. He was nothing but a burden.

"It would indeed be better if I were not alive at all," he told Maja.[13] "Only the thought that I have always done whatever lay within my modest powers, and that year in, year out I do not permit myself a single pleasure, a distraction save that which my studies offer me, sustains me and must sometimes protect me from despair."

Well, there was pleasure on the horizon. In February, with as little fan-fare and explanation as when she had abandoned the Polytechnic, Mileva announced that she was coming back. If she had indeed been trying to es-cape Albert, it turned out to be a very halfhearted attempt. Albert seized the opportunity to resume the broken correspondence. He pointed out that she had a lot of catching up to do, but he offered to let her use "our" notes, by which he presumably meant his and Grossmann's. In time, he advised, she could probably even move back into her old boardinghouse, run by Frau Bächtold. For now, however, a "certain Zurich philistine" had taken over her room. "It serves you right, you little runaway!"

By April Mileva was back in Zurich, her six-month experiment in apartness over. Perhaps, consciously or not, she was willing at last to chance the long slow process of surrendering her autonomy, a stubborn planet sighing at the first recognition of gravity and settling imperceptibly closer to an orbit around the sun.

After a short stay at her old boardinghouse, Mileva found a room on the fourth floor of the Pension Engelbrecht, one flight up from her friends Mi-lana, Helene, and the rest, whose letters home detailed Mileva's return as excitedly as if she were exiled royalty.

Mileva resumed her spartan work habits, studying all night with the win-dow open. Her friends often found her awake, still reading, at dawn. She would sleep for an hour in her coat, then grab breakfast and climb the steep slope up to the Polytechnic.

As Albert had warned her, she had missed a great deal, and in the com-ing months she would often lean on him for catching up. She was arriving back halfway through the main event of a young prospective physicist's ed-ucation: Heinrich Weber's exhaustive two-semester course entitled simply "Physics."

Heinrich Friedrich Weber stalked his physics classroom wearing a Calvinist visage worthy of the Zwingli in whose city he taught, brows knit-ted in a perpetual frown, eyes hooded, long nose standing like a rectitudi-nous prow above his straight dark beard. Born in what was now Germany, he had worked under some of the great masters of nineteenth-century German physics, including Gustav Kirchhoff and Hermann von Helmholtz. Weber was an expert on the thermal and electrical properties of matter, but his second-year course covered the entire range of nineteenth-century physics in rigorous mathematical fashion, from the nature of heat to ther-modynamics, gases, and chemical physics to electricity.

Impressed with Einstein's high science and math scores on his first en-trance exam, Weber had been one of Albert's early supporters at the Poly-technic. The respect was mutual. Albert had waited half his life to dig into the real meat of what would soon constitute twentieth-century physics. He was ready to immerse himself in the details and complexities of mysteries that he had only been able to encounter superficially in newspapers and cafe bull sessions up till now: the existence of atoms, their role in heat and thermodynamics, the properties of the aether, and the nature of light.

For the first half of the year Albert had been mesmerized by the class and looked forward to each new lecture. As Weber laid out the classical theories of heat, gases, and electricity in dense mathematical quantitative fashion, Albert took notes copiously and faithfully:[14]

"The earth in its totality can be considered as a silicate sphere which radiates heat into the low-temperature universe. The insulating effect of the atmosphere can be neglected as small. Hence, assuming that we apply the case of our sphere to the earth, i.e., if we also neglect the heat produced (by contraction and chemical processes), we can state the minimum time necessary to cool the earth by one percent."

Weber had lectured, he told Mileva, "with great mastery." Her return, however, seemed to coincide with a sudden disenchantment, as he realized that Weber wasn't going to cover contemporary topics like statistical mechanics or electromagnetic waves in his lectures. It made him angry, and he complained bitterly to Grossmann, Ehrat, and the others that Weber was more interested in the past than in the present and the future. Weber was a classicist in love with old verities and methods; he apparently figured that his students could get the more modern stuff out of books.

Albert began to skip Weber's lectures in order to stay home and study on his own. He recruited Mileva to be his accomplice and studymate, sweeping her up into his cafe-style traversals of Mach, Boltzmann, Helmholtz, and the other contending masters of modern physics who seemed to be getting short shrift from the professors at the Polytechnic.

They began by passing back and forth a textbook on electromagnetism and the aether by the German physicist Paul Drude.[15] Several of Albert's first letters to Mileva that spring were on sheets torn from the notebook he was using to learn electromagnetic wave theory, their backs covered with equations in his cramped, pointy handwriting. Formally addressed to "Dear Fräulein Maric," they were brief and businesslike. "When I came home just now, I found the apartment locked with no one at home, and had to beat a shameful retreat," he wrote in one note. "So don't be angry at me for taking Drude in this emergency in order to do some studying."

Later that spring he fell ill and was confined to his room, too weak to walk, for several days. By the time he emerged long enough to ask for visitors, he was halfway through the book and complaining that its concepts lacked a certain clarity and precision.

Soon the borrowing of notes and sharing of texts and other books had grown to near-nightly visits to 50 Plattenstrasse. Mileva also naturally came under the spell of Albert's expressive violin. She started bringing Albert around to meet her friends.

Trouble started almost immediately. Milana was excited at first to meet this "German," Einstein, who was handsome and clever and played the violin so beautifully. He was, she thought, some kind of artist, and she looked forward to playing music with him herself. That notion was soon squelched by Mileva, however, who didn't want Albert fiddling around, so to speak, in any way with other women. Milana was upset, and though she and Mileva reconciled, the incident left her less than enamored of the little German. He was, Milana thought, not good for Mileva.

Some of Mileva's other friends also had a negative reaction to the relationship. They didn't care for Albert's appearance. His hair was long and often uncombed. His clothes were often on the disheveled side, misbuttoned, and his shoelaces untied. They sometimes avoided being seen with him and Mileva in public.

Then there were the strange occasions on which he would go into some sort of trance or seizure, as if he had just disappeared into his own world. Later he would claim to have no recollection of what had happened. It was, in fact, familiar behavior. Throughout his entire life the people around him had been remarking on his ability to suddenly withdraw from even the most raucous surroundings to concentrate on some thought of his own. His coworkers at the Einstein brothers' factory in Pavia had noticed it. To the outside world and even to himself Albert looked a little unworldly.[16] People tended to tease him at first about these *Anfälle,* or fits, but later paid them no mind.

Albert's friends were likewise not particularly impressed by the unassuming Mileva, who struck them as laconic and gloomy, not to mention lame. "I should never have the courage to marry a woman unless she were absolutely sound," one of his classmates commented.[17]

"But she has a lovely voice," Albert answered.

To Mileva, Albert's casual dishevelment was probably part of his charm. She was four years older than he at an age when that distance feels like a generation. (He was eighteen, she twenty-two.) He brought out maternal instincts that were already well honed, for she had taken care of her younger sister, and was now teaching her pension mates how to sew. For his part, did Albert discern shades of his mother in the exasperated strictness with which she stamped her foot at his apparent indecisiveness and irresponsibility? He seems to have felt at home being chastised by her. From the start his letters are full of references to scoldings, real, imagined, or (almost pleasurably) anticipated. It was as if within the boundaries of her stern love he was free to be a naughty child.

Albert: "There was this old whore—"

Mileva: *"Albert!"*

He was, of course, not a child, but a man of the world who had been on his own for three years and had dropped out of a school (the Luitpold), a country (Germany), and a relationship (Marie). And she was not his mother, but a physicist and freethinker like him, and she didn't say stupid things. She could complete a thought in one sentence, unlike the distracted but sweet Marie. "We shall remain students as long as we live," he wrote her once, "and we shall not give a shit about the world"[18] It would be their anthem if he had his way.

4

THE CHESHIRE'S GRIN

IF THERE WAS ONE SUBJECT ABOVE ALL THAT WAS BURNING A HOLE IN Albert's imagination, it was James Clerk Maxwell's new electromagnetic wave theory of light. Actually, the theory wasn't particularly new—Maxwell had scribed his famous equations more than thirty years earlier—but time had only deepened their mystery. The meaning and import of those equations, their possibilities, implications, complications, and contradictions, had come to define the frontier in physics, everything that was exciting and up for grabs. Anyone who was ambitious was grappling with electrodynamics, trying to tease out the secrets of light and radiation and their relationship with matter.

The nature of light was another of those thematic controversies that had polarized science and philosophy for hundreds of years. In the nineteenth century the quest took the form of a debate about whether light was composed of particles or of waves. The great and mechanistically inclined Newton had suggested in his book *Opticks* that light rays consisted of streams of tiny particles, corpuscles of aether. Presumably, they were *very* tiny. Bernard Nieuwentijdt, in his book *The Religious Philosopher* (1718), earnestly calculated that a candle flame should emit roughly $4 \times 10^{44*}$ of these light particles every second.[1]

René Descartes, the French scientist and thinker, took the opposite tack: that light rays were some kind of disturbance—"a tendency to propagate"—in what he called the *plenum,* an invisible fluid that filled the voids in the universe. Leonhard Euler, an eighteenth-century German mathematical physicist, took Descartes a step further. Arguing from an analogy with sound waves, Euler described light as vibrations or waves in Aristotle's aether, which pervaded not only the cracks and voids of the universe but even material bodies. One of the features of the wave theory was that in the aether, as in the bathtub or the ocean, waves could combine either to en-

*Or 400 million trillion trillion trillion.

hance or to destroy each other, depending on whether their crests lined up with other crests or with troughs. When Thomas Young, in England, and Augustin Fresnel, in France, succeeded in showing that various optical phenomena could be explained as light waves interfering with each other, the wave picture seemed to be ascendant. But what were these waves? What exactly was waving?

Enter in the 1860s Maxwell, with his bushy white beard, tufted temples, and gentle, haunted eyes. Like Euler, he was one of a new breed of mathematical physicists who sought to use sophisticated mathematics to describe the behavior of physical systems. Often this model building could be done without understanding the underlying basis of the phenomena at hand. Newton, for example, could arrive at his famous equations describing the gravitational forces that controlled the falling of apples and the swinging of the planets while admitting that he had no idea how one mass could reach out and touch another with this force. "Action at a distance," it was simply called. Similarly, equations describing the spread and transfer of heat had been successfully formulated without scientists' knowing what heat itself was—a fluid, some kind of energy, jostling atoms, or divine perspiration.

The same could be said of electricity and magnetism. Technological entrepreneurs like the Einstein brothers had been using electricity for decades to send telegrams, light cities, and run streetcars and telephones, while having no knowledge of what really happens when "electrical" current runs through a wire. Was electricity a fluid, a slackening or release of some kind of tension or "polarization" of the aether, the running back and forth of positive and negative electric "masses"? Were there correspondingly magnetic "masses"?

The seminal discovery in electromagnetism, that a moving magnet could induce electric currents, had been made early in the nineteenth century by Michael Faraday, a blacksmith's son with little education. It was Faraday, an experimental genius, who first sprinkled iron filings on a piece of paper over a magnet and saw that they lined up in an arclike pattern from pole to pole. He coined the word "field" to describe that pattern. What was a field? It was just a metaphor for an invisible process, the process by which magnetic charges spoke to each other. A field was what you saw with Albert's famous compass, a condition of the aether that caused the needle placed at any place in it to point to magnetic north.

Maxwell had set out to put electromagnetism on a firm mathematical foundation and wound up making one of the boldest predictions in the history of physics. He derived four basic equations describing the behavior of electric and magnetic fields and currents. He found that by making certain assumptions he could combine these formulas into a single equation that had the mathematical form of a wave composed of oscillating electric and magnetic fields. The speed of this wave was specified by a constant, c, which turned out to have the numerical value of about 300,000 kilometers per second. That was exactly the speed of light rays that had been determined in the previous century by timing the eclipses of Jupiter's moons. The inescapable conclusion was that light rays *were* electromagnetic oscillations of a particular frequency and that other, invisible, forms of electromagnetic radiation should exist at other frequencies.

It was more than twenty years before Maxwell's conjecture was verified. In 1888, after months of attempts, the German physicist and engineer Heinrich Hertz used a spark jumping across one gap to create an electromagnetic disturbance that could somehow cross his laboratory and induce a spark to jump a second gap, demonstrating the existence of what he called radio waves. When Hertz was asked once what radio waves were, he replied, in effect, that they were a term in Maxwell's equations.[2]

This was a watershed event in the history of physics: discovery by pen and paper. Hertz hadn't *discovered* radio waves so much as he had simply *demonstrated* them, like a ringmaster dragging Maxwell's creatures into the spotlight. It was the theory itself, the mathematical machinery of deduction, that had in fact invented radio waves.

Maxwell's achievement defined a new style of research that appealed to prospective Young Turks like Albert and Mileva: Scour your habits and assumptions with doubt and cold analysis; create mathematical music; follow that siren wherever it led in search of the deeper harmonies; trust simplicity.

Albert already had a glancing acquaintance with Maxwell's theory from reading popular books, and it had confused him. During his year of reading and wandering in Italy, he had dwelled upon the subject of light waves and the aether from time to time, most presciently when he had tried to imagine what he would see if he were riding along on a light beam. One might imagine that he would see a frozen light wave, the way a surfer going along on a wave sees himself perched on a mound of water. Albert was surfing not on the water, however, but on the aether, and Maxwell's equations made a specific prediction. Light, Maxwell said, was a dance of electric and magnetic fields intertwined; according to the equations, the dance could not be frozen, therefore a cosmic surfer traveling along a light beam at the speed of light would see the light waves—those dancing fields he was traveling along with—speeding by him at the speed of light. And that seemed impossible, a paradox.

As Albert and Mileva strove to lower themselves into these mysteries under the guidance of, first, Drude's little book and, later, others, Weber noticed Einstein's absence from his classes and took offense. Looming over the pair of students the following fall was the prospect of intermediate exams, usually taken halfway through the course of study. Passage of the intermediate exam was the price of admission to the final diploma exam at the end of four years.

Despite the growing tension, both Einstein and Maric managed to get good grades from the starchy Weber that spring. In the fall of 1898 Albert passed the intermediate exams for his diploma with an average of 5.7 out of 6.0, the highest in his class, while Mileva, still catching up from her excursion in Heidelberg, elected to postpone her exams for a year.

During that year, the relation between Einstein and Weber began to deteriorate. Albert's attitude began to make his presence in Weber's classroom only barely more tolerable than his absence. As he had done at the Luitpold, Albert mocked his teacher by referring to him as "Herr Weber" instead of the standard honorific "Herr Professor."

"You're a very clever boy, Einstein," Weber grumbled, "an extremely clever boy, but you have one great fault; you'll never let yourself be told anything."

In addition to several more offerings from the grouchy Weber, the third year featured a class in experimental physics taught by the department's other professor, Jean Pernet. Albert didn't show up much for that one either. Pernet became so enraged that he flunked Albert and complained to the director of the Polytechnic, who duly issued Einstein an official reprimand for "nondiligence."

THE EINSTEIN FAMILY was faring a little better. In February of 1899 Albert's father won the concession to operate and install a lighting system in Canneto sull'Oglio, a small town in southeastern Lombardy. Perhaps the worst was over for his parents, a relieved Albert wrote to Maja. It seemed to him that he and his family had already been through enough melodrama for a lifetime. If everybody lived like him, he joked, novels would never have been invented.

It was only a month later, while Albert was in Milan on spring holiday, that his mother got her first look at the next great worry in her life, when Albert showed her a picture of Mileva. If Pauline was unhappy to see Mileva's swarthy face instead of Marie's smiling out at her, she didn't say so. Pauline simply stared, immersed in contemplation of the image, her thoughts opaque to her nervous son, while Albert, striving for empathy, cooed in her ear, "Oh yes, she is a clever creature."

Finally, Pauline told her son to convey "greetings."

The letter recounting this adventure seems to have marked a new phase in at least the recorded version of Albert and Mileva's relationship.[3] Letters that had been up till then a series of rather impersonal announcements and reflections now took a personal turn. Albert recalled sitting next to Mileva at a dinner party at Alfred Stern's house and having stomach pains from eating too much. "It was then revealed to me in harsh tints how closely knit our psychic and physiological lives are." That was also the first letter in which he addressed Mileva as other than "Fräulein Maric."[4]

Albert returned to Zurich with the teasing of his sister and relatives ringing in his ears, and his mind firmly fixed to a higher sphere. Mileva's room at 50 Plattenstrasse became their "household," stocked with goodies sent directly from Albert's mother and devoted to drinking coffee, eating sausage, and studying Maxwell's mysteries. "My broodings about radiation are starting to get on somewhat firmer ground," Albert told Mileva, "and I am curious myself whether something will come out of them."

For a guide into the electromagnetic wilderness, Albert and Mileva had traded in Drude for Heinrich Hertz, the discoverer of radio waves. Hertz was one of a group of German-speaking theorists, which included Helmholtz and Boltzmann, regarded in some circles as "mathematical terrorists." Their goal was ambitious: the mathematization and unification of physics. The triumphant enshrinement of Newton's laws as the explanation of all phenomena beckoned if they could bridge the schism between matter and waves, and explain how Maxwell's electromagnetic waves were propagated. If light was a wave, what was doing the waving? Sound waves, for example, were carried by air—no air and there was no sound. What sustained a light wave between the time it left some star light-years away until it hit our eyes on earth? The obvious answer was the one Euler had given: the luminiferous aether. "There can be no doubt," Maxwell wrote in 1878, "that the interplanetary and interstellar spaces are not empty but are occupied by a material substance or body, which is certainly the largest, and probably the most uniform, body of which we have any knowledge."

Aether had been a feature of natural philosophy since Aristotle named it as the fifth of the four elements, the substance of heaven, the formless underlayment of the rest of the physical universe. God's breath was what was left when everything else was taken away. It was the void that was not a void, the Ground of Being. Other aethers had been invented or proposed in the intervening centuries to explain various aspects of nature. In its most common incarnation the aether was virtually omnipresent, permeating every atom and suffusing every void, and it could be thought of in a way as the manifestation of Newton's absolute space itself, the elusive stage on which existence sat. As a concept the notion of an aether was of priceless metaphysical and scientific value, so useful, in fact, that by Maxwell's time, as the bushy genius once drily commented, "all space had been filled three or four times over with aethers."[5]

The positing of the aether as the seat of electrical and magnetic disturbances took God's breath out of the realm of the metaphysical and into the grubby physical realm of measurement and mechanics. Physics, as always, was reaching for territory that had belonged to the spirit. The aether was a vital missing link in physics. Understanding it, quantifying its properties and motions, was the key to the Maxwell-Helmholtzian dream of unification.

Venturing forth into the unknown by what he called the "method of analogies," Maxwell and his followers modeled the aether mathematically as a kind of fluid, using the equations that had been developed to describe the mechanical properties of real fluids. This approach tended to deflate the poetic visions that had held sway for the previous two millennia. Comparing the speed of light to the speed of sound, Euler calculated that the aether must be 100 million times less dense than air and a thousand times more elastic—qualities that made it more resilient than steel and much less palpable than air, but not entirely *impalpable.*[6] Maxwell, for example, thought that such an aether might exert a drag on the orbital motion of the earth around the sun, but could find no evidence for this drag.

Among those who took up the pursuit of aether in tangible form was a young American physicist named Albert Michelson, who worked in Helmholtz's laboratory at the University of Berlin. Michelson was an expert on measurements of the velocity of light. He proposed to detect the "aether wind" (or aether drift) caused by the earth's motion through the aether by measuring the speed of light in two different directions. Light rays, he reasoned, should appear to travel more slowly "upstream" against the aether wind, more quickly downstream, and at the classic velocity, c, in the transverse direction. His first experiments, in 1881, found no evidence for an aether wind—results that met with a certain amount of puzzlement and criticism from the graybeards of Europe. Piqued that his work hadn't gotten the attention he thought it deserved, Michelson went home to the Case School of Applied Science in Cleveland, where he and a colleague from nearby Western Reserve University, Edward Morley, endeavored to repeat the experiment with more sensitive equipment. Despite a nervous breakdown in 1885 occasioned by a seriously strained marriage, he and Morley managed during the trial run of their new experiment to carry out delicate measurements showing that light of a given color always has the same wavelength and thus can be used as a standard of length. For this discovery Michelson would become the first American to win the Nobel Prize, in 1907.[7] They failed once again, however, to find any evidence of an aether drift, and their experiments were eventually cut short by Michelson's personal problems. In 1887, on the eve of a major presentation of his and Morley's work, his housekeeper accused him of pressuring her for sexual favors and demanded money. He accused her of blackmail and had her arrested. The charges and countercharges were eventually settled and dropped, but Michelson's marriage went downhill and he never recovered his scientific stride.

By the turn of the century so-called aether-drift experiments had become a cottage industry. One experiment was even set up on a hilltop to counter the possibility that the aether in Michelson's basement laboratory was trapped and carried along with the earth. But nobody was finding any evidence of the aether wind.

In the book Albert and Mileva were studying, Hertz had tried to account for these failures by constructing a theory of what he called the "electrodynamics of moving bodies." Perhaps the luminiferous aether, Hertz argued, could be dragged along *with* moving bodies like the earth, in the same way that the atmosphere travels with the planet. In that case, the aether would appear to us on earth to be as still as a Bethlehem night, whether viewed from the bottom of Michelson's basement or from the top of the Alps. Consequently, light in the terrestrial vicinity would appear to travel in all directions at the same speed.

Unfortunately, Hertz's theory ran afoul of certain established features of light. Among them was one demonstrated in an experiment performed by a French scientist, Armand Hyppolyte Louis Fizeau, which seemed to show that light traveling along in a moving stream of water did experience a drag from the apparently nonmoving aether.[8] In Paris the great Henri Poincaré said of that experiment, "One believes one can touch the aether with one's fingers."[9]

Albert himself was already an old aether hand. During the summer of 1895, the same summer in which he had tried to envision riding on a light beam, he had written an essay in which he described magnetic fields as a deformation or strain in the aether. Although the essay was long on ideas and short on mathematics, Einstein knew enough physics to be aware that such stresses would affect the propagation of light waves, changing their velocity—an effect that could perhaps be measured. "The most interesting, and also most subtle case would be the direct experimental investigation of the magnetic field formed around an electric current," he wrote, "because the exploration of the elastic state of the aether in this case would permit us a look into the enigmatic nature of electric current."[10]

Now, four years later, Einstein admired Hertz but had grave doubts about this business of a moving aether, if not the aether itself. He resolved to take up the subject intensively during the upcoming summer vacation, using a book by Hertz's mentor, the terrorist Helmholtz.

By now Albert's professors and even some of his classmates barely knew what to do with him. Mileva often had to defend Albert from their friends' assertions that he was going off the deep end. "What he maintains he can also prove."[11]

Pernet was particularly incensed by Einstein's independent ways. According to Weber's assistant, Josef Sauter, Albert would routinely toss the chits on which Pernet had written lab instructions into the wastebasket and proceed to solve the problem on his own, in his own way. "What do you make of Einstein?" Pernet demanded furiously of his own assistant, Schaufelberger. "He always does something different from what I have ordered."[12]

The assistant replied judiciously, "He does indeed, Herr Professor, but his solutions are right and the methods he uses are always of great interest."

Occasionally Albert paid the price for his willfulness. In July of 1899 he made a mistake during an experiment in Pernet's lab, which led to an explosion and left Albert with stitches in his hand.

Pernet took him aside and gently tried to counsel him. "There is no lack of eagerness and goodwill in your work, but a lack of capability," Pernet said, suggesting that Albert had no idea how hard the physics life would be. "Why don't you study medicine, law or philology instead?"

"Why shouldn't I at least try physics?" Albert replied, contending that he at least felt he had some talent.

"You can do what you like, young man. I only wanted to warn you in your own interests."

Albert's injury forced him to set aside his violin for a few weeks, and, feeling the loss acutely, he imagined the instrument wondering why it never got to leave its case.

"It probably thinks it has gotten a stepfather," he mused to Julia Niggli, his musical friend from Aarau, who had written inviting him to visit her the coming summer.[13]

Albert, in fact, had a busy summer planned, and Julia fit right in. Hoping to escape the heat of Milan, his mother had rented rooms for the season at an inn in the town of Mettmenstetten, in the countryside outside Zurich. Albert was thus able to remain near the city's social swirl, though he still planned later in the summer to spend a couple of weeks down the road in Aarau doing research with his old science teacher Conrad Wüest. Albert wrote back somewhat leeringly to Julia that while he could think of plenty of pleasant ways to help her pass the time in her "enchanting little nest," his attendance was required in Mettmenstetten, and he scandalized her by inviting her there.

Meanwhile, Mileva was packing her bags to go home to Kac. She too had a busy time ahead of her, cramming for the intermediate exams she had postponed from the previous fall. She was nervous, however, for although her grades didn't show it, she had never recovered academically from her semester away. Geometry, in which she had gotten the lowest grades, was giving her an especially bad time. Spending so much time in the company of the confident Einstein, moreover, following him through the labyrinth of electromagnetic theory, was bound to undermine her own self-confidence.

AT FIRST OR EVEN second glance, Mettmenstetten is an unlikely-looking site for a resort. A dusty farming town, lacking even the traditional picturesque *Platz* at its center, it lies amid cornfields in a shallow valley about a half-hour train ride south of Zurich on the way to Lucerne. To the east, over a range of small mountains, lie the long, meandering Zürichsee and beyond that the Appenzell region and the towering brute Säntis, where Albert had almost been killed hiking three years earlier. The Hotel-Pension Paradise sat on the hillside above the town, overlooking it and, on a clear day, the distant Alps.

"Here in Paradise I live a very quiet, nice, philistine life with my mother-hen and sister," Albert wrote Mileva, "exactly the way the pious and righteous of the world imagine Paradise to be."[14] Within the first two weeks he had already hiked to the Zürichsee with the Paradise innkeeper Robert Markstaller and had twice been to the Säntis. What a treat to roam outdoors again, clambering across the countryside and up and down mountains. The weather was perfect. There was hardly a disagreeable moment, except for when his mother's friends visited for some empty chatter. The worst was his aunt Julie from Genoa, "a veritable monster of arrogance and dull-witted formalism."

Albert immediately began reading the Helmholtz book, out of fear that Mileva would scold him for being lazy. He promised to reread it with her. "When I was reading Helmholtz for the first time, it seemed inconceivable that you were not with me and now it's not much better."

It was amazing how just a few hours in the company of his family could sharpen his appreciation for Mileva, his *Doxerl* ("little doll"), as he called her, and their unphilistine life together. From afar Albert was the perfect lover. How he pitied and admired her, eating book dust and stuffing her fine head with gray theory. At least she was home and being properly pampered. No doubt she would prevail. "You are really great and have much vitality and health in your small little body," he wrote. "If only you would be again with me a little! We both understand each other's black souls so well and also drinking coffee and eating sausages etc."[15]

His readings were making Albert more and more skeptical about the aether. The problem was that the aether was like a Cheshire cat. In Maxwell's equations it vanished, leaving only its grin—the light waves. Whatever was waving had disappeared, at least mathematically. "I am more and more convinced that the electrodynamics of moving bodies as presented today doesn't correspond to reality, and that it will be possible to present it in a simpler way," he wrote to Mileva.[16] "The introduction of the term 'ether' into theories of electricity has led to the conception of a medium whose motion can be described, without, I believe, being able to ascribe physical meaning to it." Just because people could talk about the aether didn't mean that it or its hypothetical motions had any physical meaning. The antics of the aether, Albert was beginning to suspect, were in the same category as fairies dancing on the head of a pin, and an unnecessary complication in electromagnetic theory.

Everywhere in the world, after all, unnecessary complications and adornments were being thrown out, as if Mach had written a prescription for all of Western civilization.

Striding the trails of Appenzell, Albert wondered whether Helmholtz and the others might be working too hard. "I think that electrical forces can be directly defined only for empty space," he explained. "Further, electrical current will have to be thought of not as 'the disappearance of electrical polarization over time,' but as the motion of true electrical masses." And here he inserted part of a differential equation. "Electrodynamics would then be the theory of the movements of moving electricities and magnetisms in empty space." The answer, he went on, would have to come from experiments, perhaps the very ones he was about to perform in Aarau with Wüest. Helmholtz and the other giants of Continental physics slept unaware that they should be trembling in their smug aether beds. For the moment he was only a college student shooting his mouth off in the mountains.

In the evenings there was diversion of another sort, which he did not see fit to share with Mileva. Anneli Schmid, the seventeen-year-old daughter of an Appenzeller family from Herisau, was the sister-in-law of the Paradise innkeeper, and came to Mettmenstetten every summer. In the present-day house of Charles Schaerer, Anneli's son-in-law, there hangs an oil painting that shows her as a beautiful young woman with long reddish-blond hair and blue eyes, looking radiant in a garland wreath and holding a flower. One of her more prominent features was a high forehead, which gave her an intellectual look when she wore glasses. In later life, said Schaerer, she was very proud of that forehead because she thought that it resembled Einstein's. She was a happy, outgoing woman, he said, who enjoyed life.

In the summer of 1899 Anneli fell under Albert's spell, and Albert, it would seem, under hers. There was, at the very least, dancing and violin playing in the evenings at Paradise.

Albert inscribed this poem in Anneli's daybook, a kind of notebook calendar festooned with inspirational sayings:

> You girl small and fine
> What should I inscribe for you here?
> I could think of many a thing
> Including also a kiss
> On the tiny little mouth
>
> If you're angry about it
> Do not start to cry
> The best punishment is—
> To give me one too.
>
> This little greeting is
> In remembrance of your rascally little friend.[17]

At the end of August Albert left Mettmenstetten for Aarau to work with Conrad Wüest, who was interested in X rays. Whatever they did together those two weeks has been lost to history. Also waiting for him in Aarau was Julia Niggli, who needed both a musical partner and a father confessor. Dissatisfied with her provincial life, Niggli was eager to go abroad and was looking for a job. Albert offered to fix her up with his aunt Julie (the "monster of arrogance") from Genoa, who needed a governess for her daughter. Eventually Niggli took another job instead.

Meanwhile, Julia was agonizing over her relationship with an older man who wanted to marry her, a situation that left Albert somewhat incredulous.[18] "What a strange thing must be a girl's soul! Do you really believe that you could find permanent happiness through others, even if this be the one and only beloved man? I know this sort of animal personally, from my own experience as I am one of them myself. Not too much should be expected from them, this I know quite exactly. Today we are sullen, tomorrow high-spirited, after tomorrow cold, then again irritated and half-sick of life—and so it goes—but I have almost forgotten the unfaithfulness and ingratitude and selfishness, things in which almost all of us do significantly better than the good girls."

In the midst of this social whirl Mileva's voice sounded plaintively from Kac, where she was stuck, out in the country, because of outbreaks of scarlet fever and diphtheria, worrying that she had set a bad example by not writing enough and that was why she was now not hearing from him.[19] And she missed his letters. "Our series of shared experiences has secretly given me a strange feeling that is evoked at the slightest touch, without necessarily conjuring up the memory of a particular moment, and makes me feel, so it seems, as if I were in my room once again."

She went on to ask Albert for advice about what she could expect on the exams. Would the questions be general or specific? Could he leave his notes on Weber's heat lectures with his landlady so she could borrow them when she returned, with foreboding, in the last week of September? "Don't you feel sorry for me?" she asked pitifully, noting also that he had not bothered to tell her when *he* would be going back to Zurich.

Then, as if to underscore what pressure and uncertainty and insecurity she was really going through, as if seized by a dread thought, Mileva suddenly blurted her fear of being betrayed, perhaps to Albert's mother. "I hope you aren't letting anyone read my letters; you must give me your word," she begged. "You said once you don't like disrespect, and if I feel that this is being disrespectful, you can do it for me! What do you think?"

"You poor thing must now be cramming really awfully!" Albert responded. "If only I could help you a bit, be it merely to bring you some variety, or be it in the studies, or be it as Johann with all the lovely trifles that go with that." It would all be over soon, he assured her. But alas, he couldn't be there with her, for he had to accompany his mother home to Milan. "I would like so much to try to make the examination time more pleasant for you in Zurich if this wouldn't cause very understandable pain to my parents."[20]

He was writing from Paradise, where he had rejoined his family after his stay in Aarau. On the way back he had stopped in Zurich, where he had promised to fetch his notebooks for Mileva but instead had gone to see Weber with an idea for an experiment that would investigate the speed of light in objects moving with respect to the aether, in order to see if that speed changed.[21]

The idea, which had come to him while he was in Aarau working with Wüest, was to split a light ray into two beams, using mirrors. One beam would travel in the direction of the earth's motion through space and the other would travel in the opposite direction, and then a thermoelectric device would be used to detect any difference in the energies of the two beams. "Also a theory on this matter occurred to me, which seems to me to be highly probable," he explained to Mileva. "But enough of that! Your poor little head is full enough of different people's hobby-horses that you had to ride."

This experiment should have been to Weber's liking, for although no one was going to accuse him of being a mathematical terrorist, complex experiments were his specialty. But when Albert mentioned it to him, Weber dismissed the whole thing out of hand. The professor wasn't interested. Thus the aether went unmeasured in Zurich in 1899.

But not everywhere. Weber referred Einstein to a paper about aether theories written by the physicist Wilhelm Wien at the University of Giessen in Aachen, Germany, in 1898. In that paper Wien surveyed the results of thirteen different experiments that had been conducted attempting to detect the motion of the earth through the aether, ten of which, including the Michelson-Morley effort, had failed. It was the first Albert had heard of the Michelson-Morley work, and back in Milan he excitedly wrote to Wien about his own ideas, expecting to hear back from the great man at the Polytechnic. Wien, however, never responded, and Albert's letter to him has never been found.

"If you see there a letter addressed to me, you may take it and open it," he advised Mileva.[22] "I'll be back 'home' about the 15th. I look forward to it with great joy, because our place is still the nicest and coziest after all." First, he promised, they would climb the Ütliberg and unpack his memories of the Säntis. Then they would finally get going on Helmholtz's electromagnetic theory of light, "which 1) out of fear 2) because I did not have it, I still have not read."

But first he had to deliver Maja to the Winteler home in Aarau, where she was to start school that fall, following in her brother's footsteps with the Wintelers. He promised Mileva not to stay in Aarau this time. It seemed that Marie was going to be in town, and after all these years he was still terrified of her—or of himself. "Don't worry about my going to Aarau so often now that the critical daughter with whom I was so madly in love four years ago is coming back home. For the most part, I feel quite secure in my high fortress of calm. But I know that if I saw her a few more times, I would certainly go mad. Of that I am certain, and I fear it like fire."[23]

Alone in her den on Plattenstrasse, and so charmingly reassured, Mileva bent her dark head to the books.

5

FAMILY VALUES

SOMETIME IN 1900:

My dear Johnnie,
Because I like you so much, and because you're so far away that I can't give you a little kiss, I'm writing this letter to ask if you like me as much as I do you? Answer me *immediately.*

A thousand kisses from your Dollie[1]

At the end of July 1900, Albert, twenty-one and armed with a fresh diploma, stood nose to nose with his mother and with his future. They were high in the heart of the Swiss Alps in a hotel of subversive luxury. Outside the window, jagged snowcapped peaks walled a valley of flower-strewn meadows and rushing streams. The women of the family had plucked him from the train with kisses befitting a hero. Nevertheless it was with distinct foreboding that Albert found himself summoned to his mother's room in order to account for his relationship to Mileva.

"So," Pauline asked, "what will become of Dollie?"

The question was hardly as innocent as Pauline tried to make it sound. Albert and Mileva's intimacy had increased upward by a couple of notches during the previous year, at least judging by their correspondence. While Albert was in Mettmenstetten he had taken to addressing Mileva as Doxerl ("little doll"). Early the next year, she began addressing him as Johonzel, or "Johnnie." It became another of the personas he affected, joining company with the sensitive temperamental violinist, the loquacious cafe hound, the dog-eared poet, all of whom ringed his psyche like wagons. At about the same time the couple started using the intimate *"Du,"* analogous to the archaic "thou" in English, to refer to each other, instead of the more formal *"Sie."* It was a form of address restricted to very close family, and throughout his entire life Albert used it outside his family on a regular basis only to his friend Michele Besso. By the end of the school year Albert and Mileva were talking about marriage.

At first Albert's mother had tolerated Mileva with good cheer, keeping a quiet counsel about whatever doubts or misgivings she might have about her son's behavior and her lingering disappointment about Marie. She even acknowledged their "household" by sending packages of food to the Plattenstrasse address. For his birthday in March, Albert had received a box of "sweet delicacies," and Mileva reported that he had walked beaming down Plattenstrasse holding the box with both hands, oblivious of everybody else on the street—eloquent testimony to his sweet tooth, or of his devotion to his mother, or both.

Perhaps Pauline was hoping he would outgrow Mileva. If Albertli wanted to dally with a Serbian woman while he finished his studies at the university, that was his business, but she bridled at any suggestion that this relationship was anything more than temporary. Pauline Einstein was a middle-class German Jew, with the requisite expectations of and investments in her son. Serbs were dark foreigners; their women were reputed to be easy;[2] they were scarcely the kinds of girls a good German boy would bring home to his mother. Mileva, non-Jewish to boot, would never pass in the life to which Pauline aspired for her son.

Pauline revealed her true colors in June, however, when Mileva's friend Helene Kaufler stopped in Milan to see Albert's parents on her way to summer vacation.[3] Pauline ridiculed and mocked Mileva mercilessly. Helene reported back to Mileva that she liked Pauline and found Hermann handsome; in fact, she was so charmed by both of them that she herself had been seduced into joining in the general roasting of Mileva.

Mileva, who had been waiting anxiously for "certain reports," was suddenly pitched into despair. "Do you think she does not like me at all?" she asked Helene.[4] "Did she make fun of me really badly? You know, I seemed to myself so wretched at the moment, so thoroughly wretched, but then I comforted myself all the same, because after all the most important person is of a different opinion, and when he paints beautiful pictures of the future, then I forget all my wretchedness, or do you think that I shouldn't?"

There was little support to be had from her old friends on Plattenstrasse. In fact, that same week Milana and Ruzica announced that they were moving out, at least partly out of pique with Mileva. She had pushed them out of her life, they complained, in favor of Albert—whom they never did like. "I see little of Miza because of the German whom I hate," Milana Bota told her mother.[5]

Mileva, perplexed, wondered why her friends seemed so cross with her. "Maybe I must atone for other people's sins," she told Helene. Albert was less than grief-stricken to see the Serbian soul sisters go, and penned a sarcastic farewell poem about them.

But Pauline's disapproval was not the darkest cloud on Albert and Mileva's future. It was the stern-minded Weber, who, as their university years rushed to a close, held Albert and Mileva's professional fates more and more tightly in his fist. And it had not been an auspicious year, academically.

Rebuffed in his attempt to mount an aether-drift experiment the previous summer, Albert and his loyal accomplice, Mileva, in the fall of 1899 had turned their extracurricular interests from radiation to the newly discovered electrons and their possible role in determining the properties of matter. Until Thomson's demonstration that the so-called cathode rays were really showers of little electrically charged bits, or "electrons," physicists couldn't say for sure whether nature was particulate or continuous, or whether the fabled "atoms" really existed. Now they knew that at the very least matter contained electrons. In a gas, according to atomic theory, heat was the motion of molecules bouncing about. Perhaps in a metal, several theorists, including Weber, suggested, heat was likewise just the dancing of electrons. The longer you held a piece of metal over a flame, the faster the electrons in it would dance.

While Mileva was in Zurich taking (and passing) her intermediate exams, Albert had spent the early fall in Milan reading up on thermoelectricity. If the electron theory was true, he reasoned, there ought to be a connection between the electrical and thermal properties of matter. "I have also devised a method of great simplicity," he reported, "which permits one to decide whether the latent heat of metals is to be reduced to the motion of ponderable matter or of electricity, i.e., whether an electrically charged body has a different specific heat than an uncharged one."[6] Best of all, the experiments were simple and could be carried out with the equipment in Weber's lab.

Apparently, however, Weber was no more enthusiastic about this new scheme than the previous one. By the spring Albert and Mileva had been reduced to grinding out studies of heat conduction,[7] another of Weber's research specialties, for their dissertations, or *Diplomarbeiten*. Mileva, at least, put a good face on her thesis work, telling Helene, "I am very happy about the investigations I'll have to do for it."[8] Albert would later claim that his thesis was of no consequence and he couldn't in fact even remember what it had been about. At the last minute, Weber complained that he had not written the dissertation on regulation paper and made him copy the whole thing out again properly.[9]

Final exams, in July, caught them both in a confused state, battered by the disapproval of their parents and the defection of their friends, discouraged by the imperious Weber, enchanted by the prospects of a cozy, free life ahead, and riding the waves of lust. For Mileva, exhausted from accompanying her Johnnie on his midnight excursions to the frontiers of physics, it was her second exam hell in less than a year. In truth she was feeling only slightly more wretched than her comrades, all of whom were passing through the last dark, bitter, and required gate to a life of the mind.

"You either know it or you don't know it. This seems crystal clear," Albert remarked, dismissing one of his friend's fears.[10] Albert knew it but he hated it. His insouciance masked his loathing for all the outmoded, boring science he had resisted for twenty years and now had to swallow and parrot back.

Mileva, in the end, didn't know it. She was ranked last in the class, with an average of only 4.0 on a scale of 1 to 6, far below the cutoff for graduation. In experimental and theoretical physics her scores were very close to Albert's, but she had done worse in astronomy, and catastrophically worse in math, where he received 5.5 and she only a 2.5. It's hard not to suspect that perhaps Mileva had spent too much time helping Albert chase his will-o'-the-wisps through the aether, and not enough time cracking the books on her own account.

Albert's average of 4.91 placed him fourth in the class, behind Grossmann, Kollros, and Ehrat (who were all math majors), but was good enough for a diploma, which was duly approved on July 28, 1900. As for Mileva, there was always another year.

The polytechnic diploma officially conferred on Albert the status of *Fachlehrer in mathematischer Richtung,* which qualified him to teach math and science in secondary schools, but that career was now below his aspirations. He planned instead to get a job as an *Assistent* to one of his Poly professors, the usual next rung on the academic ladder, and write a doctoral dissertation. The Polytechnic itself didn't yet grant doctorates, but any of its graduates could get one simply by submitting a dissertation to the University of Zurich next door.

TWO DAYS LATER, Albert was in the Hotel-Kurhaus Melchtal, spending another summer vacation with the women.[11] As soon as the chattering carriage ride up the winding roads from the train station to the small lush resort valley of Melchtal had ended, Maja pulled her brother off for a short walk. She begged him not to say anything to Mama about the "Dollie affair." But all his life Albert, like a perfectly naughty little boy, had the excruciating ability to blurt out exactly that which the other person—usually a woman—least wanted to hear.

And so shortly after the pleasantries had been concluded he was summoned to his mother's room.

"So what will become of Doxerl?"

"My wife."

Pauline threw herself on the bed and buried her head in the pillow, sobbing like a baby.

Then the storm that had been building in her breast for two years burst.[12] From the recesses of her pillow, Pauline hurled her dire bolts.

"You are ruining your future and destroying your opportunities.

"No decent family will have her.

"If she gets pregnant you'll really be in a mess!"

Losing his temper, Albert denied with all the conviction he could muster that they had been "living in sin," and scolded his mother for being such a narrow-minded philistine. He turned and was about to stomp out of the room when in barged one of Pauline's friends. The conversation immediately turned social, and they all went to dinner. But as soon as they were alone, Pauline resumed her offensive. Mileva was too old, she said. "When you'll be thirty, she'll be an old hag. She is a book, like you—but you ought to have a wife."

By the next morning Pauline had relented. If Albert and Mileva were in fact not yet sleeping together, she apparently figured, the catastrophe might yet be avoided. Within days she had totally retreated from the "delicate subject," intimidated by Albert's unrelenting stubbornness—or so he bragged in unnervingly detailed blow-by-blow accounts to poor Mileva, who was sitting on the sidelines in Novi Sad.[13] His mother, he boasted, was growing reconciled to the inevitable, or at least was holding her fire momentarily. In the meantime Albert's father weighed in from Milan, announcing that he would be sending his son a letter, and promising that the two of them would have to have a serious father-and-son talk when Albert returned to Milan later in the month. That, Albert figured, would be the real battle, but he assured Mileva that he had more resolve in his little finger than his parents had between them. Victory was a certainty.

As determined as he was, Albert, with a little help from his boyhood reading of Schopenhauer, believed he understood his parents' attitude. To them a wife was a luxury affordable to a man only when he had achieved a good living. Their attitude, Albert complained, made a wife nothing more than a whore with a long-term contract. Like the overwhelming majority of people, Albert's parents' feelings were ruled by their senses. Fortunately, he and Mileva were more enlightened, but like all lonely elites, they owed their existence to the less-evolved hordes. For the latter, he declared, "Hunger and love are and remain such important mainsprings of life that almost everything can be explained by them, neglecting the other leitmotifs. I am therefore trying to spare my parents, without giving up anything I consider to be good—and that is you, my dear sweetheart!"

However prepared he might have been intellectually for his mother's reaction to his plans with Mileva, Albert had not anticipated such an emotional storm. Despite his oft-proclaimed aloofness and his rebel pose, he found himself rocked and churned by heavy tides. He should have kept his mouth shut, as Mileva was always advising him. Now he urged Mileva not to tell her parents—if she had not already done so—about their sparkling future, lest they react like the Einstein elders.

In the meantime, Albert was the good son, playing the violin for his mother's Melchtal crowd, charming them and her into good cheer. The Kurhaus was renowned for its groaning tables, and was full of indolent ladies. Thank God his Dollie was not like them. Meals here lasted forever; still, they were a welcome improvement over his coffee-and-wurst diet with Mileva. He took refuge in mountain hikes with Maja and a book of physics lectures by Gustav Kirchhoff.

"No one as talented and industrious as my Dollie, with her skilled hands, is to be found in this entire anthill of a hotel," he wrote.[14] "I long terribly for a letter from my beloved witch. I can hardly believe that we will be separated so much longer—only now do I see how madly in love with you I am! Indulge yourself completely so you will become a radiant little darling and as wild as a street urchin."

Mileva had meanwhile gone home to Kac. By now she was thoroughly cowed by Albert's mother and nearly too fearful to write, lest her letters end up in the wrong hands. But her fears were unfounded; Pauline handed them glumly to Albert without comment.[15] His father, too, said little more about Mileva after his initial sermon, but promised to take Albert to Venice because the city was near several of his company's power projects. In his newfound solicitude for his family, Albert agreed to accompany him on an inspection of the plants. It would be useful, he explained to Mileva, in case there were ever an emergency and he needed to take over the family business.[16]

In the middle of August Albert returned to Zurich for a week or so to begin looking for an assistantship for the upcoming year, but he found himself wandering around listlessly, sleeping until noon.[17] "How was I able to live alone before, my little everything? Without you I lack self-confidence, pleasure in work, pleasure in living—in short, without you my life is no life."

Albert was guardedly optimistic about his job prospects. Weber, the most logical choice, had an opening, but he didn't like Einstein, and there is no evidence that Albert ever took the idea of working for him seriously. There were no other immediate openings for which he could apply at the Polytechnic, but his classmate Ehrat, who was working for an insurance company, offered to arrange a temporary job there for Albert, at 8 francs a day. Albert turned it down, figuring he could use the time better for studying. Besides, it was best that he avoid such stultifying situations.

He returned to Milan, where the eerie calm regarding Mileva continued, and his father began to indoctrinate Albert into the family business. Albert began feeling frisky again:

Oh my! that Johnnie boy!
So crazy with desire,
While thinking of his Dollie,
His pillow catches fire.[18]

Albert spent his evenings hanging out and talking physics with his old friend Michele Besso, who had moved to Milan with Anna and their young son, Vero.[19] Besso was working as a consulting engineer, and he asked Einstein to help him figure out a problem for his firm concerning the radiation from alternating current. Albert admired Besso's mind, but as always his admiration was tempered by criticism: Michele was smart but simple; his and Anna's apartment was cozy but somewhat lacking in taste.

Albert's equanimity was shattered in the first week of August by the arrival of a registered letter from Mileva, a stratagem she had employed to make sure the letter did not fall into the wrong hands. Among its revelations was that Mileva was suffering from a goiter. This seems to have thrown Albert's parents into a whole new round of wailing and shrieking. On top of everything else this woman poised to snatch their son was *sick*—defective, unclean, diseased.

"Mama often cries bitterly," he wrote despairingly from his bed, "and I don't have a single moment of peace here. My parents weep for me almost as if I had died.

"Oh, Dollie, it's enough to drive one mad."[20]

The sight of his parents so grief-stricken, as if he had committed some great crime, broke his heart. The whole family was trapped in psychological quicksand. The more upset he got, the more upset they got, which made him feel even more wretched in turn. When he announced that he couldn't go to Venice with his father, they became so alarmed that he quickly capitulated. There were, he now knew, things far worse than exams.

"My only diversion is studying, which I am pursuing with redoubled effort, and my only hope is you, my dear, faithful soul. Without the thought of you I would no longer want to live in this sorry herd of humans."

And so he dutifully trudged along on the power plant rounds, but while his body was among wires and dynamos, his mind was among atoms, the aether, and his street urchin. "Three quarters of our stupid time apart is now over, soon I'll be with my sweetheart again and kiss her, hug her, brew coffee, scold, work, laugh, stroll, chat . . . + *ad infinitum!* This will again be a cheerful year, will it not?" he asked, declaring his intention to stay in Zurich for the Christmas holidays, instead of coming to Milan.[21]

"I cannot wait to have you again, my all, my little beast, my street urchin, my little brat. Now that I think of you, I just believe that I do not want to make you angry and tease you ever again, but want to be always like an angel! Oh, lovely illusion!

"But you love me, don't you, even if I am again the old scoundrel full of whims and devilries, and as moody as always!"

By the time he returned from his Venice tour his parents had again retreated from the subject of Mileva, and Albert eschewed talk of marriage. The fire had temporarily gone out of both sides.

ALBERT'S STUDIES had been his lifeboat during the crisis. They were taking him deeper into the nature of matter, the realm of the hypothetical atoms and molecules, of electrons, and of the mysterious forces between them. Beset and dazzled by the discovery of electrons, X rays, and radioactivity, the best minds in Europe were teetering between the continuous and discrete views of reality. The old idea of a nature with no gaps, in which particles were vortices in the all-pervading aether, was being replaced by the notion of a world that consisted of real particulate matter, atoms. The Englishman Maxwell had his feet in both camps, with his light waves resting on the aether while his theory of gases ascribed heat and pressure to the motions of masses of individual atoms and molecules—discrete entities of matter. Maxwell's version of thermodynamics had been taken up and expanded by Ludwig Boltzmann in Vienna, another of the mathematical terrorists, and the most dedicated of the atomists.

While Albert hid out from his parents, he buried himself in a Boltzmann book and reported to Mileva that it was magnificent. Though he had some quibbles with precise mathematical details, he was convinced that the theory, at least in its main principles, was correct: Gases really did consist of discrete little mass points.

In a gas these little points bounced around independently, but in a liquid, effects like surface tension testified to the presence of mysterious attractive forces between the individual atoms or molecules. What caused a drop of water or a blob of mercury to maintain its shape against gravity, or a liquid to cling to the sides of a straw or cup, producing a concave surface? In his last year at the Polytechnic, Albert had heard his math professor, Minkowski, talk about capillary phenomena, and he had remembered the lecture as the best he had heard in his four years there, the first real exposition of the kind of mathematical physics that Maxwell and Helmholtz were doing.[22] By studying these effects closely, Albert thought, it should be possible to deduce the nature of the mysterious intermolecular forces. In fact, Albert had performed some experiments on this subject during his recent stay in Zurich, and he was now anxious to get back to do more. There could be a paper in it.

He had also been thinking more about heat and conductivity, the subjects of his and Mileva's *Diplomarbeiten*. While she studied to retake her final exams the next fall, Mileva was planning to continue her previous research for a doctoral dissertation. How, she asked, were the specific properties of a body, such as its conductivity or its ability to absorb heat, related to its temperature? How, actually, *did* a body absorb heat? Einstein had thought of a novel way they could investigate an effect related to this question in Weber's well-equipped lab, so it would be essential for them to maintain good relations with him.

Returning from his excursion, Albert learned that one of the current assistants to his old mathematics professor, Adolf Hurwitz, had won a teaching job in Frauenfeld, which left a spot open with Hurwitz. Albert had been banking on exactly this sort of development. He wrote Hurwitz. Although he had neither been a mathematics major nor attended the math seminars, Albert argued, he had specialized in theoretical and analytical physics, both of which required a thorough grounding in mathematics. Hurwitz agreed to consider his application.

His spirits rebounding, Albert went off again, mountain-climbing near Lago Maggiore. While there he sent Mileva an outline of his foot so that she could knit him a pair of socks. Besso, to whom Albert confided very little regarding Mileva, saw through him nevertheless, and teased him about his eagerness to return to Zurich. The longer the summer dragged on and the more his parents complained, the more intense and florid his letters became, and the more he wanted her. "What a joy it will be when I can hold you tight once again, my little street urchin, my little veranda, my everything!"[23]

Beset by her goiter, however, Mileva was still stuck back in Kac. She announced that when she did come she was bringing her sister Zorka back with her. Albert went out and bought a pair of coffee spoons, to impress Zorka with their "European culture." He also reported proudly that he had learned to shave himself, instead of going to a barber, so now he could do that instead of burying himself in books while she cooked lunch. He painted an idyllic picture of the coming fall, promising that, no matter what happened, they could have the most wonderful life in the world, doing science, answering to no one, standing on their own two feet at last. They could save up and buy a pair of bicycles and go on bike tours.

"You don't like the philistine life anymore either, do you? He who has tasted freedom can no longer wear chains. I'm so lucky to have found you, a creature who is my equal, and who is as strong and independent as I am! I feel alone with everyone except you."[24]

"Such a boisterous fellow, and yet I love him so much," Mileva wrote Helene Kaufler.[25]

ALBERT'S SUNNY predictions were dashed almost as soon as he returned to Zurich. To begin with, his job fell through. Dissatisfied with all the candidates, Hurwitz divided the position between Ehrat and another student. At the same time, Weber was appointed an engineering student as his own assistant, leaving Einstein the only unemployed member of the class of 1900.[26] Meanwhile Albert and Mileva paid their fees and were beginning thermoelectric experiments under Weber's supervision as their thesis adviser, perhaps a less than promising arrangement.

In the meantime Aunt Julie's allowance had ceased. To help support himself Albert halfheartedly took on tutoring jobs. By Christmas his teaching load amounted to a staggering eight hours a week. He also worked temporarily for Alfred Wolfer, who had taught astronomy and astrophysics and was the director of the Swiss Federal Observatory, a small dome uphill from the Poly.[27]

Having postponed her return for so long because of her goiter problems, Mileva had finally lost her spot at 50 Plattenstrasse. When she and Zorka finally arrived in the city, they moved into a rooming house on Hottingerstrasse, a busy boulevard along the trolley line that runs up the Zürichberg to the city zoo. Mileva was subsisting on her father's allowance. Her and Albert's "household," with the addition of chubby Zorka, was running on little more than coffee, music, and physics.

By December Albert had finished writing his paper on capillarity. Whereas Maxwell had modeled electromagnetism on the mathematics of fluids, Einstein modeled the physics of fluids on the mathematics of Newton's famous law of gravity. In gravity bodies were said to attract each other in proportion to their masses and in inverse proportion to their distance apart. Albert was intrigued when an analogous proposal for molecules actually seemed to work out. On the basis of measurements of capillary effects in various fluids, the molecular force exerted by a putative atom seemed to increase with atomic weight. "The question of whether and how our forces are related to gravitational forces," he wrote, "must therefore be left completely open for the time being." The main point was that atoms seemed real. He submitted the paper to the *Annalen der Physik,* the leading physics journal in Europe. Its editor, Paul Drude, duly informed him that it would be published the following fall.[28] Einstein also proudly sent a copy off to Boltzmann in Vienna.

"Albert has written a paper in physics that will probably be published very soon in the physics *Annalen,*" Mileva chirped to Helene.[29] "You can imagine how proud I am of my darling. This is not just an everyday paper, but a very significant one, it deals with the theory of liquids."

However proud his parents might have been with his accomplishment, this threadbare existence was not the life that Hermann and Pauline had imagined for their *Schatzi.* They continued to file protests from Milan, until, despite his earlier declaration, Albert gave in and went home for Christmas. If he couldn't find a job at the Polytechnic, they argued during the holiday, he should look elsewhere. They hoped that any success would not only further his career but also separate the young lovers. Albert could stay in Zurich until he finished his dissertation in the spring, but then, they insisted, it would be back to Milan.

The news hit Mileva hard. Ever since Helene's visit with Pauline, every trip home of Albert's had thrown Mileva into a new fit of anxiety, and every letter from Milan had coiled her stomach in fright. Now, she poured out her miseries about the upcoming separation in a series of letters to Helene, who was happily married and living in Belgrade—another poignant counterpoint to Mileva's own frustrations. When Albert left, she moaned, he would be taking half her life with him.[30]

And yet how could she stand in the way of his career? "I love him too much for that, but only I know how much I suffer because of it. Both of us have had much to endure lately, but the forthcoming separation is almost killing me. I'll have a lot to tell you once we see each other, now I don't want to let filth spoil your most beautiful days. . . . When are we going to reach the point at which we'll be allowed to acknowledge our love before the whole world? It almost seems to me that I'll not live long enough to see it."

Mileva complained again to Helene just before Albert was to head home to his parents for the Christmas holidays, saying the day was not long enough to recount all the slanders and intrigues that had been perpetrated against her.[31] Although she refused to let Albert read the letter, he contributed a rather blithe postscript betraying a somewhat more detached view of the upcoming tragedy:

> *The lass cried*
> *because the lad must split*
> *but we thought*
> *we won't grumble about it!"*[32]

On the day after he mailed off his paper on capillary forces, Albert was summoned to an interview with the Municipal Naturalization Commission of Zurich. At the end of a grilling that ranged from his finances and drinking habits to whether his grandfather was syphilitic, the group made fun of Einstein's apparent lack of practical sense.[33] However, on February 21, 1901, he was finally recognized as a Swiss citizen, ending six years of statelessness. Three weeks later he reported for his physical for the obligatory induction into the Swiss army. In principle this service was less odious to him than the required German military duty, since Switzerland was a less militarized, less authoritarian society. In practice it was a moot point, for he was rejected for flat feet and varicose veins.[34] Albert, proud citizen of the Swiss Confederation, was free to follow his bliss.

In the meantime he had applied for a "practical job" in Vienna, and though Mileva started looking for a teaching position in a girls' gymnasium, the future seemed opaque.[35] "The truth is, we haven't yet the slightest idea what fate has in store for us," Mileva told Helene. They took their pleasures where and when they could, she said. When the snows came they went sledding on the Zürichberg. "Albert was always so cheerful when we went down like the Devil."

Einstein continued his investigations of the attraction between molecules into the new year, this time using very dilute salt solutions. He was now considering extending the work to gases and turning it into his doctoral dissertation. In gases, as he had surmised earlier, the forces between molecules would not be strong enough to affect gross properties such as pressure or temperature, which could be explained by atoms or molecules banging off each other like Newtonian billiard balls. But the molecular forces would, he realized, have more subtle and measurable effects—for example, on the ways in which two different gases diffused through each other. Studying these effects could reveal much about the nature of molecules themselves, the actors behind the shifting masks of nature.

"It is a glorious feeling," an excited Albert wrote to his old classmate and note taker Marcel Grossmann, "to perceive the unity of a complex of phenomena which appear as completely separate entities to direct sensory observations."[36] It was a thought that could have come straight out of his childhood reading of Bernstein, a thought that would be like a compass needle for his life.

By choosing to work on molecules and gases instead of heat and electricity, Albert had passed out of Weber's domain of expertise, finally ending the relationship between them. The new research would be performed under the supervision of Alfred Kleiner, the physics professor at the University of Zurich.

With the imminent appearance of his capillarity paper, Albert also felt ready to launch a new campaign to find himself an assistantship somewhere, anywhere. Enclosing a copy of his paper as a calling card, he wrote to Otto Wiener at Boltzmann's University of Leipzig inquiring about the possibility of an opening. He also wrote in time to Eduard Riecke at the famed University of Göttingen, to Richard Koch at the *technische Hochschule* in Stuttgart, to Heike Kamerlingh Omnes at the University of Leiden in the Netherlands, and even to Wilhelm Ostwald, the chemist at the University of Leipzig who had debated Boltzmann on the existence of atoms. As a chemist, however, Ostwald was interested in capillarity and was in fact responsible for much of the data Einstein had used. Ostwald didn't respond.

Meanwhile, the string had been played out at the Polytechnic. In late March, after five months of bohemian independence, Albert gave up, left Zurich, and went home to Milan and his family. He packed in such a state of confusion that he left behind his nightshirt, toothbrush, comb, and hairbrush.

Mileva dutifully launched those last vestiges of half of her life back to their owner, who had returned in a way to his owners, reeled back to the ones who had had him first and always. Zorka had already gone home to Serbia. The old Pension Engelbrecht, where Mileva and Milana, the Serbian soul sisters, had stayed up all night laughing at Bulgarians, singing, and studying, beckoned. Mileva packed her own things and moved back to Plattenstrasse.

PART TWO

Relative Motions

And the moral of the story
(Which one hardly ever discusses)
Is that the upper half plans and thinks
While the lower half determines our fate

Albert Einstein, 1946

6

THE WHITE WORLD

IN THE SPRING OF 1901, 50 PLATTENSTRASSE WAS FULL OF GHOSTS. MILANA and Ruzica had pulled out. Helene had moved to Belgrade with her new husband and was even pregnant. Albert had taken half of Mileva's life with him to the other side of the Alps.

Now nearly alone in Zurich, Mileva tried to concentrate on her dissertation. The plan was to extend the studies of heat conduction that she had made for her diploma thesis, but working with Weber wasn't going well. She naturally fell into brooding about what was going on back in Milan. She had become painfully accustomed to reading Einstein's detailed accounts of fights with his mother and father, yet now he was reporting none. Did that mean the fights had stopped, or did it mean that his parents were winning?

Albert kept maintaining that happiness and togetherness for the two of them were just around the corner. "If you really knew what kind of hold you have on me you little witch, you wouldn't be so afraid all the time that I'm keeping things from you."[1] His spirits, he insisted, were upbeat.

In fact, his situation was depressing, if not terrible. Albert's entreaties for employment had only rarely elicited the courtesy of even a formal rejection, and he had begun to suspect that Weber was sabotaging him from afar. As Einstein's main professor at the Polytechnic, Weber was the one who would receive the requests for recommendations from potential employers, and Albert concluded that the reports he sent back were negative. Mileva urged him to confront Weber, and when the Göttingen position seemed to be in doubt, Albert told her that he had written Weber directly to let him know that he couldn't get away with saying bad things behind Albert's back.[2] If the message was indeed sent, it has never been found. Göttingen's Riecke promptly replied with a rejection letter.

Albert wrote again to Ostwald, with the excuse that he had previously left out his address, but by now he was ready to give up altogether the thought of getting a university job in German-speaking Europe.

In addition to suspecting Weber of skulduggery, Albert was worried that his Jewishness was working against him. Maybe he would have better luck in Italy, where Michele could help. One of Besso's uncles was a professor of mathematics in Milan. Albert had met him once and concluded he was an inconsequential little man, but he had connections with important physicists. Besso was then visiting his parents in Trieste, but on his return he stopped by to visit his uncle, who offered to write to his physicist friends on Albert's behalf. Albert told Mileva, "I will soon have graced all the physicists from the North Sea to the southern tip of Italy with my offer!"[3]

At that point Albert had been in Milan for a week and a half with nobody to talk to, and the night after Michele got back, the two young men eagerly talked physics for four hours. The topics ranged from the elusive aether to atoms and molecular forces. The keenness and versatility of Besso's mind were matched only by its disorderliness, which Albert observed with a kind of dismayed fascination. A typical Besso adventure occurred when his boss asked him to inspect some newly installed power lines in nearby Piedmont. Besso decided to leave the night before, in order to have more time at the power station, but he missed his train. The next morning he remembered his assignment too late and missed that train as well. Finally on the third day Michele made it to the station on time, but had forgotten what he was supposed to do, and had to wire the home office for instructions.

His friend, Albert concluded, seemed to lack any willpower or common sense. "He's interested in our research," he faithfully reported to Mileva, "though he often misses the picture by worrying about petty things. This pettiness is a natural part of his character and constantly torments him with all sorts of nervous ideas."

One subject that their rambling discussions apparently did not touch upon was Albert's relationship with Mileva, of which Besso, having been around Zurich, could not have been unaware on some level. But Albert's gregariousness about physics was matched by reticence when it came to personal affairs.

"You don't have to worry about me saying anything to him or anybody else about you," he told Mileva. "You are and will remain a shrine for me to which no one has access; I also know that of all people, you love me the most and understand me the best." No one, he alleged, would dare say anything bad about her in front of him for fear of encountering the withering Einstein wrath. "I'll be so happy and proud when we are together and can bring our work on relative motion to a successful conclusion. When I see other people I can really appreciate how special you are!"[4]

The "work on relative motion" was presumably a reference to the problem of the relationship between light waves and the aether that had been Albert's personal obsession since he was fifteen, and which had taken up much of his time with Mileva. The subject was never far from his mind. Since arriving in Milan, however, he had become almost as engrossed with the problem of electrons and their role in matter, heat, and radiation. Drude had suggested that clouds of these particles (and equal numbers of as yet undiscovered positive charges), floating about the interiors of metals, much as molecules of a gas did, were responsible for conducting heat and electricity. On the train from Zurich it had dawned on Albert that energy or heat could be captured and stored in the dance of those electrons.

According to Maxwell's theory, when a light wave struck an electron— say in your forehead—the electromagnetic fields in the wave would force the electron to oscillate to and fro, or "resonate," in tune with the wave's frequency, absorbing the energy of the light beam and converting it into heat in your skin. Conversely, an electron that was already wiggling around because it was hot would radiate electromagnetic waves, losing energy and cooling off.

If Drude was right, it meant that these same electrons that caused a body to radiate or reflect light were also the repositories of its heat, its internal energy. Drude's theory implied, Albert realized, that there should be a correlation between the optical and the thermal properties of materials. He began pestering Mileva to look up data on the heat capacity of glass. "I'm burning with desire to work my way into this," he wrote, "because I'm hoping it will be possible to make a prodigious step forward in exploring the nature of latent heat"[5]

As it happened, Albert came upon a paper by a German physicist named Max Planck who had been trying to make a similar connection between the temperature of a body and the light it radiated. In Planck's theory the electrons comprising matter were less like a cloud of gas than like an array of springs, each capable of oscillating at a particular fixed frequency. When a light wave impinged on this array, its energy would go into jiggling the electrons, which would reradiate energy at their own characteristic frequencies. The object would then heat up until some equilibrium was reached between the radiation coming in and the radiation going back out. In principle it was a simple idea, but in practice it led to a mathematical morass: Planck hadn't been able to make the details come out right.

This was Albert's first distant encounter with Planck, who was the rising young prince of German physics, and with a problem that would become one of the defining battle fronts of modern physics, but this particular engagement didn't last too long. For one thing, Albert decided that he didn't like Planck's assumptions. It seemed to him that Drude had a better idea. "There's no doubt this Drude is a brilliant man."[6]

After a week of floundering with Drude and Planck, however, Albert gave up both. Nothing really worked. "I've retreated again somewhat from my idea on the nature of latent heat in solids," he admitted to Mileva, "because my views on the nature of radiation have again sunk into the sea of obscurity."[7]

Meanwhile Albert's job search hit a new nadir. His Italian gambit was yielding no immediate results, and Ostwald still refused to acknowledge his inquiries. Finally, to Albert's undoubted mortification, his father took it upon himself to petition the chemist behind his son's back.[8] "Please forgive a father who is so bold as to turn to you, esteemed Herr Professor, in the interest of his son," Hermann wrote.

"My son . . . feels profoundly unhappy with his present lack of position, and his idea that he has gone off the tracks with his career & is now out of touch gets more and more entrenched every day. In addition, he is oppressed by the thought that he is a burden on us, people of modest means.

"Since it is you, highly honored Herr Professor, whom my son seems to admire more than any other scholar currently active in physics, it is you to whom I have taken the liberty of turning with the humble request to read his paper published in the *Annalen fuer Physik* [sic] and to write him, if possible, a few words of encouragement, so that he might recover his joy in living and working."

The rest of family life in Milan was hardly more stimulating. Not only was Maja tormenting him in the eternal ways of little sisters, but, on the grounds that he wasn't doing anything else, his parents enlisted him as a tour guide for the various visitors who were always dropping in. Among them was Jost Winteler, Albert's old schoolmaster from Aarau. In the years since his romance with Marie had ended, Albert's relationship with the elder Winteler had cooled considerably. Winteler had let it be known that he thought that Einstein was living a life of debauchery in Zurich.[9] During his tour the conversation stayed off personal topics.

To make matters worse, Mileva got word that she was being considered for a teaching job in a girls' gymnasium in Zagreb, Croatia, perhaps the same school she had attended back in 1894. Even Besso was abandoning him. Apparently giving in to his father's wishes, Michele announced that he and his young family were moving to Trieste, where he would be working in a textile factory. "He's a frail creature without bone and marrow," Albert complained.[10]

Mileva was the one bright spot remaining in Albert's life, and he had been promising her a trip to Venice as soon as things settled down somewhat. In April Mileva proposed meeting at Lugano, on the Swiss-Italian border, but Albert was depressed and didn't show up.

It was not long thereafter, however, that Einstein's luck changed completely. First he received a letter from his old friend Marcel Grossmann, who had told his father about Albert's fruitless job search. The elder Grossmann had put in a good word with his friend Friedrich Haller, director of the Swiss Patent Office in Bern, where in fact there was an opening.[11] It sounded like a wonderful job, Albert told Marcel—at last he had something to look forward to, adding that he would have found a university position long ago if not for Weber's underhandedness. "All the same, I leave no stone unturned and do not give up my sense of humor. . . . God created the donkey and gave him a thick hide."

The next day another letter arrived, this one from Jakob Rebstein, a mathematics teacher in Winterthur, just outside of Zurich. Rebstein, a former assistant at the Polytechnic and an acquaintaince of both Albert's and Mileva's, asked whether Albert could take over for him in Winterthur for a couple of months while he went on military duty. Albert was thrilled, and it didn't even matter that he would have to teach descriptive geometry. "The brave Swabian is not afraid," he proclaimed.

Buoyed by the sudden change in his prospects, Albert decided to take the trip with Mileva after all. He proposed that she meet him at Lake Como for the journey back into Switzerland, "You'll see for yourself how bright and cheerful I've become and how all my frowning has been forgotten. And I love you so much again! It was only out of nervousness that I was so mean to you."[12]

Now she no longer had to worry about the Zagreb job. "You are a thousand times more important to me than you could ever be to all the people of Zagreb!" he assured her. "If you don't get that position, and I get the job in Bern, I hereby appoint you my dear little scientist. There's no need for you to go to a provincial backwater, dear girl—I can appreciate the value of my 'pair of old boots,' as you've always said, better than you think."

She accepted his invitation to Como and then, reconsidering, turned it down. She had, she explained, received a letter from her father, probably disparaging both Albert and her own academic lack of progress, "that has made me lose all desire, not only for having fun, but for life itself."[13] She urged to him to go to Como by himself. "I'm going to lock myself up and work hard, because it seems I can have nothing without being punished; on the other hand, I don't need anything, and will become as accustomed to this fact as the gypsy to his horse. It doesn't matter, sweetheart. Farewell, be cheerful, and if you find any pretty flowers, bring me a few."

The next day, though, she changed her mind again.[14]

Like a wishbone pointing north, Lake Como snakes for some forty-seven kilometers through northern Italy, its still blue waters walled by steep green hillsides and, beyond, the chalky limestone bones of the earth thrown up like ramparts against the great snowcapped wall of the Alps. At 411 meters, Como is the deepest lake in Europe. Its gentle waters and milky air have made it a favorite resort since Roman times. The shoreline is dotted with villas and ancient villages. The town of Como, famed for its silk industry, lies at the southern end of one of the two branches of the wishbone.

Dawn was just lightening the Alpine sky when Mileva stepped out of the train at 5 A.M. on the morning of May 5, bearing among other things a pair of opera glasses and Albert's blue nightshirt. He was already there waiting on the platform with open arms and, as Mileva later recounted in a letter to Helene, a "pounding heart."[15] For once the tentativeness and ambivalence that had marked their relationship from the beginning were absent. Albert had promised her a trip the likes of which she had never seen, and history would show that he delivered the goods. "It was so beautiful," Mileva told Helene, "that it made me forget all sorrows."

The couple spent the morning strolling through Como, past the fifteenth-century Duomo that dominates the winding downtown streets with their shops full of silken goods, to the Piazza Cavour, fronting the lake and flanked by hotels and cafes. To the right, looming atop a 300-meter-high hill accessible by cog railway from the lakeshore, is the Commune of Brunate. To the left, a few kilometers around the curving shoreline, lay Cernobbio, home of the luxurious and world-famous Villa d'Este. A winding road threads the lakeside villages, but the preferred mode of transportation is by boat. Several times a day water buses leave Como for the three-hour trip to Colico at the far north end of the lake, stopping along the way at small village landings.

Albert and Mileva boarded the boat to Colico around noon, stopping halfway up the lake at the Villa Carlotti, near the small fisherman's port of Cadenabbia. Now named after the daughter of a nineteenth-century Russian princess who received it as a wedding present, for more than two hundred years the villa had graced the lakefront with its crisscrossing marble staircase, three floors of art, and fountains. The villa was and is famed for its fourteen acres of lavish gardens, which include more than five hundred species of shrubs and trees. There are one hundred and fifty varieties of azaleas and rhododendrons alone, and on the weekend of May 5 they were all in bloom. "I have no words to describe the splendor we found there," Mileva recalled, rhapsodizing about the garden, "which I especially preserved in my heart, the more so because we were not allowed to swipe even one single flower."

We do not know where they spent the night. They might have stopped in Colico, a dreary settlement at the end of the boatride, but from there it would have been only a short train journey to the colorful old city of Chiavenna, tucked in a valley of the Alps. Chiavenna lies at the approach to two passes. To the east the road leads gently up the Breggia Valley to Maloja Pass and past the magnificent Bernese Alps and the Engadine Valley, home of St. Moritz. Albert was bent on going the other way, however, north up the Val Giacomo to Splügen Pass, 2,115 meters elevation, and down into the valley of the *Hinterrhein*.

Even this late in spring the mountain passes were still deep in snow. In the morning Albert and Mileva rented a horse-drawn sled for the journey up to Splügen. There was just enough room for two people in love to snuggle under a pile of furs. The driver rode on a plank in the rear, prattling constantly like a Venetian gondolier, referring to Mileva as *"Signora."* She couldn't imagine anything more romantic.

The road to Splügen was built by the Romans, and parts of it look as if they haven't changed much since then. Outside Chiavenna it abruptly begins to climb the wall of the valley in narrow switchback turns terraced out from the steep hillside. The first daunting course of hairpins ends at the San Giacomo Filipo, a sixteenth-century mission whose small cemetery commands an imposing view of the valley to the south. Attached is a small inn and restaurant. By the time Albert and Mileva had come this far, it was snowing gaily. The road twisted and turned higher and higher, skirting a sheer precipice on one leg, plunging through long, narrow rock caverns on the next turn.

Mileva felt enveloped in a white world that reminded her of the featureless fog of Heidelberg. "We were driving now through long galleries, now on the open road, where there was nothing but snow and more snow as far as the eye could see, so that this cold white infinity gave me the shivers and I held my sweetheart firmly in my arms under the coats and shawls with which we were covered."

In spots the snow was as much as six meters deep as they ascended to a world of rock, wind, ice, and (in the summer) tundra. On the brow of Splügen Pass, which forms the border between Italy and Switzerland, a few stone cottages mark the villages of Stütta and Montesplügen.

Albert and Mileva got out at the top of the pass, crossed the frontier, and set off on foot. Far below them in the wintery gloom they could make out the narrow rushing ribbon that would become the mighty Rhine winding from the village of Splügen through steep hills into the tight gorge of the

Via Mala. Laughing despite the weather, and clutching each other, the couple tramped downhill through the deep snow. Every now and then they stopped to roll snowballs down the slope, imagining the tiny rivulets cascading into thunderous avalanches crashing down on the sleepy inhabitants below.

From Splügen they followed the Rhine downstream through the Via Mala northeast. After a series of deeper and deeper winding gorges, the Rhine eventually leads to the old walled city of Chur, in the heart of the canton of Graubünden. By then their surrender to the sun, the snow, the fog, the hills, the torrent ceaselessly boiling downhill, and, most of all, to each other, was complete. Mileva concluded her letter to Helene, "How happy I was to have my darling for myself a little, especially because I saw that he was equally happy!" On Wednesday, most likely they took a train from Chur back to Zurich.

Albert deposited Mileva in her room back on Plattenstrasse and then went looking for a hotel. After scrutinizing him from head to toe, the manager of his first choice turned him away. The second, near the railroad station, agreed to give him a room for the night for two and a half francs. In the morning he gathered a few things from his old landlady, Frau Hagi, and took the train out to Winterthur. Rebstein, whom he was relieving, met him at the station. Hans Wohlwend, an old friend from the Aarau school, was living in the same house in a large room with a veranda, views, and an "unspeakably comfortable" sofa. "If only you could see it," he reported to Mileva.[16]

They were separated yet again, but not by a great distance. He took the train to see her that Sunday, squeezing in a visit to Grossmann as well. "Only the thought of you gives my life here a true meaning," he said afterwards.[17] "If only the thought had a little life and a little flesh and blood! How delightful it was the last time, when I was allowed to press your dear little person to me the way nature created it, let me tenderly kiss you for that, you dear good soul!"

MILEVA, of course, was back at the Polytechnic and under Weber's thumb, working on her dissertation. After four days on the trail with Albert, it was a tremendous letdown to resume her routine. And things were not going well. "I had a few spats with Weber," she recounted to Helene, "but we have already gotten used to it." Albert tried to chip in with advice, and at one point suggested that Mileva look up some old experiments that Weber himself had performed on heat transfer in cylinders, to see if she could use the data for her own purposes.[18] His suggestion probably made things worse, however, for it was precisely this sort of corner-cutting that had displeased Weber about Einstein in the first place.

With the encouragement of the head of the physics department at Winterthur, a genial man named (of all things) Weber, Albert had returned to his investigation of the vexing relationship between heat, radiation, and electricity, and to Drude's electron theory, but he missed having Mileva to talk to about them with. It would be nice to pursue this path together, he said, and then added, "But destiny seems to bear some grudge against the two of us."

It might have been on one of these cozy Sunday visits, late in May, that Mileva first voiced her concern to Albert that she might be pregnant. The exact moment is not recorded, but a new note of worry seems to have entered into Mileva's communications with Albert—and a corresponding tone of bravado in his. One Thursday night, for example, Mileva sent him a letter—"a most darling little letter"—which he felt compelled to answer before going to bed. "Be cheerful, dear Doxerl, and don't worry—you are my dear, good sweetheart, whatever may happen." That reassurance having been conveyed, he moved immediately to more pressing matters. "I am not very satisfied with my theory of thermoelectricity."[19]

If Albert was surprised by the possibility of his impending fatherhood, his breezy response suggested that he was not upset, or that perhaps he was just in denial.

Birth control—in the form of condoms, diaphragms, suppositories, douches, and abortions—was widely available by the end of the nineteenth century, although expensive. A dozen condoms cost about five marks, the equivalent of half a day's wages, while an abortion went for 30 to 150 marks.[20] (Zurich was a center of the birth-control industry, as it was home to a successful mail-order abortifacient business.) As a result, birth rates throughout Europe had fallen drastically during the nineteenth century. In Hungary there was something called the "one-child system."

In certain parts of Europe, to be unmarried and pregnant was not such an uncommon situation, however, even after the advent of birth control. Religious, economic, and class factors conspired to make illegitimate births a socioeconomic rather than a moral dilemma in many places. Half of all the illegitimate children in Prussia, for example, were born to maids and servants; university towns had two or three times the illegitimacy rates of surrounding areas owing to common-law *Konkubinat* dalliances between students and local women. Others were what the demographers call "prenuptial births" to parents who couldn't afford to get married immediately but eventually did, legitimizing their offspring later.

In Zurich, according to the 1901 official statistics, 12 percent of all births were illegitimate, but the numbers fluctuated wildly with nationality. Unmarried Austro-Hungarian residents of the city were ten times more likely than Swiss women to get pregnant. In fact, in southern Hungary, where Mileva had grown up, one of three births was illegitimate, and in this ethnic hodgepodge Serbs had the highest rate of all. Meanwhile, Jews were the tidiest people in Europe. They had been among the first to embrace birth control, and their illegitimacy rate was nil compared to that of their profligate Catholic neighbors in southern Germany.

Mileva's presumed condition, in short, left the couple straddling an awkward universe of cultural and personal chasms. Neither of their families, for example, was likely to be pleased. Perhaps, their guard lowered by job prospects at last, by springtime in the Alps, the zephyr breezes of Lake Como, and crazy ideas of electrons and light waves spinning in their heads, Albert and Mileva felt that marriage was finally just around the corner. Albert always had an exalted view of the power of his own stubbornness, and could have felt that presented with a fait accompli, their parents' resistance would melt away.

For Mileva, of course, early motherhood would be a betrayal of everything she had been working toward professionally; only a few years earlier she had scoffed at the very idea of love and marriage. In the spring of 1901 she still had hopes of earning her Ph.D. and becoming a real scientist, a goal in which her father had invested serious resources over the years.

After half a decade Albert and Mileva's relationship was at a fulcrum. If in the past he had been the one who needed her, an older woman, counselor, cheerleader, nurturer, and believer, in the future it would be Mileva who needed him.

While Mileva fretted, Albert pursued the mystery of the electron with newfound enthusiasm. By now "electron" had become a kind of shorthand word signifying an entire view about the nature of matter and the origin of electromagnetic waves, and their interaction to produce heat and other forms of energy. By spending a few hours in the Winterthur library, Albert was able to catch up on the latest developments, and they fired his brain.

One of those developments was a new article in the *Annalen der Physik* by Mileva's old Heidelberg teacher Philipp Lenard. Lenard, we recall, was an expert on cathode rays, and had just missed out on proving that these "rays" were really beams of negatively charged little particles of electricity—electrons. He had continued working with cathode rays, despite his disappointment, and now had an amazing discovery to report: He found that he could get cathode rays, that is, electrons, to stream out of a metal by shining ultraviolet light on it. Perhaps most surprising, however, the intensity of the light didn't affect the energy of the electrons: No matter how high or low he turned it, they came shooting out at the same speed. "Under the influence of this wonderful piece of work," Albert told Mileva, "I am filled with such happiness and such joy that you absolutely must share some of it. Just be of good cheer and don't fret. After all I am not leaving you and I'll bring everything to a happy conclusion. It's just that one has to be patient! You'll see that one doesn't rest badly in my arms, even if it starts a little stupidly. How are you love? How is the boy?"[21]

Albert decided to write a letter to Drude, whom he admired, but whose theory, he had concluded, had serious flaws. Among them was Drude's assumption that there were positive as well as negative particles of electricity floating around inside matter, conducting heat and electricity, whereas all the experiments in all the labs in Europe had only been able to flush out the negatively charged electrons. Einstein was sure that his arguments were so reasonable and perceptive that Drude would have little choice but to acknowledge his genius and perhaps offer him a job. (Albert made a point of mentioning that he was presently unemployed.)

No such distraction would work for Mileva as it became increasingly clear that she was in fact pregnant. Friends in Zurich noticed that Mileva had become moody.[22] She was frequently depressed and withdrew into herself. She was sure now that, despite Albert's assurances, she would be reviled in the West as the slut who had ruined his life and in the East as a fool. In her own circle Mileva's mood prompted speculation that her relationship with Albert was somehow doomed or might even have already failed. Mileva refused to say anything except that it—whatever "it" was—was intensely personal.

Einstein received an answer from Drude early in July, but it was not the response he had fantasized about. Drude simply glided over Albert's objections, noting that others of his colleagues agreed with him. Albert was furious. In his eyes Drude had suddenly gone from being a brilliant thinker to a wretched idiot. "From now on," he fumed to Mileva, "I'll not turn any longer to this kind of person, but will rather attack them mercilessly via the journals, as they deserve. It is no wonder that little by little one becomes a misanthrope."[23]

In the meantime the reality of his situation with Mileva had sunk in. He announced that he had made an irrevocable decision to take the first job he could find. He wasn't waiting any longer for Besso to scare up an Italian physics assistantship or for Grossmann's uncle to get him into the patent office, but would simply ask Besso's father for a job as an insurance clerk. He vowed to Mileva, "The moment I have obtained such a position I'll marry you and take you to me without writing anyone a single word before everything has been settled. And then nobody can cast a stone upon your dear head, and whoever dares to do anything against you, he'd better watch out!" His parents would just have to reconcile themselves to their union.

Mileva turned him down. As happy as she admitted his letter had made her, she had too much sense to want to accept responsibility for his unhappiness.[24] "But of course, dear, it shouldn't be the worst possible position, this would make me feel awful, I wouldn't be able to stand it." Then wouldn't their parents be surprised?

When his teaching term was up, Albert returned to Mettmenstetten and the Hotel Paradise. There he had another screaming fight with his mother, who was now convinced that Albert was plotting to run off with Mileva.

Mileva, incredibly but inevitably, was sympathetic. "Dear little sweetheart, how much you have to endure for me! And the only thing I have to give you for all that is the little bit of love that dwells in the human heart. But you know, this is not so awfully little after all, and it will compensate you for quite a few things, if it is humanly possible." No mother alive, she believed, could stay angry at her son. And in the fullness of time, when Mileva was more to Pauline than just a "misconception," there might even be a reconciliation between the two of them. Mileva had been thinking about how to engineer it. Perhaps she could befriend someone Pauline admired.

By then, however, the die was cast. Late in July Mileva retook the graduation exam for the Polytechnic. Not surprisingly, given the nerve-rattling circumstances of her life, she failed again, with virtually the same average as before. Things with Weber had also come to a bad pass, and she vowed never to work with him again. Mileva's science days were over.

She packed her bags to go back home to Novi Sad. Before departing, she asked Albert to send a short note to her father, and to show it to her, so that she would know what he had said. "Write to my Papa just briefly, I shall then gradually break the necessary news, the disagreeable ones included."[25]

They agreed not to tell Michele.

She proposed one last rendezvous, so that she could return some money Albert had lent her. There was a train that left Zurich in the morning, stopped at Mettmenstetten at 7:56 A.M., proceeded to Zug, the end of the line, waited there awhile, and then reversed course. If she took the train and Albert slipped away from his mother and got on at Mettmenstetten, they could have an hour together. "Would you like to take this journey with me, sweetheart?"

Albert's answer is not known.

7

IRREVERSIBLE ACTS

GIVEN THE CIRCUMSTANCES OF MILEVA'S PREGNANCY, PERHAPS IT IS understandable that Albert chose to bury himself in science. "All society necessarily involves, as the first condition of its existence, mutual accommodation and restraint on the part of its members," wrote the philosopher Schopenhauer, whom Albert had admired as a youth and had recently begun quoting to his friends. "This means that the larger it is, the more insipid will be its tone. A man can be himself only so long as he is alone; and if he does not love solitude, he will not love freedom; for it is only when he is alone that he is really free."[1]

Pregnancy may have discombobulated Mileva, but if anything it seems to have concentrated Einstein's mind. In the summer of 1901, inspired by the realization that there was a common thread to several questions about nature that were obsessing him, he set out on a new project that involved nothing short of learning a whole new way of doing physics: statistics.

Three problems in particular had absorbed Albert's attention and frustrated him during the past year. The first was his paper on molecular forces in gases, which he was trying to extend into a Ph.D. dissertation. Next was Drude's electron theory of metals, which sought to ascribe their conduction of heat and electricity to a "gas" of charged particles inside the metal. Finally there was the mysterious relationship between the radiation absorbed or emitted by a body and its internal heat. Albert had thought the latter might have something to do with electrons being vibrated by incoming waves. The German Planck had given this approach a good try but had come to grief on, among other things, the mathematical incongruities between continuous electromagnetic waves on the one hand and discrete electrons on the other.

What these subjects had in common was that they all relied on a statistical description of nature, one in which the properties and behavior of large objects were determined by the behavior of the millions of tiny atoms inside them. The master of statistical physics was Ludwig Boltzmann, the embattled and often depressed Austrian defender of atoms, currently teaching at the University of Lepzig. Boltzmann, the son of an Austrian civil servant, who had been drawn into the study of gases as a graduate student at the University of Vienna, had done more than anyone else to attempt to reformulate the laws of physics, particularly those dealing with gases and thermodynamics, in terms of the motions of little mass points—atoms. The theory, as elaborated by him and others, went by various names: the molecular theory of heat, kinetic theory, atomic theory, and statistical physics.

No aspect of this work was more far-reaching, controversial, or—as Albert was to discover—fraught with subtle difficulties than Boltzmann's interpretation of the second law of thermodynamics, one of the most fabled and mystical principles of science. Few tenets of nineteenth-century science cast a longer shadow. Rudolf Clausius, a German physicist (and Weber's predecessor at the Polytechnic) who stated the law in 1850, had expressed the idea simply in his book *Die Mechanische Wärmetheorie:* "Heat cannot of itself pass from a colder to a hotter body." But from that simple-sounding statement vast implications flowed.

In every physical transaction, it seemed, a little energy was always squandered or wasted in friction or in heating up the surroundings, producing useless heat instead of work. The result was that you could never use all the energy at your disposal. This explained why no engineer, however clever, could ever build a perpetual motion machine, why the exhaust from an engine is hot, and why ice cubes melt in a drink but never re-form. Things run down. Clausius invented a quantity called *entropy* to keep track of this useless heat energy. Mathematically, the second law of thermodynamics said that entropy always increased or, at best under very special circumstances, stayed the same. Every irreversible process—the firing of a gun, the stamp of a foot, the smudge of lipstick on a cheek—increased the entropy of the world irretrievably. Once in the world, this useless heat, like the power of the devil, could only grow.

"Things get worse" could have been the motto engraved on Clausius's coat of arms. The first law of thermodynamics stated that energy was neither created nor destroyed: You couldn't win. The second law said you couldn't even break even. (There was, in time, a third law that said in effect that you couldn't get out of the game.) Extending this reasoning to its limit, Clausius predicted that there would come a time when the entire universe would run down, everything in equilibrium and at the same temperature, and suffer a "heat death" when the stars had all burned out of whatever made them bright. Newton's clock would eventually grind down to a last feeble tick. Clausius's logic, as solid as it was gloomy, became the perfect fin de siècle idea. Entropy became a literary catchphrase of some renown. The

German sociologist Max Weber appropriated it to explain the tendency of organizations and bureaucracies toward bloating and stultification.

Boltzmann had stumbled for years trying to find a mechanical atomic basis for what he called "the second law of the theory of heat."[2] His genius was to identify entropy as disorder. Heat, he said, flowed from hot bodies to cooler ones for the same reason that aces tend to disperse themselves through a deck of cards. The laws of probability favored a situation in which a given amount of energy was spread over many particles rather than just a few. Left to themselves, assemblages of atoms tended toward the most likely configuration: Order slid to disorder; the velocities of atoms randomized.

But it turned out there was a paradox, pointed out by the physicist Johann Loschmidt in 1876. When two atoms bounced off each other, they obeyed Newtonian laws of action and reaction, in which any action can be reversed. A billiard ball coming from the right and striking another ball which goes off to the left is no more or less lawful than a billiard ball coming in from the right and hitting a ball to the left. In either case the sacred quantities of energy and momentum are conserved. Even if time itself were reversed and the planets revolved counterclockwise around the sun, all the laws of physics would still be satisfied. Now, however, imagine a bottle of perfume opened in a room and the perfume molecules drifting out until they can be smelled in the farthest corner of the room. If at the end of that process the direction of all the molecules in the room were reversed, every subsequent molecular collision between perfume and air would be in accordance with Newton's laws. But the result—all the perfume molecules reassembling in the bottle—would be an absurdity, a violation of the second law. How could microscopic correctness lead to macroscopic madness?

Boltzmann's debating partner Wilhelm Ostwald had used this paradox to argue against the atomic theory in 1895. "The proposition that all natural phenomena can ultimately be reduced to mechanical ones cannot even be taken as a useful working hypothesis: it is simply a mistake," he said bluntly.[3] The fact that mechanical laws worked equally well going forward or backward in time, he argued, meant that "in a purely mechanical world there could not be a before and after as we have in our world: the tree could become a shoot and a seed again, the butterfly turn back into a caterpillar, and the old man into a child. No explanation is given by the mechanist doctrine for the fact that this does not happen, nor can it be given because of the fundamental property of the mechanical equations. The actual irreversibility of natural phenomena thus proves the existence of processes that cannot be described by mechanical equations; and with this the verdict on scientific materialism is settled."

The reply that had eventually been dragged out of Boltzmann was that in the new statistical world of atoms, fate was a matter of probability. The second law of thermodynamics, that old pillar of classical physics, was no longer absolute, but merely extremely likely. Entropy increased *almost* all the time, but it might not. Although Ostwald was in principle correct, flukes *could* happen without Newton and Helmholtz turning over in their graves. Sometimes aces came up. The perfume could go back into the bottle, but you would have to wait forever to see it happen. For all practical and observable purposes, it was impossible.

Demoralized by the atom wars that had sprung up at the end of the century despite his theory's success, Boltzmann had drifted out of physics into philosophy and gotten depressed, and even attempted suicide in 1901.

Brushing up on his Boltzmann in the Winterthur library, Albert seized on it as the key to his own endeavors, and he began burrowing into the Austrian's work. Boltzmann's explanation of the second law sounded good at first glance, but as Albert dug into the details of Boltzmann's papers and started digesting his arguments, he was disappointed. For one thing, Boltzmann, he thought, was going about the subject backward. The Austrian had been anxious to use atoms to prove the second law, whereas Albert thought that it was more important to be able to use thermodynamic and statistical data to prove the existence of atoms. As a result, Boltzmann seemed to have fudged his proof of thermodynamics with assumptions about the nature of atoms as well as dubious statements about the nature of probability. Albert wanted a new, cleaner, more general theory of statistical mechanics that didn't have so many assumptions built in. And so he now set out to redo Boltzmann's work and reinvent what was called statistical mechanics.

"Lately I have been engrossed in Boltzmann's works on the kinetic theory of gases," he wrote to Grossmann in September of 1901, "and these last few days I wrote a short paper myself that provides the keystone in the chain of proofs that he had started. However it is too specialized to be of interest to you. In any case, I'll probably publish it in the *Annalen*."[4]

After a fast start, however, he soon bogged down when he discovered that he had unwittingly replicated some of Boltzmann's alleged mistakes. His paper, the first of what would be three, wasn't done until the following June. Thus began what would ultimately be a three-year journey deep into the thickets of probability and statistics as Albert pursued his goal of a theory that would be so transparent that he would be able to see through it the dances of the atoms themselves, the stony firmament at the bottom of reality.

In the meantime, fate intervened again to give him another job. Grossmann had just beaten him out of yet another secondary-school teaching position, this time at the Frauenfeld Canton School. The patent office was still a glimmer on the horizon, and there was no word from Besso. So once again Albert settled for a humbler position in society—this time as a tutor in a private school in the old Rhine city of Schaffhausen. His sole mission would be to coach a young Englishman for the high school *Matura* exams. Albert had been recommended for the job by an old Aarau school chum from Schaffhausen, Conrad Habicht.[5] He would be paid 150 francs a month plus room and board. While Albert would be on somewhat of a short leash financially, he would have time to continue thinking and working and wrestling with Boltzmann. Besides, it was a year's worth of security.

In September of 1901, Albert took the train north to his new home. Schaffhausen, its cobbled, walled streets arranged around an eleventh-century cathedral, a vineyard fortress whose cannons command the Rhine valley to this day, is one of Switzerland's prettier towns, famous for its rococo architecture and frescoed facades. An astronomical clock tower with gold sun, moon, and planets sweeping the zodiacal circle, capable of tracking eclipses and the changing seasons as well as the time, stands over Fronwagplatz, the central marketplace.

Albert's post, on the outskirts of town, was somewhat removed from these glories. Jacob Nüesch, the proprietor of the school and Einstein's nominal employer, was a bit of an operator. In addition to running his own school, he taught mathematics at the Schaffhausen *Realschule*. Nüesch had worked out a sweet deal with the English student's mother, who agreed to lay out 4,000 francs for a year of Albert's time with her son.[6] Out of that princely sum, Nuesch passed on 150 francs a month to Albert and kept the rest as reimbursement for his room and board. The sparseness of those accommodations led Einstein to suspect that Nüesch was making a profit on the arrangement. He spent the first two months living at the school, and taking his meals with Nüesch, his wife, Bertha, and their four children. In November Albert was moved into the home of another *Realschule* teacher, but he still had to eat with the Nüesch household.

The meal situation rankled him, but Albert put his head down during his spare time and began working in earnest, pulling together his thoughts about molecular forces and the kinetic theory of gases, including his criticisms of Drude and Boltzmann's fateful equations. He hoped it would make an acceptable doctoral dissertation.[7] Later that fall he met with Kleiner in Zurich and left him a number of papers, one of them presumably his dissertation.[8] Kleiner was gracious, and even expressed some interest in yet another experiment Albert had dreamed up to measure the earth's motion through the aether.

Meanwhile, of course, the Einstein-Maric domestic front was boiling with tension and you didn't have to be a Boltzmann to predict that a violent release of social entropy was inevitable. From the end of July 1901, when Mileva left Zurich for home, until that November, when she turned up unexpectedly back in Switzerland, there is a gap in the available record of Albert and Mileva's correspondence.

Aside from his references to Schopenhauer's heroic solitude, it's hard to know what Albert thought about the whole matter of her pregnancy, or whether he perhaps made an effort *not* to think about it. The strict angels of science, to whom he had vowed allegiance when he broke up with Marie, might have looked pretty good right now to a twenty-two-year-old, underemployed, bright, ambitious, a little emotionally immature, and in the worst trouble of his life. As he had written to Pauline Winteler five years earlier, often he appeared to himself as an ostrich with his head stuck in the sand.

Milos Maric's reaction to his daughter's "disagreeable" news has been swallowed into this historical black hole of postal negligence or kindling, but Pauline Einstein's reaction is on the record. When she discovered that her worst fears had been realized, she exploded. In October she and Hermann wrote a blistering letter to the Marics reviling Mileva, her family, and presumably the entire Serb culture, shredding whatever superficial calm might have existed along the Jewish-Serb front.[9] "That lady," Mileva said to Helene, "seems to have made it her life's goal to embitter as much as possible not only my life but that also of her son. Oh, Helene, I wouldn't have thought it possible that there could exist such heartless and outright wicked people!"[10] Mileva's parents responded in kind. It seemed almost certain now that neither set of parents would ever allow these young lovers to be legally united.

Mileva, already struggling, promptly sank beneath the waves of depression. She poured her heart out to Helene Savic in a series of letters so foul she couldn't bear to send them, ripping them up instead. It was just as well, Mileva thought. Helene, about to give birth to her own child, still did not know about their upcoming child, whom they had nicknamed "Lieserl." She should be treated gingerly, Mileva decided, in case they needed her help down the road.

At the end of October Mileva packed a bag and went secretly to Switzerland to find the only one who could console her—Albert. In order to avoid detection, she went not to Schaffhausen but to the next town upstream, Stein am Rhein, where she checked in to the Hotel Steinerhof and sent Albert flowers.[11] What she got in return was a frustrating dance.

Either busy or impoverished, Albert said he couldn't come to see her—

at least not immediately. Among other things, his cousin Robert Koch had come to visit him and then lost his ticket home. Instead Albert sent her some books, including one about hypnosis by Auguste Forel, director of the famous Burgholzi Clinic in Zurich.

"I am writing you now only a few words because I am angry with the cruel fate which ordained that tomorrow I must sit alone," Mileva responded.[12] She went on to beg Albert not to tell his sister—her lone sympathizer in that viperous family—that she was here. Maja meant no harm, she knew, but Mileva was terrified that something bad would happen.

Her secret kept, Mileva passed the time undisturbed, reading, enjoying the time away from parental tempests. She had a long-standing interest in psychology and she picked up the Forel book eagerly, only to find herself disgusted by the immorality of the hypnotic process. "Such a violent surprise attack on human consciousness," she called it.

Meanwhile Mileva's visit remained unconsummated. Albert's next day off was Thursday. When he wrote, pleading poverty, that he again couldn't come, Mileva was incredulous.

"My dear, wicked little sweetheart!

"So now you are not coming tomorrow! And you don't even say: I am coming on Sunday instead! But you'll surprise me then for sure, won't you? Let me tell you, if you are not coming at all, then I'll run away all of a sudden! If you knew how homesick I am, you would surely come.

"Very nice! The man earns 150 francs, has bed and board, and doesn't have a centime at the end of the month! But this won't serve as an excuse for Sunday, will it, and if you don't get any money by then, I'll send you some."[13]

Still, Mileva's anger couldn't stop her from being supportive when she got around to talking physics. "I am very curious what Kleiner will say about the two papers. He'd better pull himself together and say something reasonable."[14]

Shortly thereafter Mileva went back to Novi Sad. Reporting to Helene on her travails, she probably broke the news that she was pregnant. Their prospects were not bright, she explained, despite the cessation of outright hostilities. Even with a looming doctorate, Albert still lacked much chance of a real job, and Mileva added, "You know that my sweetheart has a very wicked tongue and is a Jew in the bargain.

"From all this you can see that we are a sorry little couple. And yet, when we are together, we are as merry as hardly anybody. When I was now in Switzerland, we saw each other a few times. And you know: in spite of all the bad things, I cannot help but love him very much, quite frightfully much, especially when I see that he loves me just as much."[15]

The next letter from Albert does not appear until the end of November, when he complained that his letters to her must be getting lost because none were coming back in return.[16] "Did you receive the 2 or 3 letters that I mailed to Kac and the one I mailed to Novi Sad? I almost believe that your mailmen use the letters for kindling or even . . . *horrible dictu*, but I am not saying it."

Back in Schaffhausen, Albert had moved out of the school and into the house of one of the other science teachers. Despite their common occupation, Albert found his new host insipid, and his Schopenhauerian isolation reached new heights of splendor. Aside from his student, he bragged to Mileva, he spoke to no one. Which was not quite true—he occasionally visited his old school friend Habicht and they practiced violin together. For the most part his days consisted of walking, studying, attending chamber music concerts, and waiting with mounting impatience to hear from Kleiner or Grossmann.

He reported to Mileva, "So far I have no report from Kleiner. I don't think he would dare to reject my dissertation, but otherwise, in my opinion there is nothing that can be done with that short-sighted man. If I had to be at his beck and call to become a university professor—I wouldn't want to change jobs, but would rather remain a poor private tutor."

Two weeks later he still had not heard from Kleiner nor Grossmann. It looked as if he was destined to stay in Schaffhausen for quite a while, but his indentured lifestyle had become intolerable. Albert went to Nüesch one morning and asked to be paid his meal money directly so that he could make his own dining arrangements and eat or not eat as he saw fit in order to save money.[17]

Nüesch, flushed with rage, said he'd consider it, but that evening turned Einstein down. "You know our conditions and there is no reason to deviate from them," Nüesch declared icily. "You can be quite satisfied with the treatment you are getting."

"Very well," said Albert, skating on pure nerve, "as you like, I have to give in for the time being—I'll know how to find living conditions that suit me better."

Nüesch softened and offered a compromise that would allow him to maintain financial control and Albert to escape his stifling family. He would arrange for Albert to take his meals in the Cardinal Inn, down the street. "They are now foaming at their mouths with rage against me, but I am now as free as the next man," he told Mileva.

Walking in for his last supper with the Nüesches, Albert received the last key to his freedom. There was a letter on his soup plate. It was from Grossmann. The position in the Bern patent office had finally cleared its bureaucratic hurdles and would be advertised in the next few weeks. Moreover, Albert was almost sure to get it.

He immediately wrote to Mileva. "I am dizzy with joy when I think about it. I am even happier for you than for myself," he said.

"We shall remain students (horrible dictu) as long as we live, and shall not give a shit about the world. But neither shall we forget that we owe everything to the kind Marcellus, who tirelessly looked out for me. Also, I will always help gifted young men whenever this will be in my powers, this is a solemn oath I am taking."[18]

The only problem that remained, he went on gingerly, was how they could keep their Lieserl with them. He assured Mileva that he didn't want to give the child up. Albert suggested that Mileva speak to her father, who was a man of the world, more experienced than "your impractical book-worm Johonzel." How they would care for this bundle that was turning Mileva's body—as she complained—into a "funny figure" while remaining eternal students who didn't give a shit for the world, he didn't say.

His new eating arrangement so suited Albert that he eventually moved into the Cardinal Inn. His only complaint was that his new dining partners were so stupid and commonplace.[19] "So I am sitting at the meals like a nut-cracker and am playing with knife and fork between courses, while looking out the window. The fellows must think that I have never laughed in my whole life; but, then, they have never seen me with my Doxerl."

Within a few days Einstein received a letter from Friedrich Haller, the director of the Swiss Patent Office, inviting him to apply for the new job. Albert wrote back immediately, and then went to Zurich. He'd had enough of not hearing from Kleiner; he decided he would just drop in on him and try to convince Kleiner to let him work over the Christmas vacation on his new ideas about the aether and electrodynamics.

"To think of all the obstacles that these old philistines put in the way of a person who is not of their ilk, it's really ghastly! This sort instinctively considers every intelligent young person as a danger to his frail dignity, this is how it seems to me by now. But if he has the gall to reject my doctoral thesis, then I'll publish his rejection in cold print together with the thesis and he will have made a fool of himself. But if he accepts it, then we'll see what a position the fine Mr. Drude will take . . . a fine bunch, all of them. If Diogenes were to live today, he would look with his lantern for an *honest* person in vain."[20]

In the manner of reluctant editors and advisers everywhere, Kleiner told Albert that he still hadn't read his thesis but promised to do so after the vacation. Albert told him to take his time, that he was in no hurry. They spent the afternoon discussing Albert's new ideas about what he called "the electrodynamics of moving bodies." Over the years, devising aether experiments became a sort of hobby that he turned to whenever he had a spare moment or got stuck on whatever other problem he was pondering. After being temporarily derailed by some mathematical errors, Einstein had become convinced he was on the way to a breakthrough. Kleiner was encouraging. Einstein's proposal, he said, was the simplest and most appropriate yet conceived. He urged Albert to write up his electromagnetic ideas along with his proposed experiment. Kleiner offered to give him a good recommendation whenever he needed one.

Albert concluded that Kleiner wasn't so dumb after all. "I shall certainly write the paper in the coming weeks," he told Mileva.[21] Unfortunately he never did, and so we have no idea what those ideas about electromagnetism were. By the end of Christmas he had embarked on a new round of reading and research. Besso sent him a book on the theory of the aether, written in 1885, which Albert found hopelessly antiquated. He asked Ehrat, who was working for Minkowski at the Polytechnic, to send him more modern articles by Drude and a Dutchman named Lorentz.

He professed to be amused that in the race for a doctorate he was ahead of his academically enabled friends like Grossmann, who was writing on the topic of something called non-Euclidean geometry ("I don't know exactly what it is"),[22] and Ehrat. "You see, your Johonzel has been the first to finish his dissertation, even though he is a harassed little beast. When you are my dear little wife, we will zealously do scientific work together, so as not to become old philistines, right?" he asked Mileva. "You must always remain my witch and my street urchin."

A thousand kilometers east, Albert's once and future street urchin moved into her ninth and climactic month of pregnancy, her belly swollen, complaining—no hormonal burst of maternal goddess energy for her—her fate less and less clear. Albert was still unemployed (although with better prospects) and his family was resolute. Marriage was out of the question. So, apparently, was keeping the child.

Under the circumstances Mileva's parents might already have decided to place the child with relatives or to give it up for adoption.[23] Milos Maric was a soldier, a judge, and a Serb. It had been his job to be the very personification of rectitude. The proper and customary thing to do would have been to send the illegitimate child to a distant relative.

There would have been other pressures on Mileva to give up the child. Albert's job prospects at the patent office may have depended on it, for illegitimacy was hardly tolerated in turn-of-the-century Switzerland, especially in the expensive, stuffy capital city of Bern. Switzerland was a small country, and the community of Swiss physicists was even smaller. Within this small world Albert already had several strikes against him: his career of insolence at the Polytechnic, and a Slavic girlfriend. And of course he was a Jew. He had been a citizen of Switzerland for a year, and he was up for an important civil service job. For him to show up in Bern with an illegitimate child might well have left him with his eccentricity account overdrawn.

If this was true, then Mileva was being presented with a cruel choice indeed. The prospect of Albert's patent office job was the one reed of sunshine in their lives, their best chance of eventually being united. As if in some Greek tragedy, then, the price of their life together would be their child. Mileva was too intelligent and introspective a woman not to be aware of the irony as she set about mortgaging her happiness to Albert's career. Was there anything less bohemian and more conventional, more inimical to Albert and Mileva's own ideals? Perhaps if one set or another of the parents had been willing to compromise, another choice would have been possible. But in a society in which women could still be arrested for smoking on the street, Albert and Mileva were truly alone.

In January Kleiner finally read Albert's doctoral dissertation. Its criticism of Drude, and particularly of Boltzmann, turned out to be more than he could stomach. Out of respect to Boltzmann, Kleiner asked Albert to withdraw it.[24] Albert did so, and got back the 230 francs he had paid to submit it to the University of Zurich in the first place. Then he told off Nüesch, packed his bags, and took the train to Bern.

On a clear day you can see the snowy teeth of the Alps from the terraces of the Parliament building on the south side of Bern. Nearly encircled by the mountain-green rushing waters of the Aare River, the main part of the city occupies a rocky promontory bounded by high cliffs. Bern has remained virtually unchanged since the fifteenth century, when after a fire the city was rebuilt with heavy medieval arcades and uniform red tile roofs—even the chimneys all have their own little tile roofs. Cobblestoned thoroughfares, festooned by fountains and lined with the aforementioned arcades, run the length of the peninsula from the train station to the site of the old fortress and landing in Nydegg. In a plaza across the river, brown bears, which have been the symbol and namesake of Bern since it was founded in the twelfth century by Berchtold V, cavort in a pit for tourists and children.

"An ancient, exquisitely cozy city, in which one can live exactly as in Zurich," Albert reported.[25] He was especially intrigued by the arcades, which meant that one could walk from one end of the city to the other in the rain without getting wet. Albert found a room for 23 francs a week on Gerechtigkeitsgasse, downhill in the old part of town. It came with a sofa and six chairs. He joked that he could hold a meeting in it, which he might have to. Until the patent office job came through he planned to support himself by giving private physics lessons. He sent Mileva a diagram of his chamber, with the locations of everything from the bed to the chamber pot, labeled with Greek and Latin symbols.

Mileva went into a brutal and exhausting labor. When it was over she was too sick to write. Albert, who had suspected things were not going well, got a start when he received a letter with Milos's name on it, and he was almost too afraid to open it. Inside, Milos told Albert that he was the father of a girl—a Lieserl, as Mileva had wished. For an instant, Albert said, all his own troubles seemed small. Even two more years with Nüesch seemed like a cheap price to pay if it would bring back Mileva's health.

Terror, however, turned to instant curiosity, and he showered Mileva with questions about the baby's health and her looks. Which of them did she resemble the most? "I love her so much and I don't even know her yet! Couldn't she be photographed once you are totally healthy again? Will she soon be able to turn her eyes toward something? Now you can make observations. I would like once to produce a Lieserl myself, it must be so interesting." Lieserl had been born knowing how to cry, Albert noted. But it would take her much longer to learn to laugh.

Therein, it seemed to him, lay a profound truth.

BACK IN AARAU, Maja Einstein made the mistake of mentioning that Albert was engaged to be married to Mileva. The rumor made its way around to Pauline Winteler, who wrote in turn to Pauline Einstein to ask what was going on with the son who had once jilted her daughter. Albert's mother wrote back that Maja hadn't mentioned anything to her about the purported engagement, "knowing well that in this case she wouldn't have found a sympathetic listener in me either."[26] In fact, Maja and her mother had argued violently about Albert and Mileva over Christmas in Mettmenstetten and had parted frostily.

"This Miss Maric is causing me the bitterest hours of my life," Pauline said darkly. "If it were in my power, I would make every possible effort to banish her from our horizon, I really dislike her. But I have lost every influence on Albert."

Whether she was curious about her granddaughter, she never said.

8

THE BOYS OF PHYSICS

ON WEDNESDAY, FEBRUARY 5, 1902, THE FOLLOWING AD APPEARED IN the Bern newspaper *Anzeiger für die Stadt Bern:*

PRIVATE LESSONS IN
MATHEMATICS AND PHYSICS
for students and pupils
given most thoroughly by
ALBERT EINSTEIN, holder of the fed.
polyt. teacher's diploma
GERECHTIGKEITSGASSE 32, 1ST FLOOR
Trial lessons free

By the end of the week Albert had two customers, an engineer and an architect, willing to pay two francs an hour for a private course on the mysteries of the universe, or at least on atoms, heat, and the aether.

For amusement, Einstein started spending Saturdays with an old Aarau friend, Hans Foresch, now a University of Bern medical student, at a forensic pathology class. "They questioned a sixty-year-old woman who attempted arson while senselessly drunk as well as a man accused of fraud who appears to suffer from megalomania," he reported to Mileva, mentioning that several other such "psychological swindlers" had been described in the Forel volume they had both read the previous fall during Mileva's brief stay on the Rhine.[1]

One of the first visitors in Albert's new quarters was Max Talmey, his old dinner-table mentor from Munich days. Talmey, now a full-fledged physician, had recently called on Albert's parents in Milan and found them mysteriously reticent about their son. When he saw how Albert was living, he thought he knew why. "His environment betrayed a good deal of poverty. He lived in a small poorly furnished room. He told me that he was struggling hard for a living," Talmey later wrote.[2] "His hardships were aggravated through obstacles laid in his way by people who were jealous of him." Albert later complained that he could have made more money playing his violin in the street.[3]

A couple of months later, during the university Easter vacation, Albert reran his ad, though by now he had raised the price to three francs. A restless Bern University student named Maurice Solovine came across the notice and followed it to the second floor of 32 Gerechtigkeitsgasse. There he rang Einstein's bell.

A deep voice thundered, *"Herein!"* and Albert stepped out of the door. "The hallway was dark," Solovine remembered, "and I was struck by the extraordinary radiance of his large eyes."[4]

In a photograph from the time, Solovine looks slight and swarthy with hooded eyes, slicked-back hair, and a de rigueur black moustache. The son of a wealthy Rumanian merchant, Maurice was a perpetual student. He had come to the University of Bern the year before, at the age of twenty-five, not sure of what he wanted to study. His first love was philosophy, but the discipline, he thought, was declining into a preoccupation with criticism and negativity. Solovine had therefore also taken a heavy load of mathematics, physics, geology, and medicine in order to become aware of the "paths and procedures through which the mind achieves positive results."

Unhappily, as he explained to Albert, the same disease that afflicted philosophy had also infected science. What most interested him in physics were the theories—the big picture that had been emphasized by the popularizer Aaron Bernstein, who had captivated the young Einstein. To his despair, however, Solovine's professor in Bern disparaged theoretical studies, contending that theories were little more than hypothetical constructions at the mercy of experimental facts. It is the latter that endured, Solovine's professor claimed, while theories were prone to topple and disappear. Meanwhile, Solovine's attempts to do theoretical work on his own were stymied by his lack of mathematical prowess.

It sounded all too familiar to Albert, who sensed an immediate kinship during this discussion. He confided that he too had been philosophically inclined when he was younger, but now he was strictly a man of science. The two young men talked for two and a half hours. Solovine returned the next day, and they continued the conversation. They never did get to the physics lessons and no money ever changed hands. Einstein decided that he was having too much fun to charge Solovine money; Maurice didn't need a tutor as much as he, Albert, needed a friend to talk to.

Solovine suggested that they read books by some of the main thinkers of the day so that they could talk about them together. Albert agreed. The pair was soon joined by Conrad Habicht, Albert's pal from Schaffhausen, who had moved to Bern to study mathematics. The first book they read was *The Grammar of Science* by Karl Pearson, a biologist who may have been the first scientist to compare the workings of the brain to a telephone exchange.

The "Olympia Academy," as the trio dubbed themselves, was part fraternity, part hiking group, part dinner society, and part debating club. Their weekly meetings were not exactly extravagant affairs. Dinner typically consisted of a bologna sausage, a piece of Gruyère, fruit, honey, and a cup or two of tea. Albert was a notorious teetotaler and his friends seem to have followed his lead. "The words of Epicurus applied to us," Solovine later wrote. " 'What a beautiful thing joyous poverty is!' "[5] After dinner they would begin to read from the prescribed text, but often got no further than a page or even a paragraph before arguments broke out.

When the days grew long and the nights warm the trio would often adjourn and take a midnight hike up the Gurten, a domelike hill and park on the southern outskirts of Bern. From its 840-meter summit they would gaze at the stars and talk on until sunrise colored the nearby Alps pink and the restaurants opened, and then talk some more over black coffee. On other occasions they would leave in the morning and hike to Thun, a medieval town about thirty kilometers south, on the edge of the Thuner See. There they would hang out by the lake or climb the Beatenberg, a mountain with a huge bite out of its top which overlooks the lakeshore, and take the train back in the evening.

Mileva, weighed down with more than the traditional postpartum blues, returned to a snowy, dank Zurich in the shank of the winter, presumably to the very same boardinghouse on Plattenstrasse, where by now she had lived on and off for seven years. She arrived without Lieserl. The child was presumably left behind with Mileva's family or that of a friend, the most likely candidate being Helene Savic. Whether this arrangement was intended to be temporary or a permanent adoption is not known, but it was hardly a happy situation for Mileva. Moreover, her chief source of comfort was not even around, but was off instead in Bern having what sounded like an alarming amount of fun with his new pals. Mileva had always been his one true physics partner.

With nothing to do but convalesce and wait for Albert to visit, Mileva's insecurities came to the fore, and her letters to him that spring seem to have amounted to one long petition for his attention, judging by Albert's responses. She shouldn't be jealous of his friends, he said. He longed for her every day, but it would not be "manly" to show it.

A typical reassurance prefaced a hiking trip to the Beatenberg.[6] "I'd rather be going with you to the Beatenberg than with a group of men—

I'm a man myself after all," he declaimed. "When I'm not with you, I think about you with such tenderness as you can hardly imagine; this in spite of the fact that I'm a bad boy when I'm with you."

Albert's long wait for a real job finally came to an end on June 16, when he was appointed technical expert third class to the Swiss Patent Office. In the official notices, Albert learned shortly after he moved to Bern, the director, Haller, had phrased the job qualifications in the broadest way possible, emphasizing physics, in order not to militate against Albert's lack of formal engineering training.[7] During a two-hour interview earlier in the winter, Haller had tested Einstein's mettle by springing new patents on him and asking what he thought of them.[8] Albert was no engineer, but he did have experience in his father's company as well as the gift of being able to visualize things like standing electromagnetic waves in his head. Moreover, he almost certainly knew more about electrodynamical theory than anyone else likely to walk in the door looking for a job. About half the office's business in those days consisted of foreign patents seeking licenses in Switzerland, and a large number of those were in the burgeoning electrical industry.[9] Albert's abilities and the old school tie were apparently sufficient to convince Haller to hire him—though as a third-class examiner rather than second-class as advertised—at 3,500 francs a year. Haller also prescribed technical drawing lessons for his new protégé.

The patent office was in a handsome limestone building on Speichergasse, a couple of blocks north of the main railroad station. Just across the tracks and up a flight of steps on a green grassy hill overlooking the station and the red-roofed city sat the University of Bern. On a corner down the block stood a dark, rambling cave of a restaurant known as the Cafe Bollwerk, where Albert often spent his lunch hour continuing arguments from the night before with Solovine or Habicht. At the office itself men in winged collars sat at tables along the wall under hanging electrical lamps, shuffling papers under Haller's watchful eye. Over time, most of his colleagues would conclude that Einstein was shuffling more than patents around on his desk.

Albert was in fact busy turning his rejected doctoral dissertation about statistical mechanics into a series of papers. The first one was completed just before Albert joined the patent office. In it, he used the atomic theory of gases to derive the famous second law of thermodynamics, correcting Boltzmann's alleged mistakes along the way. No sooner had he finished, however, than he set out again in a second paper to derive the law again in a more general and powerful way that would apply to mechanical systems, not just gases. Then, after one of his patent office colleagues pointed out a flaw in his reasoning, he produced yet another, even more general, paper that extended his statistical legerdemain even to electromagnetic fields.

Albert bragged to Besso that this work depended only on the "energy principle"—the assumption of the conservation of energy—and on the existence of atoms.[10] But it also hinged on a novel interpretation of the meaning of probability. Given a million perfume molecules floating out of a bottle into a room, what is the probability that they will wind up in one configuration or another? One way to find out is simply to watch the room for a long, long time and determine how much time the molecules spend in that particular configuration—plastered against the walls, say. Another method is to use statistics to count all the ways the molecules could wind up against the walls. (Solving the problem is therefore analogous to calculating the odds of drawing a pair of aces in a poker hand either by dealing out a thousand separate hands, or by calculating the number of ways five cards can contain two aces.) Albert argued that over the long run the two answers must always be the same. Time was really the unfolding of possibility. The actual was the most likely and vice versa.

It took Albert two years to finish these three papers and publish them in the *Annalen der Physik*, where the last one appeared in the spring of 1904, but they did not set the world of physics on fire. As it happened, an American physicist, Willard Gibbs, had already done the same work, laying what is generally considered the foundation of modern statistical mechanics and thermodynamics. Albert was unaware of Gibbs—those were the wages of working in isolation. But even if they did not shake up science, the papers had a lasting influence on Albert's style as a scientist. The notion that one could start with a simple fact or supposition, such as the existence of atoms, and construct, on the basis of pure logic, a grand sweeping principle such as the second law of thermodynamics would remain with him for the rest of his life as the very model of what a physical theory should be.

Perhaps even more important, his efforts at proving the second law served as an apprenticeship in the new field of statistical physics. By the end of the two years Albert had quietly become an adept at a new way of sifting the facts of nature. He was like a young guitarist who had spent years by himself playing along with old records in the basement, later to emerge with the chops of a master.

Meanwhile, Mileva waited. She moved closer to Bern in order to be near Albert, but nothing was happening. There were now no financial obstacles to their marriage, but there were still plenty of emotional ones. Albert's parents remained steadfast in their opposition to it. Albert kept his head down and did not push the issue. Obediently, he bided his time, as he and

Mileva had grown accustomed to doing. Outwardly, anyway, he seemed to be in no hurry.

Between his job, his private students, the ongoing Olympia Academy, and spinning out papers on statistical mechanics, Albert was busy, but relatively happy. He wrote to his old friend Hans Wohlwend that he enjoyed the variety of his new position, and liked the pay even better.[11] Bern was pleasant, he said, because he had no philistine acquaintances there—that is to say, family—and thus much free time.

Against bourgeois distractions, eternal vigilance was the only defense. One night on the way to Albert's apartment, Maurice, a music fan, noticed that a famed Czech string quartet was going to be in town the following week. He suggested to Einstein, who sometimes entertained his academy friends on his own violin, that they all attend the concert. Albert replied that they would be better off reading Hume, "who is extremely interesting."

Solovine capitulated, but the day of the concert he found himself walking by the theater and couldn't help buying a ticket. He hurried to his house, where the meeting was to be held that night, and set out dinner—a repast that included hard-boiled eggs in addition to the usual bologna. He left the food with a note, *"Amicis carissimis ova dura et salutem"* ("Hard eggs and a greeting to very dear friends"), and then told his landlady to explain that he had been called away on an urgent matter.

Albert and Conrad penetrated the ruse immediately and set out to teach their comrade a lesson. First they ate everything in sight. Then, knowing that Solovine hated tobacco in any form, they both began smoking like fiends, Einstein his pipe and Habicht puffing on thick cigars. When Solovine finally got home at two in the morning, the room stank so much that he could barely breathe. All his furniture and possessions were piled almost to the ceiling on the bed, and a saucer was overflowing with ashes and cigar butts. On the wall they had pinned a note: *"Amico carissimo fumum spissum et salutem"* ("Thick smoke and a greeting to a very dear friend").

Albert scowled when Maurice walked into his apartment the next week. Solovine was a wretch and a barbarian, Einstein declared, and warned, "If you ever again indulge in such folly, you will be excluded and shamefully expelled from the Academy."[12]

In such fashion was the Great Conversation begun by Plato continued. By the fall the "Academy" had read Mach's classic *Analysis of Sensations and Mechanics,* which Albert had browsed before on the advice of Besso, and moved on to works by Hume, Mills, Spinoza, Helmholtz, the mathematician Bernhard Riemann, Poincaré, Sophocles, and Dickens. The reading list was oriented toward philosophy, especially as it related to science and the burgeoning conceptual difficulties of physics at the turn of the century. Was there a real world, and could we know it? Or were we condemned in trying to understand it to studying not the world itself but merely our own bumbling sense interactions with it?

The dream of absolute knowledge had been dying for a long time, but it was David Hume, Albert maintained, who had delivered the final blow to Plato's dreams. In Hume's view, not only philosophy but physics as well was an impossible science. Hume had written in the early eighteenth century in the heyday of the empirical school, which maintained that sensory impressions—sensations—were the only basis of knowledge. The philosopher pointed out, however, that there was no purely logical pathway from "facts" to the scientific laws proposed to unify and explain them. Exact laws could not be deduced strictly from experience. Such notions as cause and effect, he maintained, were the result of the mind's expectation that patterns once learned would repeat. "Custom, then, is the great guide of human life," wrote Hume.[13]

Albert was very impressed by Hume. "Man has an intense desire for knowledge," he later wrote.[14] "Hume's clear message seemed crushing: the sensory raw material, the only source of our knowledge, through habit may lead us to belief and expectation but not to the knowledge and still less to the understanding of lawful relations." Still, he wondered whether Hume had gone too far and instilled in science and philosophy a debilitating "fear of metaphysics."

It was to answer Hume's hammer blow that Albert's boyhood hero Kant had penned his immortal *Critique of Pure Reason*. If the empiricists were right, he argued, the world was the province of the senses, and mathematics was about thinking. What could they have to do with each other? And yet the triumphs of the time in physics were Newton's laws of motion and gravitation, mathematical extravaganzas that seemed capable of explicating the course of cosmic history from the largest planets and stars down to the bumping of hypothetical atoms.

Kant *agreed* that many of the features with which we endowed the world could not be logically derived from the blizzard of impressions hitting our senses, but he turned this putative defect into a virtue. Such notions, he argued, were in fact indispensable tools that the mind used and assumed about the world in order to organize those sense impressions. Among the components of this tool kit were the ideas of absolute space, time, causality, and the axioms of geometry.

Enter the bulldog skeptic and disbeliever in atoms, Ernst Mach, as counterpoint to Kant's counterpoint to Hume. For Mach, experience and experiment were paramount. Being constantly subject to verification by experiment and endless reformulation, the laws of physics could never attain the level of axiomatic truth, he argued. At best they could be considered heuristic devices—useful fictions—for the economic ordering of facts. Atoms and the aether were heuristic devices, and so, claimed Mach in his most radical and far-reaching argument, were space and time.

Newton had conceived of space as a giant fixed stage, God's sensorium, on which atoms and people strutted and fretted, clocks ticked, billiard balls bounced, and the planets revolved, but he had left a loophole, which became known as the principle of relativity. Newton had declared that the laws of physics were the same for any uniformly moving observer—whether, say, that observer was traveling on a smoothly sailing ship or standing on a dock. But that meant there was no experiment either of them could do to tell who was actually moving and who was standing still; in fact, the ship could be anchored while the dock itself was drifting. Motion was relative.

But, argued Newton, acceleration was not, and therein lay the clue to absolute space. Newton asked us to consider a bucket of water hanging by a twisted rope. As the rope is allowed to untwist, the bucket begins to spin, and the water, initially flat, develops a concave surface due to centrifugal forces. Those forces were not due to the water's motion with respect to the bucket, Newton pointed out, because the water and bucket were moving together. Rather, they were caused by the water's rotation in absolute space.

The alternative to Newton began with his contemporary the Baron Gottfried Wilhelm von Leibniz, who famously invited his readers to explain why, if space was a kind of stage, independent of its contents, God put the universe *here* and not a few feet to the left or right. Leibniz went on to posit a mystical universe inhabited by strange entities called monads, in which the properties and identity of each part of the universe somehow depended on its relation to every other part.

Mach was more hard-boiled. In his book *The Science of Mechanics*, which Einstein and his friends read in the fall of 1902, Mach deconstructed the ideas of absolute space and time by pointing out that we don't observe the stage, only the players; we don't see space, only objects in it. "No one is competent to predicate things about absolute space and absolute motion; they are pure things of thought, pure mental constructs, that cannot be produced in experience. All of our principles of mechanics are, as we have shown in detail, experimental knowledge concerning the relative positions and motions of bodies."[15]

Physics, Mach contended, had to be based purely on relative motions. In place of absolute space, he suggested that the fixed stars might serve as a reference, or perhaps the average of all the other masses in the universe, or even a medium filling space (the aether). As for Newton's bucket, wondered out loud whether or not the same result would be achieved if the bucket stayed still and the entire galaxy of fixed stars revolved around *it* instead.

The notion of relativity was also a persistent theme in the Frenchman Henri Poincaré's writings. He agreed with Mach and stressed the point that experiments dealt with objects and not space. Empty space and absolute space were both amorphous concepts. He extolled a principle he called the "relativity of space": If everything in the universe doubled in size overnight, we would have no way of knowing it. Only relative displacements had meaning.

If anyone could resolve the epistemological knots into which the recent discoveries had tied physicists' tongues and rescue the ambitions of science from the naysaying of the Machists, that person would seem to be Henri Poincaré, professor of mathematics at the University of Paris. A portly man of medium height, with a bushy beard, pince-nez, and a legendary air of distraction about him, Poincaré was the greatest living mathematician in France and perhaps the world, arguably the greatest living Frenchman. Poincaré was the last man who seemed to know everything. He had qualified for membership in every section of the French Academy of Sciences, from geology to mathematics. As the author of thirty books and five hundred articles, he was elected as well to the literary section of the Académie Française. On top of everything else, he was an avid dancer. Every year he lectured on a different field of physics. His trademark was inventing new fields of mathematics.

"To doubt everything or to believe everything are two equally convenient solutions; both dispense with the necessity of reflection," he wrote.[16] Poincaré was nothing if not reflective.

"Now, we daily see what science is doing for us," he wrote in the preface to his 1902 monograph *Science and Hypothesis*.[17] "This could not be unless it taught us something about reality; the aim of science is not things in themselves, as the dogmatists in their simplicity imagine, but the relations between things; outside those relations there is no reality knowable."

Like Mach and Hume, Poincaré regarded sensations as the building blocks of knowledge. Like Kant, he believed that these sensations were massaged into knowledge by innate faculties of the mind. One of those faculties, he said, was the ability to invent geometries, and here he veered sharply from Kant.

According to Kant, the only possible geometry was the regular old plane geometry that Euclid and his followers had invented and that dozens of generations of schoolchildren had torturously learned. In 1827, however,

Kant had been rudely disproved when the Russian mathematician Nikolai Lobachevsky showed that it was possible to construct a new and logically consistent kind of "imaginary" geometry by simply doing away with one of Euclid's cherished but most controversial axioms: that parallel lines never meet.[18] In this new space that Lobachevsky invented, parallel lines could meet and even cross, and the internal angles of a triangle added up to less than the standard Euclidean 180 degrees. His results were met with bewilderment and derision, but other examples of alternative geometry soon followed. In fact, using a mathematical technique known as group theory, Poincaré proved early in his career that an endless number of these so-called non-Euclidean geometries could be generated.

What did these alternate geometries mean, and what good were they other than for keeping Kant rolling in his grave? In 1854 a young Göttingen mathematician, Bernhard Riemann, demonstrated that they described the surfaces of various curved three-dimensional objects. They were, in effect, what would result if you took a piece of regularly ruled graph paper, representing Euclidean plane space, and shrink-wrapped it around a sphere or a saddle, say, or a doughnut, distorting the grid lines in the process. He divided these curved spaces into three classes: flat, or Euclidean; positively curved, or spherical; and negatively curved, or hyperbolic, and laid out imaginative tests by which an inhabitant of some space could figure out what kind of space he lived in.

Riemann, a man seriously ahead of his time, died ten years later at the age of forty, in the midst of a daunting mathematical effort to unify the laws of electromagnetism and gravitation. He and Lobachevsky seemed to have left behind a universe whose very form was suddenly uncertain. "What Copernicus was to Ptolemy, that was Lobachevsky to Euclid," William Clifford, one of the other names on the Olympia Academy's reading list, wrote in 1879.[19] "And in virtue of these two revolutions the idea of the universe, the Macrocosm, the All, as subject of human knowledge, and therefore of human interest, has fallen to pieces."

It was non-Euclidean geometry in its various forms that Albert's friend Marcel Grossmann was studying at that very moment for his Ph.D. dissertation. Between 1880 and 1910 more than 3,700 popular books and articles were published about alternate geometries.[20] Among those who later pored over the writings of Poincaré and the others was the young James Joyce.[21] The apparent subversion of humankind's aspirations for absolute truth was not lost on artists, for whom non-Euclidean forms came to symbolize the rejection of tradition and authority.[22] Some of them, particularly a group of painters in Paris who called themselves the Section d'Or Cubists, began to endow their canvases with wavy lines representing the mismatched parallels and with allusions to transcendant spatial dimensions. The first cubist exhibition would be held in 1907, when Picasso finally pulled his revolutionary masterpiece, *Les Demoiselles d'Avignon,* out from under the bed where he had hidden it for a couple of years. Meanwhile, the generalization of geometries to more than three dimensions had excited spiritualists and mystics, to whom the idea of the mysterious "Fourth Dimension" conjured up visions of the Astral Plane or transcendant realms of knowledge.

A large part of Poincaré's *Science and Hypothesis* was devoted to his response to the challenge wrought by Lobachevsky and Riemann. Poincaré agreed with their math, but he challenged their philosophical conclusions. Suppose an astronomer could do the experiments that Riemann had suggested, he said, and found that a pair of parallel light rays sent off into space converged in the far distance, or that triangles drawn on large scales were curved. Would we then be forced to abandon Euclidean geometry? Or would we conclude that some force was modifying the laws of optics, making the light rays bend? Needless to say, it would be the latter.

Euclidean geometry, Poincaré concluded, was a *convention,* a standard freely chosen from all of the possible geometries of which we can conceive. What was innate in the mind was not Euclideanism, but rather the mathematical principles out of which Euclideanism and all other geometries could be constructed. This convention was a "free choice," as he emphasized over and over again. "Experiment guides us in this choice," he wrote, "which it does not impose on us. It tells us not what is the truest, but what is the most convenient geometry."[23]

In principle then, we could choose any geometry with which to analyze the world, but it would make physics and our own survival that much more complicated. There was no experiment, Poincaré declared, that was explainable in Euclidean terms that was not also explainable, albeit more complexly, in Lobachevkian terms.

"To ask whether the geometry of Euclid is true and that of Lobachevsky is false is as absurd as to ask whether the metric system is true and that of the yard, foot, and inch is false," he concluded.

Poincaré's philosophy was sometimes called "conventionalism." Another convention was the aether, which he felt was necessary to establish continuity in space as well as to transmit light waves. "Whether the aether exists or not matters little—let us leave that to the metaphysicians; what is essential for us is that everything happens as if it existed, and that this hypothesis is found to be suitable for the explanation of phenomena," he had written in 1889.[24]

Still another convention, he declared, was simultaneity, the notion of two events happening at different points in space at the same time. "There is no absolute time," he wrote. "When we say that two periods are equal, the statement has no meaning and can only acquire a meaning by a convention."[25]

To Albert, who in childhood asked questions about space and time and wondered forever about the mystery that kept his father's compass aligned, these kinds of discussion were an elixir. Poincaré's book kept him, Solovine, and Habicht engrossed and spellbound for weeks.[26]

But life on the material plane had its way of demanding attention. Hermann Einstein had for a long time suffered from a bad heart. In October of 1902 his health failed. Albert was called home to Milan. On his deathbed Hermann called his son into the room, and at last gave him permission to marry Mileva.[27] Then he sent him out. Hermann died alone on October 10.

Hermann left his family with a large debt, owed mostly to his brother-in-law and cousin Rudolf Einstein, who had financed some of his attempts to build power plants in northern Italy. Albert began sending some of his new paychecks home to help his mother weather the storm and pay off the estate's obligations. Pauline, meanwhile, moved in with her sister Fanny and Rudolf in the southern German town of Hechingen, near where she had been born. Albert's sister, Maja, who had just passed the entrance exams for the University of Bern, had to postpone her matriculation and instead took a job in Trieste, tutoring Michele Besso's younger sister, Bice.[28]

Unbeknownst to Albert, while working for the Bessos, Maja fell prey to a swindle by Hermann's former bookkeeper. Claiming that he had not been paid, the bookkeeper had approached each of the Einstein children, pressuring them to make good on their father's debt. Albert suspected the man was lying and so had sent him packing, but Maja succumbed, and had begun sending the man portions of her paycheck. As a result, she was often broke. "Such romanticism is nice but terribly impractical," Albert said, and wrote to Besso asking him to withhold Maja's pay until she came to her senses.[29] "But the whole thing is funny," he added, "and has amused me greatly, may God forgive me, it's just that I am rather level-headed and tough."

WHEN HE WAS an old man, Albert confessed to a Swiss writer who wanted to write his biography that he had felt an "internal resistance" to marrying Mileva.[30] He referred to it as a marriage he had gone into out of a feeling of duty. After six years it was perhaps natural that his ardor was cooling, and negatives were piling up in his assessment of her. "Mileva came to Switzerland because she was very anxious to learn and also very intelligent, of a certain talent for penetration, but without ease at conceptions," he recalled. "She was thoroughly not mean-spirited, but distrustful, short on words, and depressive."

By now Albert had a new life in Bern, new friends, a new job, a place for the first time in the world, a new surge of entitlement, a paycheck, and papers in the *Annalen*. In the clear light of independence, face-to-face with the matrimonial commitment, a little weary and wary of Mileva's gloomy possessiveness, he was understandably ambivalent. On the other hand, Mileva had sacrificed her whole life, her honor, her ambition, and her child for him. His new world carried with it a heavy debt—even a crushing debt— from the old world. Out of a sense of duty, he later acknowledged, he made the leap.

They were married in a civil ceremony just after the New Year in Bern. Solovine and Habicht served as the witnesses as well as the only wedding guests. The four of them spent the day partying in town. It was late night by the time Albert and Mileva returned in high spirits to the Einsteins' new apartment on quiet Tillierstrasse high over the winding Aare, and discovered that Albert had forgotten the key. He had to ring the bells of the other residents in order to gain entry to the building.

The newlyweds both promptly came down with the flu. In his subsequent report to Besso, Albert wrote from his bed, "Well, now I am a married man and am living a very pleasant, cozy life with my wife. She takes excellent care of everything, cooks well, and is always cheerful."[31]

Mileva of course made a comparable, somewhat more effusive report to Helene.[32] "I am, if possible, even more attached to my dear treasure than I already was in the Zurich days," she wrote. "He is my only companion and society and I am happiest when he is beside me."

Albert had just sent the second of his papers on statistical mechanics to the *Annalen*. In his missive to Michele he sketched out a busy agenda for the future: more work on molecular forces in gases, and then a serious assault on his old favorite and longtime obsession, the so-called electron theory, the frustrating and maddening knot into which the theories of matter, radiation, and electrodynamics had snarled themselves.

In the meantime, in addition to all his other chores, Albert decided to become a *Privatdozent,* or private lecturer, at the University of Bern. Such lecturers were the lowest rung on the academic ladder; they received no salary but were paid according to the number of their students. Einstein had recently begun to mix with physicists at the University of Bern, particularly a man named Paul Gruner, who had himself been a *Privatdozent* for nineteen years. Albert attended meetings at Gruner's house and would occasionally give talks there, and he was later admitted to the Bern Natural Science Society under Gruner's sponsorship.[33] Kleiner, Albert's putative dissertation adviser and mentor at the University of Zurich, had also probably encouraged him to get *Privatdozent* experience since it was a necessary prerequisite for a full-time university job. To be a *Privatdozent* usually required a Ph.D., but the requirement could be waived for a candidate with a substantial record of publication, such as Einstein was confident he was accumulating.

The University of Bern appears not to have agreed with his assessment, however, on what constituted a substantial record. "The university here is a pigsty," he wrote two months later to Besso.[34] "I will not lecture there because it would be a shame to waste the time." He fell ill again, this time with diphtheria, and his temperature shot to 104 degrees for a day and a half.

Mileva reported to Helene that Albert was bored with his job and was going to look around for something new. She asked if there were any possible teaching jobs in Belgrade for her or Albert, possibly an indication that they were thinking of moving to Serbia and raising their daughter, Lieserl.[35]

Nothing came from that gambit, and the dream of keeping Lieserl—if there was still such a dream—died a final, hard death. In the fall Mileva headed east to visit her family and presumably to make permanent arrangements for the child. Few of the Albert-Mileva letters have been so exhaustively scrutinized as the two that passed between them on her nearly month-long trip.

THE FIRST IS a postcard, sent presumably en route from Budapest:

> Dear Jonzerl,
> I'm already in Budapest. It's going quickly, but badly. I'm not feeling well at all. What are you doing, little Johnnie? Write me soon, okay? Your poor
> Schnoxl[36]

Three weeks later, in a letter to Mileva, who was apparently by then in Serbia, Albert wrote the last words about Lieserl that have been found to date in the Einstein correspondence.[37] The news about his daughter, clearly, had not been good, and her prospects were ominous. "I'm very sorry about what has befallen Lieserl. It's so easy to suffer lasting effects from scarlet fever," Albert said. A life-threatening malady even today, scarlet fever can leave its survivors deaf. A wave of the disease had trapped Mileva in the countryside during the summer of 1899 while she was cramming futilely for her intermediate exams at the Polytechnic.

"If only this will pass. As what is the child registered?" Albert asked, referring perhaps to the presumed adoption, perhaps to Lieserl's official ethnic ancestry. "We must take precautions that problems don't arise for her later."

And so with a few cryptic words, as if in an overheard conversation, Einstein's daughter seems to have passed into the shadows of history. For whatever reason, Mileva and Albert chose not to bring her back to Bern, which would have in effect legitimized her now that they were married. Was she adopted? Did she die?

Whether Mileva was referring in her postcard to the train ride or something else, such as an adoption process, as "going quickly" has haunted and frustrated Einstein historians. Budapest was the administrative capital of the Hungarian part of the Austro-Hungarian Empire. A Hungarian historian working for the Einstein Papers Project combed the birth and adoption records for the twelve districts that would later comprise Budapest and found no trace of Lieserl or Mileva. Robert Schulmann has concluded that whatever happened probably did not take place in Budapest.

Likewise, a search of birth certificates in Kac failed to find any trace of Lieserl. "There were nine Maritsches born in Kac in 1902," Schulmann reports, noting that the name was common in Serbia (just as the Einstein name was in southern Germany), but that none of the parents turned out to be Albert or Mileva. On a trip to Yugoslavia in the late 1980s Schulmann even walked through cemeteries looking for suspicious little headstones. Unfortunately, the ongoing Balkan hostilities, including the wars over Bosnia-Herzegovina and Kosovo and general civil unrest in Serbia, have made it very difficult and dangerous to do the kind of research that might answer the question of Lieserl's fate.

The suspicions of historians and journalists who have gone searching for Lieserl have focused—in part for lack of any better prospects—on Helene Savic, Mileva's best friend, and her family. Perhaps, the most popular speculation goes, Lieserl survived and grew up as one of them. Most of Mileva's letters to Helene were burned by Helene's daughter, Julka Popovic, and other material was lost during the Nazi occupation of Belgrade. What few letters remain are in the hands of Milan Popovic, Helene's grandson and a professor of sociology at the University of Belgrade, who made copies of some of them available to John Stachel, the original editor of the Einstein Papers Project. Popovic has recently published a book, *A Friendship—Letters from Mileva and Albert Einstein to Helena Savic* (unfortunately not available in English), in which he concludes that Lieserl died of scarlet fever. Scholars and romantics are unlikely to take it as the last word.

There was other news of a more pleasant sort coming back from Mileva's trip. While she was in the east Mileva realized that she was pregnant again, which, by stretching the calendar slightly, could explain her reference to feeling ill. She was worried that Albert would be mad at her, and there is some suggestion that she was staying away from Bern out of fear. But he responded graciously to the news when she finally told him. "I'm not the least bit angry that poor Dollie is hatching a new chick," he wrote back. "In fact, I'm happy about it and had already given some thought to whether I shouldn't see to it that you get a new Lieserl. After all, you shouldn't be denied that which is the right of all women.

"Now, come to me again soon," Albert concluded. "Three and a half weeks have already passed and a good little wife shouldn't leave her husband alone any longer. Things don't look nearly as bad at home as you think. You'll be able to clean up in short order."[38]

THE SEACOAST OF BOHEMIA

IT STARTED WITH THE RINGING OF A JESTER'S BELLS. NEXT CAME A procession of bears carrying musical instruments, crowns, and weapons; a metal cock crowing and flapping its wings; a knight in gold armor; and finally Father Time, with a scepter in one hand and an hourglass in the other. The *Zytloggeturm,* or clock tower, straddled the gate to the original walled city of Bern. The tower itself had been built in 1191; in 1530 a huge astronomical clock tracking the relentless wanderings of the moon and planets was added to its eastern, interior side. Once an hour, anyone in Bern with the price of a few minutes could get a distillation of the sum of human wisdom circa 1500 regarding the cosmos, politics, fate, and mortality, performed by puppets dancing to an eternal revolving linkage of gears, weights, cogs, levers, cranks, wheels, chains, dials, and bells, designed and assembled two hundred years before Newton would enunciate the principles that would guarantee its clockwork majesty. Among those eligible to be so enlightened were the young Einsteins.

In the fall of 1903 Albert and Mileva moved to an apartment just down the block from the clock tower. Kramgasse, which runs east from the old city gate through the heart of Bern's arcaded medieval splendor, was considered by no less than Goethe the most beautiful street in Bern and among the most beautiful in Europe.[1] Just up the street from the Einsteins' second-floor flat at 49 Kramgasse stood a fountain adorned with a statue of a noble ursine figure bearing the coat of arms of Bern's founder, Berchtold V. Just down the street in the other direction stood another of Bern's celebrated fountain sculptures, honoring Samson.

The Einsteins would live in seven different apartments in as many years in Bern, but the Kramgasse dwelling has become the most notable of them, primarily because it was later turned into a museum, the Einsteinhaus, by a local group of scholars called the Einstein Society. The rooms in the apartment were arranged railroad style. In front was a large living room with two windows looking out over the sloping arcade on Kramgasse. In the back the bedroom faced an air shaft. A small anteroom was sandwiched in between those two spaces. On the other side of the air shaft, separated from the main part of the apartment by the outside hallway and the air shaft, was a kitchen and toilet.

In what was once the bedroom there hangs today a large photograph of Albert, looking sharp in a checked sport coat and a headful of black wavy hair, bending studiously over an easel holding a large book. On the floor in front of the picture stands the same easel holding the Einsteinhaus guest book, inviting tourists to get their picture taken imitating the young physicist. In the living room stands a large square table, similar perhaps to the table where Albert spent his evenings—Mileva across from him—under an oil lamp teasing apart the equations describing the relationships between heat and energy and atoms.

Marriage had made Mileva a de facto member of the Olympia Academy, and Solovine later recalled her sitting quietly in the corner during the meetings at their apartment, following the arguments but rarely contributing. He found her reserved but intelligent, and clearly more interested in physics than housework. Despite Mileva's initial resentment and suspicion, she and Solovine became friendly.

Mileva enjoyed the intimacy and camaraderie of the group, but not the fraternity-house boisterousness to which it often gave rise. She hated it when Albert told dirty jokes, which made him, of course, go out of his way to tell them. "There was this old whore," he would begin. "Albert!" Then would come a deep belly laugh that one contemporary would later compare to the barking of a seal, a contagion of laughter sweeping the room while Mileva burned.[2]

Albert admired Mileva's resolve but he loved to tease her as well, especially on the rare occasions when she spoke up, usually to make some mathematical assertion. Then he went on the attack, needling and questioning her, making a game of his challenges. Mileva defended herself hotly, stamping her feet and turning red. Finally he surrendered the point with a laugh. He was in awe, he liked to say, of anyone who could stand her ground so stubbornly.[3]

Another occasional addition to the group was Conrad Habicht's brother, Paul, a machinist by trade, who was an inveterate tinkerer and inventor. At the time he was working on electrical relays, which Albert thought could be useful for telephones. Paul had learned the art of making Turkish coffee, which became, along with Mileva's snacks, the invaluable late-night fuel of the Academy. Afterwards, while Mileva cleaned up, the men took midnight walks through the arcades and the promontories overlooking the dark Aare. Onward they argued, through philosophy, physics, electromagnetism, atoms, Helmholtz, Plato, Hume, Dostoyevsky, Mozart, and Beethoven. Music was the other great social pole for the young couple. String quartets assembled almost weekly at the Einstein abode to play fiery renditions of the classics.

In honor of Albert's twenty-fourth birthday Solovine brought caviar to the weekly meeting and snuck it onto the plates, pretending it was just the usual fare of bologna.[4] Einstein had never tasted caviar; his usual reaction upon encountering something new, especially food, was to grow effusive and wax ecstatic. Solovine sat back waiting for Albert's reaction, but Albert was busy talking about Galileo and the idea of inertia. He shoveled the caviar down without noticing while Solovine and Habicht stared in shock.

When the caviar was all gone Solovine asked Albert if he knew what he had been eating.

"For goodness sake, it was that famous caviar," Albert said. Then he added, "It doesn't matter. There's no point in serving the most exquisite delicacies to hicks; they can't appreciate them."

The same birthday might have been the occasion for a comical drawing by Solovine commemorating Einstein and the Olympia Academy. It portrays a distinctly Greek-looking bust of Albert sitting on a swiveling pedestal like that of an office chair. His head is framed by an arch of bologna sausages; dumplings hang on either side. An inscription in Latin reads:

A.D. 1903

THE MAN OF HECHINGEN

EXPERT IN THE NOBLE ARTS, VERSED IN ALL LITERARY FORMS— LEADING THE AGE TOWARD LEARNING, A MAN PERFECTLY AND CLEARLY ERUDITE, IMBUED WITH EXQUISITE, SUBTLE, AND ELEGANT KNOWLEDGE, STEEPED IN THE REVOLUTIONARY SCIENCE OF THE COSMOS, BURSTING WITH KNOWLEDGE OF NATURAL THINGS, A MAN WITH THE GREATEST PEACE OF MIND AND MARVELOUS FAMILY VIRTUE, NEVER SHRINKING FROM CIVIC DUTIES, THE MOST POWER-FUL GUIDE TO THOSE FABULOUS, RECEPTIVE MOLECULES, INFALLI-BLE HIGH PRIEST OF THE CHURCH OF THE POOR IN SPIRIT[5]

On another, less reverential, occasion, Habicht had a tin plaque engraved with the name "Albert Ritter von Steissbein," roughly translated as "Albert, Knight of the Backside," and nailed it to Albert and Mileva's door.[6] The word *Steissbein* was also a play on *Scheissbein,* or "shit-leg." Albert and Mileva, it was said, laughed so hard they thought they would die.

By the end of 1903, however, the Olympia Academy was beginning to frazzle. Habicht had finished his mathematics doctorate and returned to his parents' house in Schaffhausen, and his brother moved back home at the beginning of January. The river town was only an hour from Bern by train, and Conrad owned a bicycle, so Albert saw no reason that he could not continue attending the Academy meetings, but Habicht didn't show up. Albert's pleadings and scoldings to him became increasingly colorful: "Hey, you miserable lazybones!" . . . "you miserable creature, you most pitiable of all the membra academia who ever inhabited this world!"[7]

Solovine had also finished school, and in the spring of 1904 he moved to Strasbourg, France.

Solovine's departure marked a changing of the intellectual guard. Only two weeks before, at Albert's urging, his old friend Michele Besso had been appointed to the Swiss Patent Office as a technical expert second class—a rank above Albert's. As Michele's route to and from work took him past Kramgasse, he and Albert walked home together, resuming the discussions about physics and philosophy that they had been engaging in since college. The Einsteins and the Bessos became inseparable.

Albert himself was also ready to shift gears. He had just finished the third paper in his series about thermodynamics. Together those papers amounted to an attempt to reformulate much of classical nineteenth-century physics in twentieth-century terms. Albert had labored to show that the laws of thermodynamics, mystical precepts that governed the engines of industry and perhaps the universe, were simply a consequence of the statistics of atoms.

By his third paper he had been ready to explore more precisely the consequences of the microscopic chaos that lay behind the facade of order and inevitability in nature. Due to the randomness of the motions of its components, even the most smoothly running systems were not completely stable but would fluctuate. Albert derived a formula describing the energy fluctuations in a system that was maintained at a constant temperature. As an example of how sophisticated his statistical techniques had become, he applied this formula not to atoms in a gas but to light waves, and obtained a result that agreed roughly with an experimental and little-understood fact of the time: The intensity and color of the glow coming from the walls of a heated oven depended on the temperature of the oven. "I believe that this agreement should not be ascribed to chance," he concluded.[8]

It would be another year before the meaning of that "agreement" would become apparent, but clearly the statistics of fluctuations was a powerful tool, and one he was eager to exploit.

He sent a copy of the latest paper to his other old buddy Marcel Grossman, who had just announced the birth of his son, Marcel Junior. Grossman had sent a copy of his new paper on non-Euclidean geometry. Albert wrote back delighted. "There is a remarkable similarity between us. Next month we are also going to have a baby," he reported.[9]

On May 14, 1904, Mileva gave birth to a boy, whom they named Hans Albert, or "Albertli" for short. As before, it was a difficult labor, and the greatly weakened Mileva required "great care" afterward.

That care was certainly not provided by the sullen mother-in-law and new grandmother, Pauline Einstein. It may have come from Mileva's mother. According to an interview twenty-five years later[10] with Mileva's old friend Milana, the birth of a legitimate grandchild softened Milos Maric's attitude toward his daughter's marriage. He visited the family in Bern shortly after Hans Albert's birth, it is said, and offered Albert a belated dowry of 100,000 kronen, equivalent to 100,000 Swiss francs—a handsome fortune indeed.

But Albert, the story continues, turned his father-in-law down, saying that he hadn't married Mileva for her money, and that she was his inspiration and guardian angel against transgressions in life and in science. Milos recounted the speech to his friends in Novi Sad, it is said, in tears. He respected Albert's views, however much he himself loved money.

In September Albert was given a 400-franc raise, and his status at the patent office went from provisional to permanent. He was on his way to becoming the bureau's reigning expert on electrical devices. Sometime that fall he was also invited, presumably on the basis of his statistical papers, to write review articles for a physics newsletter called *Beiblätter zu den Annalen der Physik.* Over the course of the next year he contributed half a dozen articles, for a few marks apiece, summarizing recent papers that had appeared in a variety of journals and languages—Italian, French, and English among them. Mileva, who had studied English, helped him with the latter.

On Sundays he could be seen pushing an ornately scrolled baby carriage around the town with a pipe hanging out of his mouth and a notepad at the ready. In photographs from the time, he appears in his shirtsleeves with a soft wary look in his eyes, bending over a pudgy frowning Albertli. Mileva looks thickened, her face coarsened and lined, as she beams at her son.

At home Albert and Mileva took to speaking almost entirely in the baby talk of their firstborn.[11] The result was that Hans Albert did not learn to speak normal Bernese German until he was around the age of five.

For Mileva, still recovering from the loss of Lieserl, it was perhaps the best of times. They were finally living the life—eternal students who didn't give a shit for the rest of the world—they had dreamed of when they were undergraduates at the Poly. Ideas flowed through the household. If it was only Albert who was the physicist, she didn't seem to mind. As she liked to tell her friend Helene, "After all we are *ein Stein* [one stone]."[12]

Nevertheless, Mileva probably experienced a pang in November, when the Nobel Prize in chemistry was awarded to the archetypal husband-and-wife science team, Pierre and Marie Curie, for the discovery of radium. Previously unknown and relegated to obscure posts, the Curies were suddenly being lionized by the French press as romantic heroes. Mysterious glowing radium was called the *"métal conjugal."*[13]

Albert and Mileva's own partnership was increasingly relegated to the corners of the day. Only after the baby had been put down for the night would there be the space and quiet to sit down at the big table in the living room and replace baby talk with the language of atoms and mathematics. In the next year—one of the most productive of Einstein's or any other physicist's life—four major papers would issue from that table. They would transform Albert's life and, indeed, eventually physics.

The first was destined to be the most revolutionary. In it, resuming where he had left off in his last thermodynamics paper, Albert pursued the curious statistics of radiation and the relationship between an oven's glow and its temperature. As the fall nights deepened and lengthened into winter and the sky became as gray and leaden as the Kramgasse arcades, Albert, like Balboa spotting the Pacific Ocean glittering in the distance from a mountaintop, came within sight of the end of physics as it had been known, although it would be twenty years before physicists completed that trek, and embraced quantum uncertainty.

Why should anybody care about such a mundane thing as ovenglow? There were only two subjects that really mattered in the world of the late-nineteenth-century physicist, two mysteries poised against each other: matter and light. According to Maxwell's theory, light consisted of electromagnetic undulations in the aether. Matter, by a growing but by no means unanimous consensus, was made up of atoms. It was in the boundary between the two, where Maxwell's aetherial waves crashed on the atomic foundations of stolidity, that radiation—light and heat—was produced and where clues to the inner processes of nature might be found, like seashells on the beach.

In 1859 a physicist by the name of Gustav Kirchhoff, from the University of Heidelberg, had pointed out that ovenglow might be such an enigmatic seashell, and challenged physicists to explain it.

For as long as there had been fire, it had been observed that any warm object, from a fevered brow to a smoldering coal or a new horseshoe plucked from the blacksmith's forge, cast off a mixture of heat and light, which Kirchhoff dubbed "blackbody radiation." Moreover, there seemed to be a correlation between the color of a body and its temperature. In fact, early potters determined the temperature in their kilns by the color of the glow of the baking clay.

Kirchhoff was a pioneer of the new science of spectroscopy, in which objects were analyzed by breaking down the light from them, such as by passing it through a prism, into the rainbow of colors of which it is composed. When light from a star, the sun, or, say, a flame was so dispersed into its component colors, dark or light lines, like cracks between the keys of a piano, would appear in the rainbow in a pattern that was characteristic of the nature of whatever had given off the light. One of the first great spectroscopic discoveries was that the sun and stars were composed of the same elements found here on earth—hydrogen, calcium, sodium, carbon, iron, nitrogen, oxygen, and so forth. But Kirchhoff argued on the basis of the second law of thermodynamics that the intensity and spectral characteristics of the broad background glow on which these lines were superimposed should depend only on a body's temperature, and not on its composition. This in turn suggested that blackbody radiation revealed something fundamental about nature, and about the relationship between matter and energy.

Kirchhoff challenged physicists to find out what that something was, first by measuring the intensity of the radiation at different temperatures and wavelengths across the spectrum, and second by coming up with a mathematical formula that reproduced those results. "It is a highly important task to find this function," Kirchhoff wrote.[14] "Great difficulties stand in the way of its experimental determination. Nevertheless, there appear grounds for the hope that it has a simple form." History would show that Kirchhoff had exquisite taste in problems.

Technically, a blackbody was an object that was in thermal equilibrium with its surroundings—that is, at the same temperature. It would therefore radiate exactly as much energy as it absorbed, a system Kirchhoff called *Hohlraumstrahlung* (cavity radiation). His challenge set a generation of physicists to building extremely precise little ovens, which could be heated and maintained at a very uniform temperature, and detectors that could accurately measure the types of light that emerged from their interiors.[15] In addition to discovering a fundamental law of nature, there was a practical incentive for whoever cracked Kirchhoff's problem: The results could be used to calibrate and optimize electrical lighting systems. In Germany much of the experimental work was in fact sponsored by the Physikalisch-Technische Reichsanstalt, the Imperial Bureau of Standards.

The nineteenth century was coming to a close by the time blackbody radiation had been sufficiently measured to meet the first half of Kirchhoff's challenge. Humans and other room-temperature objects whose absolute temperature is about 300 degrees Kelvin were brightest at wavelengths of a few thousandths of a millimeter, deep in the infrared part of the electro-magnetic spectrum, invisible to the eye but readily manifest as heat. At 3,000 degrees a "blackbody" would no longer appear so black but would glow a very dull red like an underpowered light bulb; at 6,000 degrees, roughly the surface temperature of the sun, it would be yellow. Above 10,000 degrees or so the blackbody would look blue, but most of its energy would actually be radiated as invisible ultraviolet rays at even shorter wave-lengths and higher frequencies than visible light.

Although only one color dominated at any given temperature, a black-body actually radiated a whole spectrum of hues. On a graph of intensity versus frequency, the blackbody spectrum looked like an asymmetric hump or a cresting ocean wave. The higher the frequency (or shorter the wave-length) of the light, the more intense it was, up to a point that depended on the temperature of the blackbody. Beyond this so-called characteristic frequency, which corresponded to about 6,000 degrees for the sun, the in-tensity declined steeply. The hotter the body, the higher was the frequency of its maximum emission and the more energy it radiated overall. So as the temperature went up, blackbodies got bluer and hotter.

The second half of the challenge—understanding what this curve was saying—proved harder, however. The great breakthrough of nineteenth-century physics had been Maxwell's electromagnetic theory of light. If any-thing could unlock the mystery of the blackbody, it was surely that theory. After all, according to Maxwell the myriad varieties of blackbody emission were only electromagnetic waves of differing wavelength. But in 1899, when Lord Rayleigh, a.k.a. John William Strutt, an English expert on ther-modynamics, and James Jeans went to the effort of actually trying to use Maxwell's theory to compute the blackbody spectrum, they got a shocking and nonsensical answer—that such objects should not shine at all (no heat and no visible light) except for a deadly infinitely intense flash at the ultra-violet and X-ray wavelengths.

The reason was simple: To be in thermal equilibrium the energy in a Kirchhoffian oven had to be evenly distributed among all the wavelengths that would fit in the oven; in effect, each possible "color" was entitled to its own separate but equal dollop of energy. The problem was that there was an endless number of possible different colors or wavelengths that could fit inside the oven, and while there was no limit on how short the wavelength of the light in the oven could be, there was a firm limit on how long it could be, namely the size of the oven itself. Just as a violin string could vi-brate with one fundamental note or higher harmonics of it, but none lower, so too a blackbody could contain light waves smaller than itself but not larger.

One implication of this was that it would take an infinite amount of energy to attain equilibrium, because there were an infinite number of ever shorter wavelengths in which to deposit it. Another was that whatever energy a blackbody did have would rapidly drain to the short-wavelength high-frequency end of the spectrum and be radiated only as ultraviolet or X-ray radiation—invisible, heatless, and potentially deadly—and the world would consequently freeze. Physicists termed this the "ultraviolet catastrophe." The fact that the world was not frozen and dark suggested that there was something seriously wrong with the combination of electromagnetic theory and thermodynamics that passed, as of the year 1900, for standard physics.

Kirchhoff died in 1889 without having learned what the blackbody spectrum was trying to tell him about nature. The University of Berlin, where he had been working, offered his chair to Boltzmann, who declined, and then to Hertz, who also declined, before finally approaching a young theoretical physicist named Max Planck, who had just been rejected for a job at Göttingen. It was Planck who would finally decode Kirchhoff's blackbody puzzle.

The descendant of a long line of scholars, pastors, and lawyers, Planck, then thirty-one, embodied all that was admirable in German culture. A photograph from his graduate student days shows a slender, handsome, clean-shaven youth in formal dress and a pince-nez staring off into space with almost angelic intensity. Later he grew a trademark drooping moustache. Shy and modest, yet possessed of an inner confidence and unshakable resolve, Planck had progressed upward through the ranks by steady diligence as much as by native brilliance to become one of the world's experts on classical thermodynamics, which to him represented the kind of universal unchanging truths to which humans had aspired since the time of Aristotle.[16]

To call Planck a conservative would be an understatement. He still didn't believe in atoms, contending that Boltzmann's statistical approach to the world undermined the majesty of thermodynamics. The second law—that entropy should always increase—was, for Planck, just that: a law, and not a statement of probability. "Despite the great success that the atomic theory has so far enjoyed, ultimately it will have to be abandoned in favor of the assumption of continuous matter," he wrote in 1882.[17]

The conflict between atomism and thermodynamics continued to tear at him for the next two decades, long after most of his compatriots had accepted the atomic worldview. A historian of the time joked that there were only four physicists left in the world who troubled themselves about that problem.[18] In 1897, as he embarked on a typically intense study of the so-called cavity radiation, Planck was one of them.

With Rayleigh's colossal failure and constant revisions in the blackbody curve staring him in the face, Planck sat down again in October of 1900 to tackle the blackbody problem. In his approach he made use of the new theory—made popular by Drude, Thomson, Lorentz, Lenard, and others—that matter is full of bits of electricity called electrons. Planck envisioned the walls of the oven as filled with electrons, which incoming electromagnetic waves would cause to vibrate back and forth. The oscillating electrons would radiate waves back out into the cavity until the entire cavity radiation was vibrating like the springs of a mattress and a balance—thermal equilibrium blackbody radiation—was reached.

In a heroic act of mathematical legerdemain, Planck finally found a formula that worked. But in his desperation to invest his symbols with physical meaning he had been forced finally to embrace Boltzmann's statistical concept of entropy and to make two additional assumptions that seemed as mad as they were inspired: The first was to treat the energy of the radiation mathematically as if it were divided into discrete bits to which he could apply the techniques that Boltzmann had developed for the molecules in a gas. The second was a peculiar restraint on the behavior of his little electron oscillators. Planck was forced to assume that they could only emit or absorb energy in the form of discrete bundles he called *quanta*. The effect of this bizarre adjustment was to make the oscillators in effect like fussy bank tellers who refused to make change, and wouldn't accept it either.

The response from the scientific community to these quanta was mostly bewilderment. What did they mean? Were they real or just a convenient calculational device? While Planck's formula worked to predict the blackbody curve perfectly, he himself was a little embarrassed at what he had wrought and hoped that its implications wouldn't be taken too far. "I had to obtain a positive result, under any circumstances, and at whatever cost," Planck later explained.[19] For the next four years his formula was regarded as a miraculous mathematical contrivance without a clear physical meaning. In the meantime, swayed by his success with Boltzmann's statistics, Planck became a belated convert to atoms.

ONE OF THOSE who had been bewildered by Planck's paper was Einstein, who first read it in 1901 while he was teaching at Winterthur. He hadn't liked it then, feeling as if the ground had been pulled out from under him.[20] At the end of 1904, as he took up the subject again with his newly gained mastery of statistical physics, Albert disapproved of it even more. Planck's calculations, he concluded, were full of errors. He wrote to Planck, presumably to offer criticism in his usual tactful way. A correspondence ensued, but the letters were lost when Planck's house was bombed during World War II.

In early 1905 Albert set out to redo Planck's calculation correctly. By now he had the full power of statistical mechanics at his disposal. If he had some hope while commencing the calculation of restoring the ground under his own feet as well as those of other physicists, he was wrong. By March, when he sent his results in to the *Annalen*, the ground was clear out of sight. While Planck had been willing to consider the notion that the exchanges of energy between light waves and matter happened only in certain discrete amounts, Albert found himself arguing—seemingly in the face of Maxwell himself—that the light waves themselves behaved as if they were made of particles, bits of energy that under the right circumstances seemed as particular as marbles.

In his paper, titled "On a heuristic point of view concerning the production and transformation of light," Albert began by noting that there was a basic contradiction between the conceptions of matter and electromagnetism.[21] Matter was treated as a collection of a large, but finite number of discrete elements—namely, atoms—while light was represented as a continuous wave. It was easy to see that there might be problems mating these ideas mathematically.

Traditional optical experiments supported the Maxwell wave picture, Albert pointed out, but these experiments involved averaging the behavior of light waves over periods of time that were long compared to their frequencies, with the result that any nonwave properties that might exist would not show up. Indeed, he maintained, recent experiments, such as the studies of blackbody radiation and Lenard's demonstration that electrons could be bounced out of metal by shining ultraviolet light on it, were more easily understood by assuming that light was *not* continuous, but granular in form. "According to the assumption to be contemplated here," he wrote, "when a light ray is spreading from a point, the energy is not distributed continuously over ever-increasing spaces, but consists of a finite number of energy quanta that are localized points in space, move without dividing, and can be absorbed or generated only as a whole."

Albert then redid Planck's calculation of the blackbody formula and showed that while Planck might have gotten the right answer (or at least one that fit the experimental data), he had been able to do so only because he had used the wrong definition of probability. Which just went to show how miraculous Planck's derivation of his formula really was. If he had used the correct definition, Albert contended, Planck would have wound up with the same answer as Rayleigh, namely, that the blackbody intensity would be infinite at high frequencies.

Planck's formula, Einstein concluded, could therefore not be trusted. Neither could Rayleigh's, which worked fine only for frequencies below the turnover frequency where maximum intensity occurred. But there was another formula, he pointed out, worked out by trial and error by Wilhelm Wien, an Austrian physicist at the University of Würzburg, which successfully described the other half of the blackbody curve where the intensity dropped off at high frequencies. This formula was a perfect fit to the most baffling and troublesome part of the blackbody spectrum, but nobody, including Wien, knew what it meant.

Albert set out to deconstruct Wien's formula, finding that it led to an equation for the entropy of the blackbody radiation, and that it bore a striking resemblance to Boltzmann's formula for the entropy of an ordinary gas made up of molecules. The blackbody radiation, Einstein concluded, was behaving thermodynamically as though it were not continuous, but rather made up of separate individual bits, or quanta, of energy.

It was a subtle argument, a brilliant reading between the lines of nature, but there was more. The notion that light consisted of little bullets of energy, Einstein went on to point out, could also explain recent experiments like Lenard's. One of the curious features of that experiment was that increasing the intensity of the light increased the number of electrons coming out of the metal but did not increase their energies. Suppose, Albert theorized, that each electron could absorb only a quantum of energy, or none at all. In his and Planck's theory, the energy of a quantum of light was the mathematical product of a numerical constant—denoted by h and later named "Planck's constant"—multiplied by the frequency of the radiation. Increasing the intensity of the light merely increased the number of quanta and thus the number of electrons that could be bumped out. The only way to increase the energy of the quantum, and thus of the resultant electron, therefore, was to turn up the frequency. Einstein proposed a simple equation for Lenard's experiment, relating the frequency of the illumination, the properties of the metal plate used, and the energy of the cathode rays, by which his explanation of the "photoelectric effect" could be tested and verified. It would be nearly ten years before Robert Millikan, a Caltech physicist, confirmed those predictions on the way to winning a Nobel Prize.

In the meantime, physics stood precariously poised on the edge of a chasm. What did the quantum mean? Was it only, as the title of Albert's paper implied, a heuristic, "a useful fiction" practical for calculating, or did light quanta, *Lichtquanten,* really exist?

Albert liked to explain the quantum principle by comparing nature to a drinker who can guzzle beer only by the pint. Other than the fact that this characteristic made the universe a warm and habitable place, why nature should choose to guzzle energy in gulps rather than sip it in a continuous flow from the tap was a mystery. Neither Newton's mechanics nor Maxwell's electromagnetic wave theory, it was now clear, could claim the absolute validity that had been vouchsafed them in his youth. Some new principle was needed to serve as the bedrock of physics.

Nobody agreed with Einstein, and his paper was largely ignored. Even Planck, the father of the quantum, thought that he had gone too far. Planck and the rest of his colleagues found it easy to resist the idea of quanta as anything more than a necessary device to be reluctantly employed in a poorly understood calculation. From the standpoint of the established physics community, Albert was an outsider, a dabbler who worked at the patent office and browsed the journals on the side. From his own standpoint Albert had no expectations and nothing to lose. He had been raised in no school that obliged his allegiance. He had no mentors and no favors to pay back. Seemingly he had no fear. He could afford to be radical.

The publication of his quantum paper marked the beginning of an extraordinary period for Albert. In the next two months he performed a densely mathematical analysis of viscosity and diffusion of sugar in water in order to estimate that the size of a putative sugar molecule was about a millionth of a millimeter in diameter. The same calculation also yielded a value for an important constant known as Avogadro's number. If there was anything that conveyed the mind-dizzying smallness of scale of the so-called atomic world, it was this number, which represented the number of molecules of some substance you had to assemble in order to equal its molecular weight in grams. For example, one gram of hydrogen, 18 grams of water, or 56 grams of iron, according to chemical and gas theory, should all contain the same number of molecules. But what was that number that represented the statistical complexity of the world? Albert got a value of 210,000,000,000,000,000,000,000—more conveniently written as 2.1×10^{23}—about one third of the modern accepted value. Even the smallest ink flecks of material you could perceive under a microscope contained literally uncountable billions of particles.

After four years immersed in thermodynamics, statistical mechanics, and kinetic gas theory, Albert was closing in on the fabled and controversial world of atoms. "My major aim in this was to find facts," he explained, "which would guarantee as much as possible the existence of atoms of finite size."[22] Although the majority of physicists and chemists had by now surrendered to the logic of atoms as the organizing principle of the world, there were still doubts among many of them about whether atoms, like quanta, were actually real, or just a convenient (to use a favorite word) heuristic, a useful fiction.

What would turn out to be the coup de grace for the skeptics arose from Albert's interest in the physics of liquids and solutions, which had formed the basis of his first published paper, on molecular forces, and had been the subject of numerous letters between him and Besso, as well as of the work he had just completed. It had been known ever since the invention of the microscope that even the cleanest water is full of microscopic particles, which appear to be dancing around like dust motes in a sunbeam.[23] The phenemenon became known as Brownian motion, after the botanist Robert Brown made careful observations of it in 1828. In the intervening years a variety of explanations for Brownian motion had been tried and found wanting: convection currents, electrical and solar effects, evaporation, capillary forces, simple thermal kinetic energy, and even the possibility that the particles were alive. The French physicist Louis-Georges Guoy argued that Brownian motion violated the second law of thermodynamics, and his ideas were mentioned by Poincaré in his book *Science and Hypothesis*, which Einstein, Solovine, and Habicht had read.

Albert hadn't set out to solve the puzzle of Brownian motion; he was looking for some visible, measurable, and unequivocal manifestation of atoms.

According to Boltzmann's kinetic theory, the velocities of the molecules in a gas or liquid were not uniform but rather deviated randomly from some average that depended on the temperature. Einstein reasoned, therefore, that the pressure felt by a small particle suspended in such a fluid would fluctuate as it was hit by molecules of differing velocities. In less than the blink of an eye the particle would be randomly kicked this way and that at high speed. Each kick would last only a fraction of a second, during which time the particle traveled—like the oxygen molecule that Lenard had described in a lecture to Mileva a decade before—only a fraction of a hair's width and would thus be invisible. Over time, however, the effects of these tiny random kicks would add up and the particle would appear to execute a kind of drunkard's walk, darting here and there. Using the statistical machinery he had developed to study fluctuations, Albert calculated that a typical particle would be displaced by about six microns (a thousandth of a millimeter) over the span of a minute. That distance could in theory be measured with a good microscope, and for Albert it was the next best thing to seeing atoms themselves. Such jumps, he suggested, might be what people were observing in Brownian motion, although he claimed that he did not know enough about the phenomenon to be able to judge.

Confirmation of this effect, he concluded, would allow an exact determination of the size of atoms. But, he acknowledged, "Conversely, if the prediction of this motion were to be proved wrong, this fact would provide a weighty argument against the molecular-kinetic conception of heat."[24]

The paper was published in the *Annalen* in July. Unlike his article on light quanta, this one was met with a flurry of attention and correspondence. Later in 1905 a professor of forensic medicine at the University of Zurich, one Heinrich Zangger, who had done research on Brownian motion, came to Bern to visit Albert. Some spark was struck; next to Besso, Zangger would become Einstein's best friend in the world.

Four years later Einstein's prediction—and with it Boltzmann's—would be confirmed with stunning precision by the Frenchman Jean Perrin. (The honor came too late for Boltzmann, who committed suicide in 1906.) Perrin later wrote: "The atomic theory has triumphed. Until recently still numerous, its adversaries, at last overcome, now renounce one after another their misgivings, which were, for so long, both legitimate, and undeniably useful."[25]

THE SPRING OF 1905 bloomed along an Aare swollen with alpine melt. Tulips and lilies sprouted on the terraced south-facing slope of the fortress rock of Bern. Albert and Mileva, in search of more room for themselves and their scrambling, active one-year-old, packed up their belongings and moved from beautiful Kramgasse across town to Besenscheuerweg, midway between the university and the Gurten, which Albert loved to climb with his friends.

Two weeks later, toward the end of May, he wrote to the perpetually absent Habicht to summarize what had been an extraordinary spring for any physicist, let alone Albert.[26] "What are you up to, you frozen whale, you smoked, dried, canned piece of soul, or whatever else I would like to hurl at your head," he began, and went on to describe his spring harvest of four papers, which Habicht would soon receive. The first, which he described as "very revolutionary," was about radiation and light, while the second and third were about atomic physics, the sizes of molecules, and their random movements in liquids.

Finally, he continued, there was another paper that was not yet written. It was still an idea ticking like a time bomb in his head. "The fourth paper is only a rough draft at this point, and is an electrodynamics of moving bodies which employs a modification of the theory of space and time; the purely kinematic part of this paper will surely interest you."

IO

SIX WEEKS IN MAY

IF THE YEAR 1905 REPRESENTED A SCIENTIFIC EXPLOSION IN ALBERT'S
life, then the month of May in that year was an explosion within an explosion. It was during a six-week burst of ferment, desperation, and invention
that Albert resolved the questions and paradoxes about light waves dancing
on the aether that had been bedeviling him for ten years. When the fog in
his brain finally cleared, the universe stood revealed as a simpler and
stranger place than Albert or any other physicist had ever dared to dream—
a more modern place, shorn of unobservable metaphysical baggage and ab-
solutes of space and time, a world with neither an aether nor a center, and
with no landmarks except change. Few so-called revolutions in science are
truly revolutions, but relativity was one.

It was a revolution fed by many sources. Albert had hardly been alone in
his preoccupation with light and how it fit into the other mysteries of the
universe. In 1894, two years before Albert entered the Polytechnic, Heinrich Hertz, the discoverer of radio waves, had defined the agenda for an
entire generation of physicists as the unification of mechanics and electromagnetism. Eleven years later, in the spring of 1905, that vision was being
pursued on terms strikingly different from those Hertz had envisioned.
Whereas Hertz had sought to explain light waves in terms of Newton's mechanical laws, as vibrations of the aether, some proponents of the new
emerging worldview attempted instead to explain mechanics as an effect of
electromagnetism. According to the electron theory, as it was called, matter
was constituted of little balls of electrical charge whose interactions with
one another and the aether could be made, by adroit mathematical manipulations, to account for almost all physical phenomena, including the existence of mass and momentum.

The greatest minds of Europe had contributed to this edifice, but some
of the most influential ideas leading to it were sketched out by a genial
Dutch physicist, Hendrik Antoon Lorentz, while sitting at a writing table

at his home in Leiden with a cigar and a glass of wine, in the bosom of his family. Revered by his colleagues as much for his cosmopolitanism and engaging personality as for his scientific genius, Lorentz seems to have glided through life. The son of a gardener, he had finished his courses at the University of Leiden in only a year and half and then spent the next seven years working and studying at home. He emerged with a doctoral dissertation on Maxwellian optics that was celebrated throughout Holland and that won him—at the age of twenty-four—one of the first chairs of physics in Europe explicitly designated for *theoretical* physics.

There, in bourgeois splendor, Lorentz pursued a dualistic theory of the universe. In contrast to predecessors like William Thomson, a.k.a. Baron Kelvin, and Hertz, who theorized that atoms were comprised of knots or vortices in the aether, Lorentz argued that atoms and aether were separate and distinct elements of nature. The aether was continuous and all-pervading. Molecules (or atoms—he used the terms interchangeably) were discrete entities and contained electrical charges, both positive and negative, which he called ions. It was the flow of these charges that produced electricity; they were also the source of force fields that rippled out across the aether. "Molecular forces," he wrote, "act by intervention of the aether."[1]

In Lorentz's scheme, the aether was immovable and was not, as some other physicists had speculated, dragged along with moving matter. Unfortunately, this contradicted the results of the Fizeau experiment, the only vaguely tangible evidence scientists had for the aether. In order to correct for this seeming discrepancy, Lorentz postulated a strange mathematical relationship between light waves in a moving substance and those in a stationary one. One of the features of this "theorem of corresponding states," as he called it, was a fictitious coordinate called "local time" that applied to moving objects and was slower than real time.

It was in the context of this theory that Lorentz proposed in 1892 what he called a desperate solution to the failure of Michelson and Morley's experiments to detect any effect on the speed of light by the earth's presumed flight through the aether. Their results suggested that either the earth was the one celestial body in all the swirling heavens that was dead, solid still with respect to the aether—a rather un-Copernican notion, at this enlightened stage of history—or that something else entirely was going on. The "something else" is what would become known as the Lorentz (or the Lorentz-Fitzgerald) contraction. Lorentz theorized that it was not unreasonable to presume that the forces—electrical and otherwise—responsible for holding together a moving material object would be affected by motion through the aether. The "wind" would push the molecules close together, causing the body to contract. Lorentz therefore speculated that the arm of Michelson and Morley's measuring apparatus had shortened when it was pointed into the "aether wind," which changed the resulting measurement of the velocity of light along that direction just enough to compensate for the earth's velocity. (Unknown to Lorentz, George Francis Fitzgerald, an Irish physicist, had proposed an identical hypothesis three years earlier[2] in a letter to *Science*, which he mistakenly thought[3] had never been published.

Lorentz explained this contraction hypothesis more fully in the last chapter of an 1895 treatise. It was one of many hypotheses that he would adopt and add to his precious theories during the following decade in order to ensure that Maxwell's equations, and other laws of optics and light, would be the same for resting and moving observers. This notion, as he developed it, went by the cumbersome rubric of the theorem of corresponding states.

In Lorentz's scheme the aether had replaced Newton's absolute space as the ultimate but unknowable universal reference system, God's elegant little stage. It was there, but you could never detect it, never know if you were at rest in it or moving through it. It was this quality of invisibility in the aether that preserved an important tenet of Newtonian mechanics: namely, that the laws of physics should be the same for any freely moving observer. That meant—as Mach had stressed—that the only motions that could be detected and have physical significance were those relative to other "ponderable matter." Poincaré called this notion the principle of relativity.

In his role as the central clearinghouse of ideas, Poincaré grudgingly endorsed Lorentz's theory as the "least defective" of the ones then in circulation. "An explanation was necessary," he wrote drily, referring to the Michelson-Morley experiment, "and was forthcoming; they always are; hypotheses are what we lack the least."[4]

All Europe began to embrace Lorentz's theories in 1899 when the Englishman J. J. Thomson finally showed that the so-called cathode rays were in fact electrically charged particles, particles that were indeed the basis of all matter. Electrons, enthused Walter Kaufmann, a Göttingen experimentalist, were the long-sought primordial atoms.[5] At a meeting in 1900 to celebrate the twenty-fifth anniversary of Lorentz's doctoral dissertation, Wilhelm Wien, the German physicist to whom Albert had written in 1899 about his own proposed aether-wind experiment, predicted that electromagnetism would replace mechanics as the foundation of physics.

Temporarily caught up in Wien's spirit, Lorentz even attempted an electromagnetic explanation of gravity. He speculated that Newton's grand but unexplained law might arise from a slight mismatch in the attractive and repulsive forces between equal electrical charges: Perhaps unlike charges attracted slightly more strongly than likes repelled.*

Max Abraham, a young *Privatdozent* at Göttingen and former student of Max Planck's, took Wien's suggestion the most seriously of all and proclaimed in 1902 the first Electromagnetic Theory of Everything.

The centerpiece of Abraham's theory was the so-called rigid electron, a spherical field of negative or positive charge. It had to be rigid, Abraham believed, to avoid being blown apart by its own electrical repulsion. This electron, including its mass, was nothing but energy—that is to say, an electric field.[6]

Abraham was regarded as one of the brightest physicists in Europe as well as one of the most acerbic, and his tongue kept him in trouble his whole life. On his death it was said of him that "he loved his absolute aether, his field equations, his rigid electron as a young man loved his first flame."[7]

One of the stranger implications of Abraham's and other electron theories was that the mass of an "electron" would increase with its speed relative to the aether, since moving charges created additional forces and fields in the aether around them. In fact, at Göttingen, Kaufmann had already set out to measure the mass of real electrons as they moved. Between the poles of a strong electromagnet inside a small box the size of a cigarette pack he set a grain of radium. He had already demonstrated that so-called beta rays emitted by radioactive materials were identical to the cathode rays studied by Thomson and Lenard—they were, in other words, little charged particles, only faster. The radium spat out electrons at nine-tenths the speed of light, which was more than fast enough for weird effects to show up. Kaufmann recorded the amount by which electrons emerging from one end of the box were deflected by the magnetic field, and concluded that his preliminary results favored Abraham's theory.

*Lorentz applied his theory to the outstanding gravitational anomaly of the time, a puzzling motion of the planet Mercury. Like all planets, Mercury traveled in an ellipse around the sun. However, in their increasingly detailed studies of the cosmic clockworks, astronomers had noticed that Mercury's perihelion (the point of its closest approach to the sun) was not fixed in space but itself rotated about the sun. Over time—measured in thousands of centuries—Mercury would therefore trace a kind of rosette pattern through space. Most of this behavior could be accounted for by the gravitational influences of the other planets and a slight oblateness in the sun's shape, but a tiny fraction—a slight but stubborn discrepancy amounting to 43 seconds of arc per century—remained to mock Newton and the astronomers. Every astronomy student of the time, including Albert at the Polytechnic, was acquainted with this mystery, a riddle hovering on the margins of perception. With his theory, Lorentz got an answer that went in the right direction but had the wrong magnitude. His theory was the first of many to stumble at Mercury.

Lorentz countered in 1904 with a full-blown electron theory of his own. Unlike Abraham's rigid sphere, Lorentz's electron was soft and deformable. At rest it was spherical, but in motion the resistance of the aether squeezed the electron into a pancake shape. The faster the electron went, the flatter it became, until at the speed of light it had no thickness at all in the direction of flight. As the electron went, Lorentz argued, so went the atoms and the material bodies they comprised. Lorentz's original hypothesis that moving bodies contracted as they sped through the aether now could be attributed solely to the contraction and flattening of the "electrons"—positively or negatively charged—of which they were constituted.

He produced a comprehensive set of equations, known now as the Lorentz transformations, that described how the electron's properties would change as it went faster and faster relative to the aether. These effects were too minuscule to have been noticed on the scale of everyday life. Traveling at one-seventh the speed of light—42,000 kilometers per second—a meter stick made of Lorentz's electrified particles would be compressed by a mere 1 percent—a single centimeter—by the force of the aether. At that velocity time would also slow, and mass would appear to increase by a similar fraction.

Lorentz's theory worked, but in its final form it embodied eleven different fundamental assumptions. "It need hardly be said that the present theory is put forward with all due reserve," Lorentz wrote sheepishly. He went on to add: "Our assumption about the contraction of the electron cannot in itself be pronounced to be either plausible or inadmissible. Whatever we know about the nature of electrons is very little, and the only means of pushing our way farther will be to test each hypothesis as I have here made."[8]

Abraham attacked Lorentz for betraying the electromagnetic worldview—the Dutchman's squishy electrons would need additional mechanical forces to stay intact. Poincaré, for his part, was wary of Lorentz's rickety edifice, while acknowledging that it seemed to work. He complained at the Congress of Arts and Sciences in St. Louis, Missouri, in September 1904 that his principle of relativity was getting "battered" by recent events. "Perhaps," he wondered out loud, "likewise we must construct a whole new mechanics, that we only succeed of catching a glimpse of, where inertia increasing with velocity, the velocity of light would become an impassable limit."[9]

Poincaré proceeded to recast Lorentz's theory into a more mathematically elegant scheme. Among other things, he recognized that the Lorentz transformations comprised what is called in mathematical terms a *group*: any number of transformations could be added together to produce a new transformation that also obeyed the Lorentz rules. A physicist could hop from the ground onto a 100-km-per-hour train and then from that train to one going twice as fast; the effect on his clock or his ruler was the same as if he or she had hopped straight onto the faster train.

There was one curiosity that Poincaré did not fail to notice: The Lorentz transformations could be inverted. Since all motion was relative, the mathematics predicted that just as a physicist on the ground would see his speeding counterpart shortened and slowed, so the physicist on the train would perceive his earthbound colleague to be the shortened one with the lagging clock. This, Poincaré thought, was nonsense. He regarded these inverse transformations as mere mathematical artifacts with no physical significance. The aether, he knew, did not move.

ALBERT, working away at his patent office desk or under the oil lamp at home, counted himself a partisan of neither the mechanical nor the electromagnetic worldview. The first seemed too gimmicky. As a student he had given up on Hertz's book when the old master began to populate the aether with "hidden masses" to replace force fields with mechanical connections.[10] Albert's patent office colleague Sauter had written a paper about mechanical models of light waves, but when he once tried to discuss it with Einstein he was cut short. "I am a heretic," an uninterested Albert declared.[11]

Lorentzian electrodynamics seemed to have more going for it aesthetically, but Albert was bothered by the fact that it was a dualistic theory—in other words, it allowed discrete points and continuous fields to coexist in the same space. Reality, he thought, should be either discrete or continuous, but not both. Moreover, Albert had not followed Lorentz's work very closely. He had read the 1895 book in which Lorentz discussed the contraction hypothesis, but not the later papers in which he generalized that transformation to other quantities such as mass, in the electron theory. Nor had Albert read Poincaré's own version of the electron theory, which was published in a French journal on June 5, 1905, just two weeks after his triumphant letter to Habicht.[12] Where then, did Albert get the idea to overthrow the reigning concepts of space and time?

What was a blessing for Albert has proved to be something of an obstacle for historians. With his best friend Michele walking him to and from the patent office, and Mileva waiting hungrily at the writing table at the end of every day, there was no need for him to develop his ideas in correspondence, which could later be rifled for anecdotal details of this, the climax of the Einsteinian spring. Neither Albert, Mileva, nor Michele is known to have kept a diary. Moreover, it would be another fifteen years before Einstein would become famous enough to be hounded by journalists wanting to know how he had done whatever it was he had done. The lack of detailed documentation from this period has led to a general misapprehension in the public mind that the origin of what we now call the special theory of relativity is a mystery. Two generations of scholars scouring Albert's later correspondence, memoirs, old textbooks, and school curricula for clues to the trajectory of his thoughts have proved, however—to paraphrase Poincaré—that hypotheses are what historians lack least.

One trail that Albert might have followed to relativity, as ingeniously and doggedly reconstructed in Gerald Holton's classic *Thematic Origins of Scientific Thought* (1973), leads to an obscure German, a civil engineer by training and a philosopher by temperament, named August Föppl. In 1892, apparently bored while teaching a course on agricultural machinery at the University of Leipzig, Foeppl wrote a textbook on Maxwell's electromagnetic theory.[13] In its fifth chapter he exposed the deep fault line running through physics, and challenged physicists to consider whether they should change their concepts of space.

In Newtonian and Machian mechanics, Föppl explained, there was no absolute reference frame, and only relative positions and motions had physical significance. Electromagnetism, however, *did* have one, namely the aether, the medium that filled space and transmitted waves and other force fields. This led to conflicts of principle.

Consider, for example, the simple case of a magnet and a loop of conducting wire. The electromechanical transformation of Europe had sprung from Michael Faraday's discovery in 1832 that moving the magnet near the wire would induce an electrical current in it. According to the principle of relativity, it should not matter whether the coil or the magnet was moving. The same current would ensue. According to Maxwell's theory, however, which of the two was moving made all the difference: If it was the magnet, its very motion through the aether would create an electrical field around itself, which in turn would drive a current around the wire; if it was the wire that was moving, however, electrons inside it would feel an "electromotive force" from the magnetic field and respond accordingly. In each case, a very different mathematical equation would be required to determine the resulting current. Were both guaranteed to give the same answer? Was it relative or absolute motion that counted?

Föppl answered that question by imagining what would happen if a coil and a magnet were moving together, at the same speed, through the aether. In that case, as an experiment showed, no current would result, just the same as if the two objects were resting immobile on a table. It was, therefore, relative motion that counted, at least in this case. But Föppl's thought experiment (*Gedankenexperiment*) left no role for the aether. He compared the idea of space without an aether to a forest without trees—not a very Machian concept—but warned that physicists might soon be forced to accept that strange and unpalatable concept. "The decision on this question," he wrote, "forms perhaps the most important problem of science of our time."[14]

Published the same year that Albert was fleeing the gymnasium, Föppl's book was eagerly taken up by students trying to find a foothold in Maxwell's world. It has been suggested that among those students was Albert, who may have read Föppl when he was boning up for his patent office interview or wandering in the academic wilds to supplement Weber's teachings at the Polytechnic.[15] Besso, too, later recalled having suggested to Albert that he investigate Föppl's work.[16] Being his uncle's nephew, Albert was familiar with dynamos, and the idea that something deep about the universe could be revealed by thinking very hard about an experiment that could be conducted on the kitchen table would have appealed to him greatly.

Albert had already encountered firsthand the paradoxes of electrodynamics through his own *Gedankenexperiment* during his sojourn in Italy, trying to imagine himself surfing the aether. When is a light wave not a light wave? When you're traveling alongside it at the speed of light. That was the riddle that had taken root in his mind like a Zen koan for the last decade. If the purpose of a koan is to drive a student slightly mad enough to be able to make a perceptual leap, it had at least part of the desired effect on Albert over the years. Newton's mechanics or Maxwell's? The relativity principle or aether waves? His repeated failure to solve the problem had left him increasingly frustrated and had created in his soul what he later called a "psychic tension."[17]

In the spring of 1905 that tension was exacerbated to the breaking point by Einstein's adventures with Planck's quantum. This brilliant but disturbing reading between the lines of nature indicated that blackbody radiation behaved not like waves but like little particles of energy. In this case, at least, Maxwell's equations failed. Neither Newton nor Maxwell could, as Albert liked to put it, "claim exact validity."[18] Suddenly it seemed that physics had no trustworthy foundation. "It was as if the ground had been pulled out from under one, with no firm foundation to be seen anywhere, upon which one could have built," he remarked.

Albert took the failure of physics to heart. His recollections of the period paint the picture of a man in a daze, nerves ajangle, beset by the trances (*Anfälle*) about which Mileva and her friends teased him, his thoughts racing and caroming in unpredictable directions.[19] He would "go away" for weeks in a state of confusion. In a way, he found himself back where he had been at age twelve, when the religious paradise of his youth had been suddenly collapsed by an orgy of hard thinking.

"By and by I despaired of the possibility of discovering the true laws by means of constructive efforts based on known facts," he said later.[20] "The longer and the more despairingly I tried, the more I came to the conviction that only the discovery of a universal formal principle could lead us to assured results." What did it take for a law to pass as universal? As Albert saw it, there were two kinds of theories in the world. One was based on grand, overarching principles or laws learned the hard way from experience. Examples of that kind of thinking were the law of conservation of energy, or the second law of thermodynamics: Thou shalt not be able to build a perpetual motion machine. Much power and wisdom and verifiable physics flowed from such principles, but the explanations for the principles themselves remained mysteries. The other kind of theory was constructed from more elementary concepts, like the idea of atoms, and built up with logic and mathematics into an elegant edifice that included prediction and explanation. For example, if you began with the humble foundation of the notion of heat as bouncing atoms, and accepted Boltzmann's idea that entropy was just the probability of different arrangements of these atoms, the laws of thermodynamics stood revealed as the necessary and inevitable consequences of simple statistics. You now knew *why* there was a second law and *why* you could not build a perpetual motion machine.

Such bottom-up explanations were much more satisfying intellectually, but in this case none was available to help Albert solve his dilemma. In his previous work that spring on Brownian motion and blackbody radiation, Albert had marched through the traditional truths deliberately and systematically—in the opinion of scholars like Holton—to show that physics had no clothes. The facts alone explained nothing; that was just mental stamp collecting. The only alternative was to make a stab at truth by inventing some new principle that would restore order. Maybe someday, by studying it like a clever architect, he could figure out the nature of its foundations.

But where was he going to find another principle as general and as grand as the second law of thermodynamics? Mach's keen precepts could tell him how to analyze and criticize an idea, but they couldn't tell him how to have the idea in the first place.

It was in such a quicksand-grabbing mood that he considered the problem of the magnet and the coil that Föppl had posed. Did it matter which one moved? Mechanics said no; electromagnetism said one thing empirically and another theoretically. Under the circumstances, that was no choice at all. Because electromagnetism was already suspect due to quantum theory, Albert chose relativity. The lesson of the magnet experiment was that what held true in the realm of mechanics would be equally valid in the arena of light and electromagnetism: Only the relative movements of real things—that is, so-called ponderable bodies—had any significance. The aether went out the window. Relativity would become Albert's new commandment, a new principle, the foundation of his own desperation to set his lands in order again.[21]

Albert's first instinct was to continue the line of thought he had developed in his quantum paper—that light behaves like particles dashing through empty space—and abandon the constancy of the speed of light. In this case, *c,* the constant in Maxwell's equations that represented the speed of the light wave, would be a kind of muzzle velocity, the speed of the light quanta relative to the source that had emitted them. This approach quickly led to a muddle, however, and he gave it up.

Back where he had started, Einstein took another look at Maxwell's equations. How, indeed, was it possible for a boy riding a light beam and a man puffing his pipe on the bank of the Aare to get the same answer for the velocity of the same light wave? "The difficulty to be overcome then lay in the constancy of the velocity of light in vacuum, which I first thought would have to be abandoned," he later recalled.[22] Only after groping for years, he went on, had he finally realized that the problem lay in our concepts of space and time themselves—especially time.

Compared to the dapper French geometer Poincaré and the sophisticated, genial Dutchman Lorentz, Einstein was a mechanic, an *elektrotechnischer* brat with a Machian yearning for economy and, conversely, a seemingly biological need for fundamental principles, universality, order, law, God—call it what you will—*logos.* Faced with a choice between two of those principles—Mach's insistence that only relative motions and positions have physical significance, and Maxwell's seeming commandment that light waves move with the absolute velocity *c* in the aether—Albert ultimately chose both.

It would be pretty to think that the breakthrough to this conundrum came as he and Besso were walking and arguing past that great clock tower straddling the old city gate at the top of Kramgasse. In fact, Albert might have had to look no further than his desk at the patent office. Recently, work by the Harvard University historian of science Peter Galison has established that the synchronization of clocks over long distances and far-flung networks was one of the key technological goals of the late nineteenth and early twentieth centuries in Europe. How could the trains run on time without a way for everybody to agree on what time it actually was? The "unification of time" was considered to be of great practical and symbolic significance, especially to Prussian military leaders, who hoped that the master clock for Europe would be located in Berlin, not Paris or Vienna.

Like many cities, Bern in fact had a master clock, which had been set up in 1890 and with which all the clocks in downtown Bern were synchronized. As a nation of clockmakers, Switzerland was at the center of the push to unify time, and patent applications for various schemes, based on telegraph lines, radio waves, or other devices, flowed into the Bern patent office, Galison reports, and probably onto Albert's desk.

Nevertheless, at some point Albert asked himself in his best patent-examiner Machian manner: What did it really mean to measure a motion in space and time? Time and space were only metaphysical abstractions; in the empirical world of sense data there were only measuring rods and clocks.

The simple act of determining the time at which some event occurred—such as the passing of a light wave or of a locomotive—Albert realized, was fraught with epistemological problems and assumptions. We are used to thinking of things happening simultaneously with our seeing them, as if the transmission of light were instantaneous. Light is fast, but not infinitely fast, and in its speed is masked the weird grace of the world.

Suppose that light were as slow as sound, and that you are a young student standing on, say, the hillside behind the Polytechnic in Zurich looking out over a sea of clock towers stretching infinitely into the distance. How do you tell what time it is? The first obstacle in obtaining a reliable answer would be figuring out how to synchronize all those clocks. You might think you could simply go around town and set each clock against your wrist-watch, but that would violate the take-no-prisoners stance of fin de siècle science and philosophy by assuming that moving clocks (your watch) keep the same time as stationary ones. Another way would be to bounce light beams off the clock towers, measure the delay time, and use that to correct each reading, but that approach—the one favored by nineteenth-century physicists—involves its own set of assumptions, namely, that space is homogeneous and that light travels at the same speed in all directions.

Imagine, nevertheless, that some genius has managed mysteriously to synchronize all those clocks to standard time. The nearest to you, then, might read 6 P.M., but as they grow more distant the clocks will each read an increasingly earlier time: 5:58, 5:45, and so on, the news that it is six o'clock delayed by the time interval that it takes for light to travel from each clock to your hillside perch. Alternatively, the clocks might have been synchronized in a such a way that to you, standing on the hillside over Zurich, they *all* read six o'clock—in which case it will be six o'clock for you and only you among all the inhabitants of the entire city. In any event, you will be in a certain amount of epistemological trouble if the police come and ask you what time you saw a flaxen-haired beauty tossed by some dark figure from a high balcony of a distant building. Should you report the time on the clock nearest the unfortunate incident, or your own time? What if there is no clock nearby? Would other witnesses be able to agree on the time of the woman's death?

And there, at the end of this long, winding argument that physics was having with itself and nature, at the fulcrum of the young twentieth century, was the slippery metaphysical worm at the center of the apple: the concept of simultaneity. All temporal measurements, Albert realized, were observations of simultaneous events—the position of the falling maiden combined with the positions of the hands on a clock. Moreover, any such measuring process relied on a hidden assumption, namely that events that appeared simultaneous to one observer would appear simultaneous to all. But absolute simultaneity—as commonsensical as it sounded—was a diseased concept. Poincaré had dissected it in *Science and Hypothesis,* arguing that simultaneity was a convention, a choice imposed upon nature. "There is no absolute time," he had written.[23]

Nobody appreciated that argument better than Besso, Albert's little Machian enforcer and sounding board. Years later Albert recalled that he had gone to see Besso one fine spring day and laid out the entire problem, the schism between mechanics and Lorentzian electrodynamics, that he had been vainly struggling with for the last year.[24] They talked for hours, rehashing every aspect of the dilemma. At the end of the day, Albert announced that he was giving up the entire quest.[25]

That night, however, everything changed. The next day Albert went back to Michele's house and without even greeting him blurted, "Thank you. I've completely solved the problem." With that Albert returned home and began to write.

It took Albert six weeks to complete what is arguably the most famous scientific paper in history. Its title, "On the Electrodynamics of Moving Bodies," echoed the Hertz book he had studied long ago in the Hotel Paradise. Most of his argument is presented verbally; what little math is involved is algebra that any high school student could follow. Indeed, the whole paper is a testament to the power of simple language to convey deep and powerfully disturbing ideas. Reading it is like following the writer, Albert, into a deceptively straightforward-looking maze, taking one obvious, even boring, step after another, until all of a sudden you are standing on your head and there is no way home.[26]

"It is well known that Maxwell's electrodynamics—as usually understood at present—when applied to moving bodies, leads to asymmetries that do not seem to attach to the phenomena," he began, recalling the example of the magnet and the coil. When the first is moved, it produces an electric field; when the second is moved the ions inside the conducting coil feel an "electromotive force," although *relative* motion in each case was the same. Whereas Föppl had been more concerned about whether the same current was induced in each case, Albert emphasized the ugly asymmetry of the ascribed causes of that current: an electric field in one case and not the other. In this, perhaps one of the first appeals in modern science based on an aesthetic principle, Einstein was applying a new and most un-Machian standard: A theory must not only explain the data, it must be in some mathematical sense beautiful.

This asymmetry, together with the well-known failure to detect any motion of the earth relative to the "light medium," Einstein suggested, meant that in electrodynamics, as in ordinary mechanics, all motions were relative. There was no state of "absolute rest." The laws of physics did not depend on who was moving and who was still. He declared, "We shall raise this conjecture (whose content will be called 'the principle of relativity' hereafter) to the status of a postulate and shall introduce, in addition, the postulate, only seemingly incompatible with the former one, that in empty space light is always propagated with a definite velocity V which is independent of the state of motion of the emitting body." These two postulates were all that was needed, he wrote, to produce a simple and consistent theory of electrodynamics, once proper care had been taken to understand the relationships between the clocks and measuring rods that formed the underpinnings of scientific fact. The concept of aether would prove superfluous. In effect, Albert was turning electrodynamics upside down: Instead of taking space and time for granted, he took Maxwell's equations and the constancy of the speed of light for granted and asked what they revealed about the nature of space and time—that is to say, the behavior of measuring rods and clocks.

"We have to bear in mind that all our propositions involving time are always propositions about *simultaneous events,*" he explained as he launched into a discussion as much philosophical and epistemological as scientific. "If, for example, I say that 'the train arrives here at seven o'clock,' that means, more or less, 'The pointing of the small hand of my clock to 7 and the arrival of the train are simultaneous events.'" But what, he asked, did it mean to say a train arrives someplace else at seven o'clock? He then extended this definition of time to more distant realms by imagining a set of clocks extending throughout space—all at rest with respect to each other—that could be synchronized by bouncing light beams between them. Since the speed of light was a basic and universal constant, half the round-trip time between the master clock and any distant clock gave the delay for that clock. (For example, if it was 3 P.M. here and the round trip to some other clock was thirty minutes, that other clock would read 2:45 to an observer standing at the first clock.)

The centerpiece of this elemental discussion was another innocuous-sounding thought experiment: There are two people—one at rest, the other moving at a constant speed—each of them with his own set of synchronized clocks. Moreover, there is a rod traveling along with the second person, pointed in the direction of motion. (In the future no physicist would be a physicist who had not had to play Albert's deceptively subtle little game of moving clocks and rods.) To begin with, Albert asked what what would happen if the two men both tried to measure the length of the rod. "The commonly used kinematics tacitly assumes that the lengths determined by the two methods mentioned are exactly identical," he drily noted. In fact, he promised, this would not be the case.

Before explaining why, however, Albert went on to examine the very weird behavior of the clocks in this game. Assume, he wrote, that the moving and stationary clocks are initially synchronized. The moving man then shoots a light beam from the trailing end of the rod up to the front end and back. From his point of view the beam travels up the rod at the speed of light, bounces off the end, and comes back; each leg of the trip takes the same amount of time. The "stationary" person, however, sees something quite different while watching the moving man's light beam. To him, the two legs of the trip along the measuring rod are not equal and take different amounts of time. From his point of view the beam also travels at the speed of light, but on the initial leg of its journey it must traverse not just the length of the rod but also the extra distance that the rod has moved while the light was traveling down it. Similarly, on the return leg of its trip the light beam has a shorter distance to go along the rod, which is still moving forward. As a result, the two men will not agree on when the light hit the end of the rod and bounced back. They will not even agree, Albert declared, on whether their clocks are still synchronized.

The source of this smoke-and-mirrors magic, of course, was the commandment that the speed of light was always the same. Recall from high school algebra that the distance anything travels is simply given by multiplying its velocity by time. If our two men were therefore compelled to agree on the uniform velocity of the light beam they are watching, then by simple mathematical logic they would have to disagree on how far and for how long it had traveled. In other words, if velocity was fixed, then space and time—that is, rods and clocks—had to be elastic. "Thus we see that we must not ascribe *absolute* meaning to the concept of simultaneity," Einstein wrote, "instead, two events that are simultaneous when observed from some particular coordinate system can no longer be considered simultaneous when observed from a system that is moving relative to that system."

From there the argument proceeded with logical inevitability. Detailed calculations followed of exactly what value that observer *did* measure for the length of the moving rod, and the consequences for space, time, and electromagnetism. In his 1895 paper, which Albert had read, Lorentz had already provided a sort of mathematical template for transformations that preserved the sanctity of the speed of light. Mathematically, Albert's results were identical to those of Lorentz and Poincaré. To a person at rest, moving rods seemed to contract along the direction of their motion, moving clocks appeared to tick more slowly, objects appeared to gain mass as they got faster, and velocities did not strictly add up: a bullet fired at 3,000 feet per second out of a gun that was on a train traveling 100 feet per second did not fly at 3,100 feet per second relative to the ground, but slightly, imperceptibly, less than that. At the speed of light, the rod would contract to zero, the clock would stop, and a body's inertial mass would appear to become infinite, preventing further acceleration.

In Albert's hands, however, the meaning of these equations had changed completely. Lorentz believed the transformations were real electrodynamic effects, caused by forces created by the passage of objects through the aether. In the new relativity theory, however, they were purely intrinsic to the nature of motion, a consequence of nature's presumed desire to keep the speed of light constant. In this new universe, exceeding the speed of light was not so much impossible as meaningless. If we could exceed the speed of light, Albert later remarked, we could send telegrams to the past. Moreover, since there was no aether, there was no absolute rest frame, just as there was no real time. Any observer could view himself as being at rest and everybody else moving. Two physicists sailing past each other could look out and each see the other as shortened and moving in slow motion, and they would both be right.

But in Einstein's formulation did objects actually shrink? In a way the message of relativity theory was that physics was not about real objects; rather, it concerned the *measurements* of real objects. And each of those measurements included time as well as space. Relativity was not an explanation of nature at all, but of how we know about nature. In that sense it has sometimes been claimed that relativity is not a theory at all, but a language or a convention, a set of rules for how to talk about the universe. After relativity, no law could claim to be a law of nature that did not speak its language, that could not be expressed in a form that was true for observers moving at any constant speed—so-called inertial observers. Although Albert spent the last half of his paper solving problems in optics and electromagnetism, he knew that the relativity principle, as he called it, transcended any particular problem.

No such declarations of grandeur, of course, intruded on the flat and somewhat brisk tone of the paper. Albert simply presented his argument and in many cases left it to the reader to fill in the gaps and to realize the implications. Unlike most scientific papers, it did not specifically refer to any other scientist or body of experimental data and contained no footnotes. This, remarks Galison, may be a reflection of Einstein's experience in the patent office, since footnotes, suggesting that somebody else has been there first, are anathema in a patent application. At its end Einstein listed no references, but only a brief acknowledgment. "In conclusion," he wrote, "let me note that my friend and colleague M. Besso steadfastly stood by me in my work on the problem here discussed, and that I am indebted to him for many a valuable suggestion."

The paper was finally finished at the end of June. We are told in Peter Michelmore's biography, which was based on conversations with Hans Albert, that Einstein went straight to bed for two weeks while Mileva checked and rechecked his work. He had planned to submit the paper to Kleiner at the University of Zurich as his doctoral dissertation,[27] but it was too weird for them, so it was duly offered to the *Annalen der Physik* instead, where it was published on September 26, 1905. (Copies of that volume of the *Annalen* fetched $15,000 at auction in 1994.)

Shortly after sending in the paper, Albert had another flash of insight, a little piece of unfinished electrodynamic business. "Namely, the relativity principle, in association with Maxwell's fundamental equations," he told Habicht, "requires that the mass be a direct measure of the energy contained in a body; light carries mass with it."[28] The idea that mass was associated with electromagnetic fields was not new to physics—it formed the basis of the electromagnetic worldview promulgated so vigorously by Max Abraham. Einstein, however, was now arguing that all energy contained mass and vice versa. He suggested that it might even be possible to measure the mass lost by radium when it underwent radioactive decay. "The consideration is amusing and seductive; but for all I know, God Almighty might be laughing at the whole matter and might have been leading me around by the nose."

Nevertheless Albert rushed off a three-page sequel to his first article and submitted it to the *Annalen*. "If the theory agrees with the facts," he concluded, "then radiation transmits inertia between emitting and absorbing bodies." It is here that Albert first implied in words, if not actually setting down, the famous formula $E = mc^2$.[29]

Meanwhile he had dusted off his earlier paper on the size of molecules and a month later sent it in to Kleiner as his long-awaited dissertation. His doctorate was in the bag by now, and he knew it. That night he got drunk with Mileva for one of the few times in his life. They sent a postcard to

Habicht announcing that they were under the table and jokingly signed it "Steissbein and wife."[30] A week later, July 27, the faculty of the University of Zurich voted to approve his petition for the doctorate.

In August Mileva finally made a triumphant return home with Albert and Hans Albert in tow to the "little land of bandits" that she had bragged about and promised to show Albert long ago. The tour began with a week near Belgrade to visit their old friends Helene and Milivoje Savic, and Milana Bota, who had also married and was living there.[31] Albert, it is said, was fascinated by the geography of the region.

From there they proceeded on to Novi Sad, where they received a warm welcome from Mileva's family and the rest of the community. It was a great moment for both of them. Whatever bad feelings may have been engendered by Albert's previous adventures with Mileva and with Lieserl seem to have been forgotten. The autodidact Milos could not have helped being charmed by his son-in-law's style, for Albert paraded through the streets carrying his son on his shoulders. "Not long ago we finished a very significant work that will make my husband world famous," Mileva told her father in a conversation widely repeated through the years.[32]

To the villagers and relatives who remembered her as a childhood genius in mathematics, Mileva had a heroic aura, the local girl who had gone out into the world and made good. Now she had brought back a handsome, adoring husband. Albert knew how to play the crowd. "I need my wife," he is reported to have said. "She solves all the mathematical problems for me."[33]

The villagers nonetheless did find Albert a bit eccentric. The "crazy Maric son-in-law," as they called him, looked like a clown with his long, messy hair as he ran down the street with Hans Albert. The students, however, clustered around him in the local coffeehouse. One of them later recalled a strange conversation, in which Albert was asked about alcohol.[34] "I don't believe in doctors, medicine, or anti-alcoholism," Albert answered. "A Serb drinks from birth unto his grave, during growth, on trips, at weddings and funerals, and yet the Serbs are a brilliant nation. I base my opinion on what I know about my wife."

"We were stunned," the student said, "and the result was that we started drinking and smoking."

PART THREE

The Revolutionary at Home

Marriage is the unsuccessful attempt to make
something lasting out of an accident.

Albert Einstein, quoted by Otto Wathan

II

THE DISCREET CHARM OF
THE BOURGEOISIE

BY THE TIME RELATIVITY APPEARED, ALBERT HAD LONG SINCE PROVED himself a capable bureaucrat. In March of 1906 Haller promoted him to technical expert second class and raised his salary to 4,500 francs per year, praising him for his adept handling of the most difficult cases. Two months later he and Mileva and little Hans Albert moved again, from the university district across the river to Aegertenstrasse, a high tree-lined street of wood-frame houses overlooking the Aare and downtown Bern in the distance.

Albert watched bourgeois life closing in on him a little warily and joked that he was about to become the kind of sterile old man who complains about revolutionary youth,[1] a "federal ink-pisser" who played the violin and rode his "physical mathematical hobbyhorse" within the increasingly narrow confines allowed by his growing "Bubi."[2] At least, he allowed, the improvement in his circumstances gave him something to think about. To occupy himself scientifically, Albert spent time at the Bern Natural Science Society, which met at his friend Paul Gruner's house once a month. He continued to miss his old Olympia Academy, however, and felt the loss of its camaraderie. Solovine had moved to Paris, and Habicht rarely came around. Even the walks home with Besso had ended because of his move.

His only regular socializing during this period was on the Sundays when he and Mileva hosted casual get-togethers at their home. Among the frequent Sunday visitors were the Bessos and Maja, who after two years at the University of Berlin had moved to Bern to study for a Ph.D. in Romance languages. She soon became engaged to Paul Winteler, who was studying law at the University of Bern. Her eventual marriage would further entwine Albert with the Winteler family, in whose home he had lived, and whose daughter he had courted. Albert had always had Edenic memories of his year in Aarau, so what happened next must have been doubly disturbing. On the first of November, 1906, Paul's brother Julius came home from

America, where he worked as a cook, shot and killed his mother and his brother-in-law Ernst Bandi, who was married to his sister Rosa, and then turned the gun on himself.[3]

Albert probably heard the news from Maja or Michele, and it reawakened his guilt over Marie. "The dear departed has shown me so much kindness while I caused her only sorrow and pain," he wrote to Jost Winteler. "That distresses me all the more at this hour!"[4]

The subject of Marie, of course, had never ceased to distress him. Their relationship, he once told Besso, remained a dark stain on his life.[5] The very mention of her seemed to be sufficient to send his mind back to wander the caverns of what-might-have-beens. Was it she, or the feelings she had inspired in him—the craziness that he had admitted even to Mileva he feared like fire, that still lurked after all these years—that had caused him to run from her?

The fact is that Albert was always more comfortable longing for his lover, his family, or his great mentor from a distance than being embraced, being consumed, up close. The gauzy memories of Aarau life that he spun around Mama Winteler suited Albert perfectly. If he was going to be alienated or an outlaw, it was much more tangy and poignant to be separated from this boisterous, beautiful clan of cocoa-sipping philosophers and bird-watchers than from the dead weight of his own mother's philistine attentions. But what if his beautiful memories were a myth? In the course of Albert's young life the events of November 1 stand out like an iceberg, signifying a submerged mountain of grief and madness. The threat of the latter, it seemed, was always lurking.

Julius, it was generally agreed, was deranged. And then there was Michele's brother Marco, who had been a suicide at eighteen.[6] A terrible blow for his family but better than a miserable life, Albert had concluded about Marco when he heard. What if happiness in general was a myth? The Wintelers, apparently, were a fragile race. The dark side of their version of what we now call family values had ultimately claimed Marie, as well. She had lapsed into depression after the affair with Einstein had ended and was consigned for a while in Waldau, a mental hospital affiliated with the University of Bern. A synopsis of her case appears in the official history of the hospital. According to a statement by one of her doctors, "a student named Albert Einstein had turned her head."[7]

If one era of science was ending, with the revelation of the relativity principle, as Albert called it, its successor was slow to take its place. In the months following the publication of his relativity theory, Albert eagerly scanned the journals for reaction to it, like a playwright awaiting opening-night reviews.[8] Finally a short note in the *Annalen* by the Göttingen physicist Walter Kaufmann alluded to "a recent publication by Mr. A. Einstein on the theory of electrodynamics which leads to results which are formally identical with those of Lorentz's theory."[9] Otherwise, there was only a disappointing silence, a stinging reminder of his rejections by Drude and Weber. In fact, most physicists couldn't tell the difference between his approach to electrodynamics and that of Lorentz, for their mathematical consequences, the Lorentz transformations, seemed to be identical. Only abstract philosophical principles distinguished the two approaches—the classical electron theory of Lorentz and Poincaré maintaining the aether and a preferred reference frame to the universe, or Einstein's hypothesis, which jettisoned the aether and apparently common sense as well in favor of a purer but more slippery notion of reality. Even those who did understand the difference didn't always approve of it. Arnold Sommerfeld, an influential physicist and electron theorist at the University of Munich, confided in a letter to Lorentz that Einstein's work contained "unhealthy dogmatism" and exemplified "the abstract-conceptual manner of the Semite."[10]

Lorentz himself admired Einstein's work and appreciated the logical clarity of his theory, but he was uncomfortable with the notion of relative time. Moreover, he was committed to the reality of the aether. "I cannot but regard the aether, which can be the seat of an electromagnetic theory with its energy and its vibrations, as endowed with a certain degree of substantiality," he explained once, "however different it may be from all ordinary matter."[11] Poincaré, as well, had trouble abandoning the idea of the aether. In his publications over the next few years he simply pretended Einstein didn't exist. When he finally addressed the issue of relativity squarely, in 1912, it was clear that he didn't understand it.[12]

One scientist who both comprehended Einstein's theory and liked it was Max Planck, the reluctant inventor of the quantum, who was now working at the University of Berlin. Einstein's critique of Planck's derivation of his famous radiation law and the dubious expansion of it to embrace the notion of atomistic light beams had naturally caught his attention, and they had, by some accounts, already corresponded.[13] When the *Annalen* with Albert's electrodynamics paper landed on his desk, therefore, in September of 1905, just as the school year was beginning, Planck was prepared to take it seriously. He scheduled a lecture on Einstein's new theory, which he dubbed *Relativitätstheorie,* for the first weekly physics colloquium that fall. He later wrote to Albert asking for clarification of some points.[14] It was the first substantial response Albert had received, and the fact that it came from a major figure like Planck lifted his spirits considerably.

Planck soon joined the very select few who had published papers on relativity. The theory resonated with his notion that science was a quest for absolute truth. "For me its appeal lay in the fact that I could strive toward deducing absolute invariant features following from its theorems," he explained. "Like the quantum of action in quantum theory, so the velocity of light is the absolute, central point of the theory of relativity."[15]

Meanwhile, Planck's relativity seminar caught the imagination of young Max von Laue, Planck's new assistant, who had recently arrived from Strasbourg, and he too began writing to Einstein. The following August, in 1906, he arranged a trip to Switzerland so that he could visit Bern. On his arrival, von Laue went straight to the patent office. Albert came down to the lobby to greet him, but von Laue at first let him walk right by, not expecting this swarthy young cherub with a helmet of frizzy black hair and smiling eyes to be the father of relativity.[16]

Von Laue's impression of him was perhaps best expressed by advice he later gave the physicist Paul Ehrenfest. "You should be careful that Einstein doesn't talk you to death. He loves to do that, you know."[17] Of his visit, von Laue said to another physicist, Jakob Laub, "He is a revolutionary. In the first two hours of the conversation, he overturned all of mechanics and electrodynamics, and this on the basis of statistics."[18]

MILEVA WAS QUICK to sense that a change was coming over her and Albert's lives. By now she was almost completely out of her husband's scientific life. The shift was changing, passing over to young men like von Laue with their own handlebar moustaches and something of the same ruthless glint (though less of the twinkle) in their eyes as Albert. She found herself unexpectedly wistful for the old days of student poverty and oppression by Weber, when it was a triumph for her and Albert even to be together. In December of 1906 Mileva sat down to write her annual Christmas letter to Helene. "Did you find us very different aside from the fact that we've become older?" she asked. "I often have the feeling that I'm sitting in Zurich in a certain little room spending my most beautiful days. It is too strange, in my old days I like to think of this time too much."[19]

With the passage of time, Mileva was like a hiker who had been ascending through a fog and then finally emerged into the clear to discover that the vista beneath her bore no resemblance to what she had so long anticipated. At thirty-one she suddenly found herself a middle-aged woman, with two strikes against her. She had already lost a physics career and her Lieserl. At least there was Albertli. It was time to start thinking about his development and planning his education. Seemingly without her noticing, little Hans Albert had metamorphosed from a baby into an independent being with his own droll (the same word they had used to describe Albert at that age) personality, so roguish that she and Albert had to bite their tongues to keep from laughing at his antics. "I don't know if it's generally this way," she told Helene, "it was as if suddenly his spirit began to awaken; all of a sudden he became completely different."

As for "my husband," well, she reported, he was incredibly busy; what little free time he had, he spent playing with the boy. "To his honor, however," she wrote, as if she were his mother and very far away indeed from his scientific life, "I must note that that is not his only occupation outside of his office position, the papers he has written are piling up terribly." Perhaps the fog had lifted to reveal him across the valley climbing another mountain altogether, by himself.

BY 1907 A FLURRY of letters was being exchanged between Einstein, Planck, von Laue, Lorentz, and others. Much of Albert's correspondence found its way to him addressed to "Albert Einstein at the University of Bern." Paul Ehrenfest, an unemployed physicist then living in St. Petersburg, wrote to ask how relativistic contractions would affect the internal forces holding together the electron. Albert realized that Ehrenfest had misconstrued the meaning of relativity, and his answer allowed him to draw a clear distinction between Lorentz's thinking and his own. Lorentz's was a theory about the structure of matter, in which matter was composed of electrons that were flattened by the resistance of the aether when they moved. Albert's theory, in contrast, concerned motion, space, and time. Relativity had nothing to say about the structure of matter at all, except to add one more hurdle that any prospective theory of matter would have to clear if it staked a claim to universality: Whatever laws governed the structure of matter would henceforth have to be Lorentz-invariant; that is, they would have to appear to be the same for any freely moving observer. Ehrenfest understood the distinction and became a fan and defender of relativity, but he kept pestering Albert with embarrassing paradoxical questions about the effects of contraction on so-called rigid bodies.

Between 1905 and 1909 Albert would publish more than two dozen papers elaborating on relativity and responding to criticisms. "As long as the proponents of the principle of relativity constitute such a modest little band as is now the case, it is doubly important that they agree among themselves," Planck wrote approvingly at the resolution of one such problem.[20]

Planck unwittingly added to the confusion at one point. In a sort of turnabout to what Albert had done to his quantum calculation, Planck criticized Einstein's derivation of the famous mass-energy relationship and then redid the calculation to show that heat as well was associated with mass. Planck's paper was read by Johannes Stark, a physicist at the Technical University of Hanover, who had been corresponding with Albert about quantum theory and had even recruited him to write a review paper about relativity for the *Jahrbuch der Radioaktivität und Elektronik,* which he edited. Stark's knowledge of relativity was sufficiently superficial that in a subsequent paper he credited *Planck* with having invented the mass-energy relationship.

One of his new friends, Jakob Laub, urged Albert to read the article in the latest issue of the *Physikalische Zeitschrift* by a J. Stark, and to take steps to protect his status in the relativity field.

"I find it somewhat strange," Albert promptly wrote to Stark, "that you do not recognize my priority regarding the connection between inertial mass and energy."[21]

Stark apologized by return mail, explaining that he had misunderstood one of Albert's earlier papers. "You are greatly mistaken, esteemed colleague, if you think that I have not been doing sufficient justice to your papers. I champion you whenever I can, and it is my wish to be given the opportunity to propose you for a theoretical professorship in Germany quite soon."[22]

Albert was sufficiently mollified by Stark's letter to write back contritely, "If I regretted even before the receipt of your letter that I let a petty impulse goad me into making that remark about priority in the matter in question, your detailed letter showed me all the more how misplaced my sensitivity was. People who have been granted the privilege of contributing to the progress of science should not let their pleasure in the fruits of joint labor be spoiled by such things."[23]

The most serious challenge to relativity was not philosophical, however, but experimental. According to the transformation equations that were the mathematical core of the so-called Einstein-Lorentz theory, objects should appear to gain mass as they move faster and faster relative to the observer. But that's not the way it was working out, according to the Göttingen physicist Walter Kaufmann. The Lorentzian prediction of mass increase was not unique; competing theories, such as Abraham's rigid electron, had made similar predictions, but with a different dependence on velocity. In fact, J. J. Thomson, the discoverer of the electron, had speculated as early as 1881 that the mass of a charged particle might depend on its velocity.[24]

Kaufmann, as mentioned, had been looking for this effect since 1901 by studying so-called beta rays, electrons that came shooting out of radium salts at up to nine-tenths the speed of light. In his apparatus a narrow beam of electrons passed crosswise through electric and magnetic fields that pushed them in mutually perpendicular directions before they splashed onto a photographic plate, where, owing to their initial spread in velocity, the electrons painted a curve. Making certain simplifying assumptions about the dynamics of the electron and its behavior in his apparatus, Kaufmann was able to calculate the velocity and mass of each electron that made a point along the curve.

Abraham's rigid electron model had been looking promising to Kaufmann all along, but so had Lorentz's model. In 1906, however, Kaufmann published a new set of results. These compared the curved line painted by his electrons to the predictions of Max Abraham's rigid electron model and to those of the so-called Einstein-Lorentz theory, which, Kaufmann argued, was now ruled out, along with the principle of relative motion. He suggested, therefore, that experiments to determine the "absolute resting ether" be resumed.[25]

Lorentz was discouraged by the news. "Unfortunately my hypothesis of the flattening of electrons is in contradiction with Kaufmann's results, and I must abandon it," he told Poincaré. "I am therefore at the end of my Latin."[26] Poincaré advised him to be patient. Although Kaufmann's data looked threatening, it would be a mistake to abandon fifteen years of work on the basis of one experiment; someone should repeat the measurements. The normally imperturbable Lorentz nevertheless went off to New York and proclaimed the impending collapse of his theory with one breath while continuing to propound it with the next.

To Planck, Einstein's theory was too beautiful to die so soon. At a meeting of German scientists in Stuttgart in September of 1906 he challenged Kaufmann's data analysis, arguing that there were too many uncertainties to verify or refute either theory. Kaufmann stood his ground and defended his calculations. He pointed out that though neither theory was supported by his data, the Lorentz model was off by more than 10 percent, a "strong" deviation, while Abraham's was only wrong by 3 or 5 percent.

Abraham began his own talk in his usual cutting fashion by noting that the deviations of the Lorentz theory from Kaufmann's findings were twice those of his own rigid spherical electron theory. "You may say that the sphere theory represents the deflection of beta rays twice as well as the *Relativtheorie*." The assembled physicists laughed loudly, if nervously. Trying to defuse the tension, Arnold Sommerfeld joked that there seemed to be a generation gap: Scientists under the age of forty seemed to prefer Abraham's more radical electron theory, while those over forty were inclined to Lorentz and relativity.[27]

Fortunately, more experiments were already under way. In Kiel, August Becker had measured the masses of cathode rays moving roughly at one-third the speed of light. In Göttingen, Adolf Bestelmeyer, in the course of measuring the speeds of cathode rays produced by bombarding a heavy metal with X rays, had made measurements of electron masses that overlapped with and contradicted the low-velocity end of Kaufmann's data, raising the prospects for relativity. Bestelmeyer had used a different geometry of magnetic and electric fields that was borrowed from Thomson's groundbreaking experiments twenty years earlier, and turned out to allow a cleaner measurement. Accordingly, Alfred Bucherer, a Bonn University researcher whose own electron theory had been denounced at the Stuttgart meeting, adopted this technique to begin his own series of, presumably, definitive measurements.

Emboldened, meanwhile, by the criticism against and apparent inconclusiveness of Kaufmann's results, Albert offered his own judgment of Abraham's electron theory: It simply wasn't very interesting. To Albert's taste, a theory without ambition—one that only solved the problem it was posited to solve—was hardly worth a fuss. Perhaps the data *did* slightly favor Abraham's or Bucherer's theories, Albert allowed in a fifty-one-page review article for Stark's *Jahrbuch*. So what? "In my opinion, however, a rather small probability should be ascribed to these theories," he wrote, "since their fundamental assumptions about the mass of a moving electron are not supported by theoretical systems that embrace wider complexes of phenomena."[28]

RELATIVITY, on the other hand, was an idea with legs. Without bothering to wait for confirmation of his theory, Albert had already begun wondering how to expand its domain.[29] Among the things that bothered him, as he sat preparing his *Jahrbuch* review in the fall of 1907, was that relativity seemed to apply to nearly every physical phenomenon except gravity. This was a tantalizing anomaly, because he was fairly certain that the standard model of Newtonian gravity would have to be modified to fit into a relativistic universe. For one thing, according to Newton's law the gravitational force between, say, two planets was proportional to the product of their masses divided by the the square of the distance between them. According to relativity, however, the precise numerical value of those masses and that distance would depend on the velocity of the observer, meaning the force of gravity would be embarrassingly relative. Moreover, in Newton's world the gravitational forces were transmitted instantaneously from distant bodies, in contradiction to the principles of relativity.

Albert was at work one day, probably in November of 1907, when he had a revelation, which he later described as the happiest thought of his life. Perhaps he had tipped his chair too far back and fallen over. "I was sitting in a chair in the patent office at Bern," he later recalled, "when all of a sudden a thought occurred to me: 'If a person falls freely he will not feel his own weight.' I was startled."

It was an obvious insight, although not one that a person in that situation usually had much time to contemplate in depth. The reason a falling person did not feel his own weight, Albert realized, was that every molecule and fiber in his body was being accelerated by gravity at the exact same rate, an experimental fact known since the time of Galileo and for which there was no theoretical explanation. As a result, a falling man did not feel gravity, and there was no experiment he could conduct on himself, Albert realized, to tell that he was in a gravitational field. If he jumped off a rooftop with an armful of oranges, he could toss them sideways from hand to hand, because the oranges would accelerate at the same rate right along with him. If he dropped one, it would float beside him. All the other laws of physics would work fine. The faller, Albert concluded, would have every right to conclude that *he* was the one at rest and that his surroundings were accelerating *upwards* at 9.8 meters per second per second. Gravity, in other words, seemed to be relative.

There is no record of what he said to Mileva that night, and no "Eureka!" letters went out to friends. He had not formulated a theory of gravity, but had perceived the barest glimmer of an idea for such a theory. Nevertheless, in the last section of his *Jahrbuch* article Albert set out to exploit this new notion. "So far we have applied the principle of relativity, i.e., the assumption that the physical laws are independent of the state of motion of the reference system, only to *nonaccelerated* reference systems. Is it conceivable," he asked in his introduction, "that the principle of relativity also applies to systems that are accelerated relative to each other?"[30]

The key to this notion was what Albert called the principle of equivalence. There had always been two different conceptions of mass in physics: *inertial* mass, which was a measure of a body's resistance to being moved; and *gravitational* mass, a measure of how strongly gravity tugged on it. There was no guarantee that these quantities had to be the same. In the case of electrical or mechanical forces, for example, the kick that an object felt was determined by its electrical charge, but its response was governed by its inertial mass. In the case of gravity uniquely, however, it seemed as if the force on an object and its response to that force were both due to the same property. The falling physicist could only feel weightless if every atom in his body, regardless of its composition, charge, or inertial mass, fell at exactly the same rate. Gravity seemed to be able to reach out and tug each individual particle of matter with exactly the right force necessary to overcome its innate inertia. That was the result of a series of experiments that the Baron Roland von Eotvos had carried out in 1899, and it was too much of a coincidence, Albert argued. Inertial mass, he concluded, was gravitational mass.[31]

Albert used this fact as a springboard to a more radical proposition, namely, that there was no physical difference between a gravitational field and uniform acceleration; mathematically one could be replaced by the other. This allowed Albert to use the tricks he had developed for relativity—imagining little men with clocks and measuring rods zipping around with respect to each other—to formulate an idea of what the properties of a relativistic gravity might be. Most famously, he would place them in a pair of elevators, one hitting on the ground in earth's gravitational field, the other being jerked upward through space by a cable. If Albert was right, what held true in one elevator had to hold true for the other, and from that presumption vast consequences would flow, for gravity and for the universe.

Two preliminary and radical conclusions immediately leapt out. The first of these was that gravity (or acceleration) should affect time. The farther our jumper was from the center of the earth or whatever gravitational field he was in, the faster his clock would run. On earth the effect would be less than minuscule, but Albert pointed out that there already existed very precise clocks scattered through space at different places of gravitational strength—namely, atoms, which radiated light at precise frequencies in the form of spectral lines. Atoms at the surface of the sun, where gravity was much stronger, would vibrate slightly slower than those on earth; the light produced by those solar atoms would have wavelengths longer by about one part in two million, he calculated, than light from the same atoms on earth. Proving this prediction was a worthy but not impossible challenge for an ambitious spectroscopist.

The other effect, Albert admitted, was beyond the reach of experimental science: Gravity would affect light as well, causing a light beam to bend in a tiny arc around the earth, as if its electromagnetic energy was indeed (as Albert had already suggested) equivalent to mass and weight. The amount of deflection by the earth was so small, however, that he didn't even bother to quote a value.

Albert hoped this new theory for a theory would yield further insights into the other mysteries of gravitation, particularly the failure of classical celestial mechanics to explain a small peculiarity in the orbit of the planet Mercury. After submitting his paper to Stark, he wrote to Habicht, "At the moment I am working on a relativistic analysis of the unexplained secular changes in the perihelion of Mercury."[32] But he soon bogged down on this subject and would not mention it again for years.

IN THE SPRING of 1907 Jakob Laub wrote asking if he might come to Bern and work with Einstein for a few months. Laub, another of the Young Turks of physics, was an experimenter at the University of Würzburg. He had been a graduate student working for Wilhelm Wien, the editor of the *Annalen,* when Albert's relativity article appeared, and upon Wien's urging had become a student and proponent of relativity and a frequent correspondent of Albert's. Like the rest of Einstein's new scientific friends, none of whom (with the exception of von Laue) had actually met him, Laub was under the impression that Albert worked at the University of Bern. Only now, apparently, did he learn that Albert was a patent clerk.

"I must tell you quite frankly," Laub wrote, "that I was surprised to read that you must sit in an office for 8 hours a day. History is full of bad jokes."[33]

Laub arrived at Einstein's flat in April to find Albert there alone, kneeling in front of the oven trying to stoke the fire; Mileva and Hans Albert were away visiting her family back in Kac. For the next couple of weeks, Albert and Jakob led a bachelor existence, brainstorming over lunch and dinner every day in a cheap restaurant until both their stomachs rebelled. Laub was Albert's first collaborator, and over the course of the next year they published several papers on the behavior and properties of electrons dancing a relativistic jig in electromagnetic fields. While Albert put in his hours at the patent office, Laub stayed home and ground out calculations. "Laub is a very nice, however very ambitious, almost greedy person," Albert wrote to Mileva ("Dear Treasure"), "but he's doing the calculations, for which I would not have the time, and that's good."[34] He was lonely, he added, and the apartment was very dirty.

BY NOW, sitting eight hours a day, six days a week in the patent office "pissing ink" had begun to seem to Albert, too, a bad joke. In June of 1907 he had applied again to be a *Privatdozent* at the University of Bern. This was a strategic move on Albert's part: If he wanted to ascend to a university professorship and have time to be a real physicist, like his friends von Laue and Laub, he would need the credential of having taught as a *Privatdozent*. With men like Planck and Lorentz taking him seriously, this was no longer such a wild fantasy. Alfred Kleiner, the University of Zurich physics professor who had shepherded through Albert's doctorate, had also been scheming for years to get a second physics chair approved at Zurich.[35] He had his eye on Albert as one of the possible candidates and had been coaching him in his career choices from afar.

Albert's first encounter with the University of Bern, five years earlier, had foundered on his lack of a Ph.D., and ended with his dismissing it as a pigsty. This time, his application scarcely went any smoother. One of the requirements was the submission of a *Habilitationschrift,* a sort of postgraduate dissertation of unpublished original work. Albert, however, in his usual blithe way, merely attached a collection of seventeen already pub-lished articles.[36] Not surprisingly, this caused a row when the full faculty met in October to consider his candidacy, and they duly voted to reject his petition pending the completion of the required paper.

Albert's response was not recorded, but he was no stranger to disap-pointment and the maliciousness of petty men, and so he looked for an-other avenue into teaching. Tipped off by an old friend that a position was likely to open up at the Technikum Winterthur, where he had taught as a substitute back in 1901, he wrote to his old friend Marcel Grossmann for advice on how to campaign for the job. Grossmann was now a mathemat-ics professor at the Polytechnic, having succeeded their old teacher Fiedler.

Albert explained that he needed the work in order to have more time to devote to science. "So let me ask you: how does one go about doing that?" he asked. Should he seek out someone to impress? How could he avoid triggering anti-Semitic feelings? "Furthermore, would it make sense to sing the praises of my scientific work on this occasion?"[37]

The vacancy, however, failed to materialize. To Albert, it must have be-gun to seem like 1900 all over again, when he had bombarded all of Europe with job applications to no avail. Still, when another position—this one to teach math at a gymnasium in Zurich—opened up, he dutifully pursued it.

A week later he received a message at the patent office from Kleiner wondering where to write to him at home about "a matter of impor-tance."[38] Either Kleiner had heard through the grapevine about his im-broglio at Bern or had news about the impending position at Zurich. In any event, he advised Albert to swallow his pride and submit a *Habilitationschrift* to Bern.

By the first week of February 1908, he had completed a paper on black-body radiation and sent it to Bern's dean, Gustav Tobler. Kleiner was im-pressed by Albert's speed, but cautioned him that the evaluation of the paper might be dragged out as a delaying tactic. "I must be kept informed about this matter, for if necessary, i.e., if you do not get to lecture next se-mester via *Habilitation*," said Kleiner, "I must see to it that you be given an opportunity of another kind to prove yourself as a lecturer."[39]

Kleiner's concern proved unfounded; three weeks later Albert was ap-proved as a *Privatdozent,* the lowliest rung on the academic ladder. The po-sition allowed him to hang out a shingle as a lecturer under the auspices of the university, his pay, such as it was, coming directly from his students. Which is to say that it was a not very lucrative part-time job, and another drain on his precious nonoffice hours. Three students enrolled for his first class, that April 1908, on thermodynamics.[40] Although he tended to talk above the students' heads, the number increased to four in the next term.

One regular attendee was Michele Besso, whom Albert knew could always be counted on to ask provocative and fruitful questions in the discussions afterward, which often trailed into coffeehouses and home. Another was Maja, who reports that, during this period, Albert's immersion in his work began to affect his appearance.[41] Long stubble covered his face, and his hair was left to the mercy of the wind. "Indeed," she said, "he was not even aware of what he was wearing on his body." At the time the low end of the sartorial ladder around Bern was occupied by poor Jewish students from Russia, "often very intelligent slobs," in Maja's words, who got no respect from the orderly Swiss. When Maja asked the university's janitor once in which auditorium her brother held his lectures, he replied, astonished, "What, the Russian is your brother?"

In the meantime the much-lamented Olympia Academy had undergone a rebirth, with a surprising consequence. As a result, in another part of town a backroom in one of the local gymnasiums was filling up with wires, meters, and batteries running to a squat, ugly cylinder vaguely resembling a wastebasket. Protruding from it were arms, spouts, and nipples, while its innards were an intimate maze of sheet metal, grids, wires, and spinning disks. Night after night Albert trudged in from the patent office, fortified by coffee, sleeves rolled up, and hooked up wires and set switches as the room filled with pipe smoke and then ozone as the *Maschinchen,* or "little machine," cranked and sparked and usually fizzled. Albert muttered imprecations and oaths in which the name "Habicht" frequently appeared.

This ungainly device was in fact the product of the second coming of the Olympia Academy (sans Solovine), although in its latest incarnation the Academy resembled less a bohemian debating and drinking society than a gathering of garage mechanics. "I have found yet another new method of measuring very small amounts of energy," Albert had announced to the Habicht brothers several months earlier, on a July day in 1907, as he summoned them to meet him and Mileva and Hans Albert at a resort in the Oberland, where they would be staying during the next ten days, so that he could explain his new brainstorm.[42]

After five years in the patent office scrutinizing and visualizing other people's inventions, and a boyhood spent among his father's and uncle's gadgets, it was hardly surprising that Albert should turn to tinkering himself, especially since there might be a good profit in it. The *Maschinchen* he envisioned was an electrical device that would amplify minute electrical charges so that they could be measured. Very small amounts of energy were the essence of the statistical fluctuations, the signatures of random molecular movements, with which Albert was attempting to plumb the Boltzmannian atomic nature of reality. And, if the *Maschinchen* worked, every lab in Europe—or maybe the world—that wanted to do state-of-the-art science would require one.

As he sketched it, the instrument resembled eighteenth-century machines used to generate high voltages electrostatically. In principle its operation was simple. When two conducting plates are moved very close together, an electrical charge on one will induce an equal and opposite charge on the face of the other. In Albert's machine a series of metal strips would rotate past a fixed plate holding the initial charge, each picking up that charge and transferring it to a second plate, thus building up a multiple of the value of the original charge. Another series of strips would rotate past the second plate, multiplying its charge onto a third plate, and so on. The initial charge would thus in a few stages grow exponentially from a small fraction of a volt to a quantity easily measurable with conventional electrometers.

In practice the devil was in the details. By the time the group returned to Bern in mid-August, Paul, a talented machinist, had already built a prototype of one stage of the machine, based on Albert's description. But this first version didn't work, inaugurating a two-year bout of trials and tinkering that kept Albert in a constant state of anxiety. September 2, 1907: "Dear Habichts: I am driven by *murderous* curiosity to ask what you've been up to."

And it is here, some accounts suggest, that Mileva had her last fling with science. According to Seelig's biography, Mileva occasionally pitched in on the *Maschinchen* project. One can almost imagine a return to the comfortable rituals of the past: Paul firing up the Turkish coffee machine; Albertli asleep in the next room; Albert squeezing out ribald jokes in his cellolike voice; Mileva one of the students again, hunched over circuit diagrams and electrical formulas in the lamplight. In her biography, Trbuhovic-Gjuric suggests in fact that Mileva was the main inventor of the *Maschinchen:*

> She began work with Paul Habicht on an influence machine [*Influenzma-schine*] for measuring small electric tensions by way of multiplication. It took a long time, not only because she had a lot to do, but mainly because of the thoroughness she applied toward perfecting the instrument. When they were satisfied with it, they left it to Albert to describe the machine as a specialist in patents.[43]

Mileva's name, however, is never mentioned in the correspondence between Albert and the Habicht brothers. In the fall of 1907, while he was pondering the relativity of gravity, Albert did prepare a patent application for the device, but then decided not to send it in, because no one seemed interested in manufacturing it commercially. In the meantime, Conrad managed to be run over by a sled, casting a pall over Christmas, and it was too cold for Paul to work in the basement workshop in his parents' house.

That February Albert wrote a short paper describing the device and sent it to the *Physikalische Zeitschrift*. On the basis of this he found a potential sponsor for the *Maschinchen*, Joseph Kowalski, a physics professor at the University of Fribourg. In June Albert spent a day there working in one of the laboratories, but that particular collaboration came to naught.

In October of 1908 Paul, who had moved to Basel, announced that he had finally built the whole machine, but it would not charge, perhaps because the insulation was bad. "So something is wrong. I have no way of knowing what, because I have no measuring instrument," he explained.[44] The following weekend, on his regular visit, he brought the machine to Bern.

With the combination of borrowed and homemade equipment, and help from a local mechanic, Albert commandeered room in a lab in one of the local gymnasiums and cobbled together a small laboratory. By November he had managed to rebuild the *Maschinchen* and get it up and running and began to test it using an electrometer and a voltage battery that he had built. "You wouldn't be able to suppress a smile," he told Laub, "if you saw the magnificent thing that I patched together myself."[45] The finished product had six stages of amplification. A photograph of a later version shows the *Maschinchen* as a sealed cylinder roughly the size and shape of a water heater with electrical leads coming out of it.

Albert reported that winter to Kowalski's Fribourg colleague Albert Gockel, "So far I have found that with the *Maschinchen* one can measure voltages lower than 1/1000 volt, which is an unparalleled sensitivity for the measurement of electrical quantities."[46] Another factor of 100 lower would allow Albert to test electrostatic aspects of molecular theory. In 1910 Albert and the Habichts published a detailed description of the machine in the *Zeitschrift*. The following year Paul demonstrated it to the German Physical Society with great success, but by then Albert seemed to have drifted away from the project. The prospect of becoming a tycoon had melted away under the pressure of teaching, theoretical science, and economic reality.

Along the way, however, a patent for the device was obtained under the names of Einstein and Habicht. According to Trbuhovic-Gjuric, Paul Habicht asked Mileva why she hadn't put her own name on the patent. Mileva is alleged to have punningly answered, "Why? We're both *ein Stein* [one stone]."[47]

Habicht manufactured and sold various versions of the *Maschinchen* for the next twenty years, but it proved to be too temperamental and complicated, and never really caught on.

WHILE ALBERT and his friends were obsessed with building their gadget, relativity was taking giant steps on its own. In September 1908, Alfred Bucherer wrote to Albert: "First of all I would like to take the liberty of informing you that I have proved the validity of the relativity principle beyond any doubt by means of careful experiments." Bucherer had been weighing speeding electrons with his new apparatus in Bonn. Not only did his new results support the Lorentz-Einstein electron, but his research also superseded Kaufmann's old findings, suggesting that his "pioneering work" had been flawed by an unhomogeneous electric field.[48]

When physicists from around northern Europe gathered in the shadow of the great cathedral in Cologne for the 80th *Naturforscherversammlung,* or Congress of Natural Scientists, Albert was not present, but he was the star of the show. The meeting would amount in a way to the erection of a new sort of mathematical cathedral in which future physicists would genuflect to the principle of relativity.

Bucherer's news was greeted with relief and joy by the partisans of Einstein and Lorentz at the convention, although it would be years before further experimentation established the phenomena beyond reasonable doubt. Hermann Minkowski, Albert's old Polytechnic math professor, is said to have abandoned his usual reserve and waxed enthusiastic with joy, and then definitively dismissed Abraham's discredited theory: "The rigid electron is in my view a monster in relation to Maxwell's equations, whose innermost harmony is the principle of relativity."[49]

Abraham and Kaufmann were not there to respond.

It was Minkowski's subsequent lecture, however, for which the 80th Congress would be mostly remembered. Minkowski had left the Polytechnic in 1902 for Göttingen, the Valhalla of mathematicians. It was here that Gauss and his student Riemann had explored the non–Euclidean geometries of curved surfaces and multidimensional spaces, and here that young geniuses like David Hilbert were reinventing mathematics. Minkowsi, a former child prodigy as a geometer, stepped into this pantheon like a man walking into his living room.

Among his interests had been Lorentz's electron theory. Like Poincaré, Minkowski was impressed by the strictly mathematical properties of the Lorentz transformations that described how the aether compressed length and expanded time to make the speed of light appear to measure the same for all observers. Einstein's relativity paper had been a revelation for him in two respects. The first was that this once "lazy dog" had amounted to something after all. "Oh, that Einstein, always cutting lectures," he is said to have commented. "I really would not have believed him capable of it."[50]

The more important revelation was that in doing away with the aether

Einstein had unleashed the true power and symmetry of the Lorentz transformations. They held true whether backward or forward: Two travelers
going in opposite directions could, with complete consistency and logical
clarity, each see the *other* as compressed and slowed. Each could regard himself at rest and the other in motion. The beauty of the theory was still obscured, however, behind the scrim of Einstein's awkward math.

Relativity became Minkowski's life. The story is told that, after he and
David Hilbert went to an art show, Hilbert's wife asked them how it was.
They answered that they didn't know, explaining, "We were so busy dis-
cussing relativity that we never really saw the art."[51]

Minkowski set about to reinvent relativity in a language better suited to
mathematicians obsessed with elegance and symmetry—in the language,
that is, of geometry. Minkowski's space had four dimensions—the three
usual spatial coordinates of length, breadth, and depth, and one of time. So
far, his formulation was only common sense; things happen at a time, after
all, as well as a place. A plan to meet at the Rebel Cafe wouldn't work without an appointed time. Only four numbers—x, y, z, and t—on the other
hand, sufficed to specify any happening in cosmic history. Points in this
four-dimensional space-time Minkowski called events; the distance between them, which was a combination of their temporal and spatial dislo-
cations, was called an interval. The trajectory of a particle or a person
through it was called a world-line.

One way to visualize this idea was to imagine a two-dimensional graph
on which the vertical axis represented time and the horizontal axis represented spatial position. A stationary rock or a lamppost would be represented by a line drawn straight upward; even sitting still we are moving
forward in time. A light beam traveling as fast as anything ever could was
another straight line inclined at 45 degrees to either axis. Minkowski himself would be represented by a line encompassing roughly forty-five years
of time and one thousand kilometers of space that wandered and curved a
bit, with one end in Russia marking his birth in 1864 and the other, alas, in
Göttingen in 1909.

It was Minkowski's genius to see that in this weird mathematical space,
the Lorentz transformations amounted simply to rotating and skewing these
space-time axes. All the mysteries and apparent paradoxes of relativity could
be explicated and solved by simple geometry on this graph. One result of
the finding was particularly fundamental, as any reader can check by drawing a number of dots on a piece of paper and spinning the paper. As the
space-time axes spin, the coordinates of the event-points change, but the
relation between the points—the interval between them—remains fixed. In
some cases that interval might be composed of different mixtures of space

and of time. Space and time could melt back and forth into one another depending on how you looked at them, similar to the way mass and energy did. In this context, the mysterious relativistic "shrinking" of a moving rod was an artifact of seeing it, as it were, at an angle in space-time; the whole thing was still there, the same way a whole couch is still there when you turn it on an angle to get it through a doorway. This invariance was the magic at the heart of Minkowski space and relativity.

"Gentlemen!" began Minkowski in Cologne. "The views of space and time which I wish to lay before you have sprung from the soil of experimental physics, and therein lies their strength. Their tendency is radical. From now on, space by itself and time by itself are doomed to fade away into mere shadows, and only a kind of union of the two will preserve an independent reality."[52]

Albert, however, was not impressed by this great leap forward, regarding it as "superfluous learnedness." He liked to joke that, since the mathematicians had taken over relativity, he didn't even understand it himself. But in time he would learn to love it.

Three months later Minkowski suffered a fatal attack of appendicitis, allegedly bemoaning that he had to die while relativity was still in its infancy.[53] He was only forty-five when his own world-line was snipped off.

Though it soon became the rage in physics and relativity, Minkowski's new four-dimensional space-time had no effect on the popular quest for the fourth dimension that had sprung from the discovery of non-Euclidean geometries in the nineteenth century and the subsequent writings by Charles Hinton and others. Despite H. G. Wells's *The Time Machine,* which in 1895 did indeed treat time as a fourth dimension, the concept by then had become associated with mystical insights and transcendence, an extra direction in which to *see* if only one could know how to look. "When one reaches the country of the fourth dimension," wrote the Frenchman Gaston Pawlowski in 1912, "when one is freed forever from the notions of space and time, it is with this intelligence that one thinks and one reflects. Thanks to it, one finds himself blended with the entire universe, with so-called future events, as with so-called past events."[54]

In 1909 *Scientific American* announced an essay contest for the best explanation of the fourth dimension. Entries came in from all around the world. Not one of them mentioned time.

12

QUANTUM DOUBTS

IN THE SPRING OF 1908 MARX AND MACH STRUGGLED FOR THE SOUL OF Friedrich Adler. For three years Adler had been the assistant to Alfred Kleiner, head of the physics department at the University of Zurich, as well as its new rector. Now, after a seven-year campaign, Kleiner was on the verge of winning approval for a second chair in physics. Adler, an expert on heat and mechanics, was the logical candidate. Moreover, if he really wanted the job, he had the clout to get it. His father, Victor Adler, had founded the Austrian Social Democrat Party and was well-known and admired in liberal circles throughout Europe, including the Zurich Board of Education, which was dominated by Social Democrats. Kleiner, however, had his eyes on a different candidate.

"It is a man named Einstein who studied at the same time I did and with whom I attended some lectures," Adler reported to his father. "Our development was rather parallel. He married a woman student at the same time I did, had children, but no one who supported him, was for a while starving, and was treated by the professors at the Polytechnic outright contemptuously; he was not admitted to the library, etc. He had no understanding of how to relate to people."[1]

Pale and thin, with a stubborn romantic streak and doctrinaire tendencies, Fritz Adler had more the temperament of a philosopher than a physicist. His father had initially sent him to Zurich to get him away from politics, which proved a rather futile gesture. Fritz inherited his student lodgings in the radical-infused student quarter from Rosa Luxemburg.[2] He entered the Polytechnic the year after Albert and became one of his admirers and coffeehouse companions. In 1903 Adler married a fellow physics student, a Lithuanian-born Russian, Katya Germanischskaya.

In physics Adler was a staunch Machist. During their student days, Fritz had once described to Einstein his dissertation project on the specific heat of chromium. What, Albert inquired, was his hypothesis? "None," Fritz answered, "I want to prove it purely experimentally."[3]

Ten years later, the class struggle had captured Adler's imagination and physics was becoming more and more oppressive to him. If forced to choose between science and the party, he told his father at one point, he would choose the party. "I believe I must abandon theory, seek some kind of practical occupation, and that can *only be activity in the party . . .* that is the only thing that is important to me after epistemological theory."[4]

Fritz's romanticization of Einstein's own professional struggle knew no bounds, so that in his eyes it was both inevitable and fitting that Einstein get the appointment in Zurich. "They have a bad conscience over how they treated him earlier," he wrote. "The scandal is being felt not only here but in Germany that such a man would have to sit in the patent office." Indeed, Adler predicted that Einstein would not be long for Zurich, but would certainly soon be called to Germany.[5]

It was shortly thereafter that Kleiner traveled to Bern to see his fabled protégé in action, and Albert's chances for the post almost ended right there. Albert only had one student by then, and was not in the habit of preparing his lectures, and in addition, the prospect of being on trial made him unaccountably nervous.[6] Kleiner arrived in time to watch a disheveled young man ramble disconnectedly on material over the head of his own student.

Kleiner went away disillusioned, and word eventually got back to Albert that Kleiner had complained publicly that he was an unsatisfactory lecturer. Albert consoled himself with the fact that he had never asked to be a University of Zurich professor.

At that point it seemed as if the hapless Adler had the job he only half-heartedly wanted. Subsequently, however, Albert had second thoughts about the post and did the sort of thing Mileva was always advising him to do: Namely, he wrote back to Kleiner directly and scolded him for criticizing him behind his back. At this rate, he complained, he'd never get a teaching job.

Kleiner apologized and said he'd be happy to get Albert a professorship if he could be convinced that Albert was capable of teaching. Albert suggested the Zurich Physical Society as a venue.

Meanwhile, Adler continued to campaign for Einstein. "If it is possible to obtain a man like Einstein for our university," he wrote the governing board in November, "it would be absurd to appoint me." Legend would subsequently have it that Adler had valiantly stepped aside in favor of Einstein, but it remains unclear whether he was ever formally offered the job. Kleiner seems to have been noncommittal with both men and had to know that Adler was going through the motions to please his father. The final turndown, Adler reported, felt like a great weight off his chest. Einstein, for his part, felt that Adler was a self-sacrificing fool with a martyr complex, but all the same, he wasn't about to rebuff him after having waited so long for a teaching job.[7]

Albert's second audition took place before the Zurich Physical Society in January, and as he reported, "I was lucky. Contrary to my habit, I lectured well on this occasion."[8]

In February 1909, Kleiner formally recommended Einstein for a professorship, which had come to be specified for theoretical physics—a branch of the discipline that had not existed officially eight years prior when he had begun to campaign for the existence of the chair. In secret deliberations held in March, the faculty allowed that Albert, despite being a Jew, did not share the unpleasant characteristics—such as "intrusiveness, impudence, and a shopkeeper's mentality"—usually attributed to that race.[9] Accordingly, by a ten-to-one vote they elected to offer him the job.

Albert promptly refused; the salary they had proposed was less than he was making at the patent office. Finally the university agreed to match the 4,500 francs he was getting in Bern, and on May 7, 1909, the Governing Council of the university approved Albert's appointment as extraordinary professor of theoretical physics. His long exile from academia was over.

"So," he said to Laub, "now I too am an official member of the guild of whores."

A FEW DAYS LATER a housewife in Basel by the name of Anna Meyer-Schmid read of Einstein's appointment in the local newspaper and was transported ten years back in time to the summer when she had frolicked at her brother-in-law's Hotel Paradise with a flirtatious physics student named Albert. They had not seen each other since. Anna, who had been happily married to George Meyer, a Basel bureaucrat, for the past four years, sent off a congratulatory note to Bern.

Albert's heart leaped when he received Anneli's letter, and he wrote back immediately. "I probably cherish the memory of the lovely weeks that I was allowed to spend near you in the Paradise even more than you do."[10] He might now be old, and famous enough to have his name in the papers, but he assured her that he was still the same simple soul who expected nothing from the world. Nothing except, perhaps, that she would visit him in Zurich, to which end he gave her his office address on Ramistrasse.

Anneli's reply was apparently intercepted by Mileva, who immediately became furious. Who was this woman writing so affectionately to her husband? Mileva made Albert send the letter back, claiming that he didn't understand it. This letter, in turn, may have been intercepted by George Meyer, who inquired of Einstein exactly what was going on. Mileva replied with an angry letter to George Meyer, complaining about his wife's inappropriate behavior.[11] Anna, she claimed, had read too much into what was in reality a simple friendly exchange of greetings. "We don't know for certain what misled her to write another, rather inappropriate letter."

When he discovered what Mileva had done Albert was mortified. This was not the first time that what Albert called his "gypsy blood" had gotten him in trouble with her, particularly during their school days, but it was the first time she had lashed out so publicly. Characteristically, Albert felt as if he were the innocent victim in this fracas. Mileva seemed to have the knack of going into crisis at critical moments in his career—first the baby when he was up for the patent office, and now a noisy explosion of jealousy on the verge of his university professorship.

Albert wrote to Meyer apologizing for causing so much fuss. He had overreacted to Anneli's card, he explained, which had reawakened old feelings. His intentions were pure, however, and Anna, he said, had behaved totally honorably.[12]

"It was wrong of my wife—and excusable only on account of extreme jealousy—to behave—without my knowedge—the way she did," he went on. He promised not to bother them again and said he hoped Meyer wouldn't be angry with Anna, who, he repeated, was blameless. As if to underscore his authority on this account, he signed the letter "Professor Doktor Einstein," one of the few times in his life that he deigned to pull rank.

The incident left a bad aftertaste with Albert. From his point of view the intensity and devotion that, coming from an older woman, had enticed and sustained him as a teenager was beginning to cloy. The passion that was once a lifeline, and a welcome alternative to the smothering love of his mother, was now feeling more like a millstone. As Mileva aged, the darker Slavic parts of her soul were becoming more pronounced.[13] She neither forgave nor forgot. In fact, now that he thought about it, as the bloom of youth faded from her face, Mileva was actually growing ugly.[14] The angrier and more unattractive she became, the more haloed became Albert's memory of a sweet mountain maiden who wanted nothing from him but a good time.

"IN MID-OCTOBER, on the 14th, we leave Bern, where I have now spent seven years, so many beautiful and, I must say, also bitter and difficult days," Mileva wrote to Helene late that summer. In 1909 Albert was thirty years old and Mileva was thirty-four. In photographs from the time, the saucy light in her eyes has been extinguished, replaced by a fragile, hunted stare. She has broadened, and her face has coarsened and thickened. It was in

Bern that she had lost her Lieserl; in Bern that she had been allowed to be one of the boys sitting quietly in the corner of the Olympia Academy, taking what succor she could from the raucous display of wit; in Bern that Albertli had been born and the air was full of baby talk; in Bern that she had sat across from her sweetheart under the oil lamp wandering on the edges of physics. It was in Bern that she had gained Albert's attention and his company at last; in Bern that young Germans with savage gleams in their eyes and their moustaches foamed around words like *Relativitätsprinzip* and *Lichtquanten* began finding their way to the door with questions, offers, propositions, arguments, papers, work, and more work to pull him away from her.

She was proud of and happy about Albert's success, Mileva told Helene. He had earned it, literally burned the midnight oil for it. "I only hope and wish," she added, "that fame does not exert a detrimental influence on his human side."[15]

EARLY THAT SAME SUMMER an engraved Latin document had arrived at the patent office from the University of Geneva. Albert promptly tossed it in the wastebasket. Swiss science being a small world, the university's authorities were able to make inquiries after a few weeks of silence wondering why Herr Doktor-soon-to-be-Professor Einstein had not responded to their invitation of an honorary doctorate at the big upcoming 350th anniversary celebration of the university.

A friend was enlisted to make sure Albert appeared at the festivities. Albert duly traveled to Geneva on the appointed day, checked into the designated inn, and found himself in the company of numerous Zurich academics, to whom he claimed he didn't know why he was there. Whereupon he was unofficially enlightened—too late, of course, to be able to provide for the proper wardrobe. In honor of its founding by Calvin, the university had decided to indulge in an orgy of degree giving. Some 110 luminaries, ranging from Marie Curie and Wilhelm Ostwald to a drug magnate named Ernest Solvay to a local restaurateur and folk poet, were to be bestowed with honorary doctorates under the aegis of the president of the Swiss Confederation. Present as guests of honor, resplendent in the robes of their homelands and home institutions from throughout the world, were 210 delegates from universities and other learned bodies.

It rained throughout the three-day festival, but Albert had the time of his life. Wearing in a straw hat and an ordinary suit he marched in the official procession shoulder to shoulder with the berobed dignitaries. The festivities ended with a lavish feast in a grand hotel. The general opulence inspired Albert to lean over to the Genevan patrician who was sitting next to him at the banquet and ask if he knew what Calvin would have done had he been present.

The patrician was puzzled and asked what he meant.

Calvin, replied Albert with relish, would have erected an enormous stake and had them all burned for sinful extravagance.

"The man never addressed another word to me," Albert later recalled.[16]

Later in the summer a more significant invitation appeared in the mail, asking him to come to Salzburg, Austria, in September to speak at the annual congress of German scientists, the same society whose meeting the year before in Cologne had been the occasion for Minkowski and Bucherer's anointing of relativity. Founded in 1822 to promote the unity of the sciences—a typical Enlightenment dream—the Gesellschaft Deutscher Naturforscher und Ärtze, as it was officially known, was one of the oldest scientific organizations in the world, predating by half a century the German Empire itself. With three thousand members it was one of the largest.

Albert was to be the honored guest of the Deutsche Physikalische Gesellschaft, which met under the umbrella of the larger congress. Perhaps the organizers expected a star turn on relativity from their young phenomenon, a masterly response to Minkowski's new four-dimensional space-time. What they got, however, was something more disturbing and prophetic. For all the praise that he was getting for relativity, Albert thought more and more than he hadn't really solved any of the fundamental questions plaguing physics. Relativity, he recognized, was based on a foundation of sand. Albert had been brooding about this for at least five years, a period during which relativity was ascending from incomprehensible dogma to acclaim. The more relativity succeeded, the more he brooded about the real nature of light.

It was Planck's weird quantum, of course, that was undermining Maxwell's equations, the ones that described the interweaving of electric and magnetic fields to make light waves, and thereby also undermining relativity and most of the rest of physics. Relativity was based on the principle that the speed of light was the same for any observer. That was what the equations said, although it had taken the famous null results of the Michelson-Morley experiments for physicists to take them literally.

But nowhere in Maxwell's equations, magnificent as they were, or in the work of Faraday, Hertz, Lorentz, Helmholtz, and Boltzmann that preceded and followed them, did it specifically say that electromagnetic energy spurted, in the tiny goblets that Planck called quanta, rather than in continuous waves. Nowhere did that mathematical apparatus predict that light was made of little grains, atoms of energy, *Lichtquanten,* or that one of them could punch an electron out of a piece of metal.

At best, it seemed to Albert, Maxwell's equations were incomplete; at worst they were simply wrong. Something new was going to have to replace the traditional concept of electromagnetic fields and explain the existence of the quantum. Maybe that something would also explain the nature and origins of electrical charges themselves (Why were there just two kinds, for example? Why not, say, three?)[17] and what role they played in the structure of matter. By the time of Salzburg, the quest for this "something" had already left a four-year trail of tears, arguments, and incomplete ideas running through the written record of his thoughts.

In 1906 Albert had gone back to reexamine Planck's original derivation of the radiation law. In his first quantum paper, he had shown that Planck, using classical electromagnetic theory, should have gotten the wrong answer for the spectrum of radiation from a so-called blackbody. Instead Planck had derived a strange mathematical expression that fit his meticulously collected data. How was it, Albert asked now, that Planck had gotten the right solution? The answer was that at a crucial point in the calculations Planck had made a crazy (at first Albert thought simply wrong) guess about how to count the different possible ways that energy could be distributed among the putative electrons that jiggled back and forth producing electromagnetic waves. Albert proceeded to demonstrate that Planck's premise was equivalent to assuming that only multiples of a certain energy were allowed, leading to the existence of the same quanta that Albert had written about in his 1905 paper.

"In my opinion the above considerations do not at all disprove Planck's theory of radiation," Albert wrote, "rather, they seem to me to show that with his theory of radiation Mr. Planck introduced into physics a new hypothetical element: the hypothesis of light quanta."[18]

Not even Planck agreed with this conclusion, for he had no intention of being a revolutionary and he wasn't about to accept responsibility for destroying physics. As far as he had been concerned, this "quantum of action," as he technically called it, was merely a calculational device with no physical meaning. In fact, Planck was already trying to argue his way out of his own quantum hypothesis by suggesting that one of Boltzmann's key statistical laws, on which it was based, might not apply in the case of radiation.[19]

Although Planck's young assistant, von Laue, was willing to give Einstein and the quantum more credence, he wrote back to Albert after reading a preprint of the 1906 paper almost urgently to warn him that he was going too far.[20] Quantization, von Laue maintained, was a property of the way light was emitted or absorbed, not a characteristic of the light itself. He concluded firmly, "and hence radiation does not consist of light-quanta as it says in . . . your first paper; rather it is only when it is exchanging energy with matter that it behaves as if it consisted of them." Beer might be sold in pints, but in the barrel it was all one.

Far from backing off, however, Albert was wondering if the quantum might be the answer to another problem he had been pondering. The issue in this case was heat. Back in 1901, during his teaching stint in Wintherthur, when he was rummaging through the library, reading Drude and Planck, Albert had wondered if the internal energy of a body (the heat you felt when you touched it with your hand) was due to resonating electric charges. If that was true then there ought to be a connection between the optical properties of a substance—since those vibrating electrons also transmit electromagnetic waves—and the thermal ones, in particular a quantity known as the specific heat.

The specific heat measured the amount of energy it took to raise the temperature of a certain number of atoms of some substance by one degree Kelvin. In 1819 a pair of Frenchmen, Pierre Dulong and Alexis Petit, had concluded after measuring various materials that this number was the same for all elements and temperatures. Half a century later Boltzmann had provided a tidy explanation for this phenomenon: Atoms only had three dimensions in which to vibrate: up-down, back-and-forth, and sideways. So any atom, whether hydrogen or iron, had the same limited number of ways to spend its energy. Moreover, according to Boltzmann, it had to spend, on average, equal amounts in each dimension, or "degree of freedom."

The so-called Dulong-Petit law worked fine until experimenters got to carbon, which turned out to have an exceedingly low specific heat. Among those who investigated this puzzle was Heinrich Weber—later Albert's starchy tormentor at the Polytechnic—who took extensive measurements of diamond while he was at the University of Berlin. In 1875 Weber had concluded that carbon was anomalous only at low temperatures and that its specific heat rose with its temperature; upwards of a thousand degrees or so Kelvin, carbon behaved like all the other elements—a finding that represented a temporary victory of sorts for the theory. In fact, Weber, an avowed classicist who twenty years later wouldn't even teach Maxwell's equations, was laying one of the foundation stones on which his student would one day erect the worst affront to tradition of all: quantum theory.

By Albert's time the anomalies in the specific heats of other elements were piling up again. Albert asked himself what would happen if he applied

Planck's quantum assumption to thermal motions: Instead of bouncing back and forth or up and down at just any frequency, might the atoms, or molecules (or whatever was doing the vibrating) be constrained to vibrate only at multiples of some fundamental frequency? If this was true, it would mean that their energies could only have certain discrete values.

When Albert folded these assumptions into the theoretical calculations, he got a very different picture of the behavior of specific heat. At high temperatures the specific heat was constant, just as the Dulong-Petit law had said, but below a critical temperature (about 1,300 degrees for carbon) his equations predicted that it should begin to fall, and vanish when the temperature reached all the way down to absolute zero. The reason for this was that as a material cooled and changed from a gas to a liquid to a solid, its atoms or molecules became more rigidly attached to one another and so gradually lost various possibilities for motion; once locked into a crystal lattice, for example, they could no longer rotate. As a result they gradually had fewer and fewer ways to spend their energy. Einstein's result suggested that it should be impossible to ever cool something to absolute zero, because the colder something got, the more susceptible it would be to having its temperature nudged upward by some stray quantum of radiation. A graph of his predictions for carbon matched Weber's thirty-year-old diamond measurements almost exactly. This graph, it has been said, was the beginning of the field of modern solid state physics.[21]

Einstein's paper, published in early 1907, was another triumph for the quantum viewpoint, but it came, typically, at a terrible price. The notion that bouncing atoms could have only discrete amounts of energy violated the laws of Newtonian mechanics—the laws on which the atomic kinetic theory of gas and heat were based. The more successful quantum theory was, the more mysterious it became.

Albert redoubled his efforts to understand the quantum, to no avail. "I am incessantly busy with the question of radiation," Albert told Laub at one point. "This quantum question is so uncommonly important and difficult that it should concern everyone."[22]

Planck, meanwhile, having lost his statistical arguments to eliminate his quantum, had adopted a minimalist approach. A photograph from the time shows him in profile, sitting stiffly in his office chair staring intently ahead, looking hawklike with his bald forehead and hooked nose, his drooping moustache imparting a stern frown to his visage. "Does the absolute vacuum (the free aether) possess some sort of atomistic properties?" he asked Albert rhetorically in a letter, noting that "you seem to answer this question in the affirmative, while I would answer it, at least in line with my present view, in the negative. For I do not seek the meaning of the quantum of action (like quantum) in the vacuum but at the sites of absorption and emission, and assume that the processes in the vacuum are described *exactly* by Maxwell's equations."[23]

In the fall of 1907, having gotten nowhere, Albert put the problem aside to work on his review paper about relativity and gravitation. Von Laue, doubtless relieved, misinterpreted his sudden silence on the matter and wrote that he was pleased that Albert was abandoning his theory of light quanta. "As you know, I never had much use for it."[24]

Von Laue rejoiced too soon. Albert's dissatisfaction with the philosophical origins of relativity made him even more determined to get to the bottom of the quantum. Relativity, he had to admit to his admiring and not-so-admiring colleagues, would never do as a satisfactory theory of nature.

The reason, he explained in an exchange with Arnold Sommerfeld, went back to his old dichotomy between "constructive" theories and theories of principle.[25] The former were logical edifices built up from simple, elementary but far-reaching ideas, such as the kinetic theory of heat, based on the idea of atoms. Theories of principle were based on sweeping commandments, such as the second law of thermodynamics. Relativity was a theory of principle, which meant it lacked an elementary foundation. So, for that matter, were Maxwell's equations and electromagnetic theory in general.

"I feel, by the way," Albert wrote, "that we are still far from possessing satisfying elementary foundations for electrical and mechanical processes." He had come to this pessimistic view, he explained, primarily as a consequence of his "endless vain attempts" at interpreting Planck's quantum in any sort of intuitive way. "I even seriously doubt," he ventured, "that it will be able to maintain the general validity of Maxwell's equations for empty space."

In 1908, these were provocative words indeed. The roll call of famous scientists opposed to light-quanta was substantial, and included even the agreeable and judicious Lorentz.[26] Like Planck, he thought it might be the result of inappropriately applied statistics. He had set out to derive Planck's radiation formula but by beginning with different assumptions and eliminating the quantum requirement. The effort backfired, however, and Lorentz obtained the old classical answer for the blackbody spectrum, which didn't agree with the measurements.

When Lorentz called for a new round of experiments to decide the issue, he was shouted down in the pages of the *Physikalische Zeitschrift*, whereupon he conceded, admitting in 1908 that "a derivation of the radiation laws from the electron theory will hardly be possible without a profound modification in the foundations of the latter."[27]

Albert was trying to reinvent electromagnetism, and his ambitions were nothing if not profound. Early in 1909 he toyed with the idea of modifying the fundamental electromagnetic field equations into a more complicated and powerful form. He was hoping that the solution of such an equation would yield not only light-quanta, but also the "quantum of electricity," that is to say, the value of the charge of the electron, for which until now physicists had no explanation. He sent yet another light-quantum paper to Lorentz, groaning that it was "the insignificant results of years of reflection. . . . I have not been able to attain a real grasp of the matter."[28]

Lorentz responded with a long letter arguing against the existence of individual light-quanta, this time on the grounds of optics. A single quantum, he calculated, could be at least 5,000 square centimeters across—and would therefore be too big to be captured whole by the eye or even by a large telescope. That meant that fragments of thousands of quanta would have to merge simultaneously in the eye for vision to occur. "The individuality of each single light-quantum would be out of the question," he concluded. "It is a real pity that the light quantum hypothesis encounters such serious difficulties, because otherwise the hypothesis is very pretty, and many of the applications that you and Stark have made of it are very enticing."[29]

Johannes Stark, the man who had misattributed the mass-energy relationship to Planck but had solicited and published Albert's major review paper on relativity in his *Jahrbuch,* was indeed the orphaned quantum's only friend on the entire Continent, it seemed, besides Einstein. Stark was studying X rays, and he had concluded that they seemed to travel through his laboratory more like bullets than like waves that dispersed with distance.

Albert was looking forward to meeting Stark at Salzburg. "I am curious to learn the reasons that moved you to ascribe to the quanta a spatial existence in the aether," he said. "I am carrying on a lively correspondence about this subject with Planck; he still stubbornly opposes material (localized) quanta. You cannot imagine how hard I have tried to contrive a satisfactory mathematical formulation of the quantum theory. But I have not succeeded thus far."[30]

In England the quantum hypothesis was taken more seriously. In Cambridge the Trinity College professor Geoffrey Taylor attempted to isolate light-quanta by photographing an interference pattern through a series of smoked-glass screens.[31] Taylor hoped that with enough smoked glass he could cut the intensity of the light down to the point where too few quanta would survive to interfere with one another. At that point the characteristic dark and light fringes of interference would disappear. Taylor eventually piled on so much smoked glass that it took 2,000 hours to complete an exposure. Even in the feeblest light, however, the interference pattern remained. In extremis, light continued to cling to its waveness.[32]

IN HIS MORE REFLECTIVE moments, Albert found it a little spooky how quickly the European physics community was willing to confer on him the status of genius, especially since, as far as he was concerned, he was failing to solve what he regarded as the most important problem of the time. The quantum was eating away at the foundations of physics, but he seemed to be the only one that felt the threat. Having given them relativity and the Brownian motion proof of atoms, Einstein, the community thought in its collective tolerant wisdom, had earned the right to be wrong once. Albert felt that they had missed his point and he resolved to set them straight.

Salzburg, an architectural and historical gem set amid the foothills of the Austrian Alps astride the Salzach River, and the birthplace of Mozart, was a fitting spot for Albert to meet his colleagues at last, and he held nothing back in his speech. What they saw was a young man in his prime, riding a wave of confidence. He was now thirty years old. Age had removed the callowness from his cheeks and thickened his dark moustache. His wavy hair was still black, his eyes level and sardonic. He was not so much beautiful anymore as handsome, and in photographs he exudes a new air of confidence, like a gambler who knows the cards are going to fall his way.

Albert's address tied together explicitly for the first time the different strands of thought that had occupied him for the last nine years—statistical mechanics, electrodynamics, and the quantum hypothesis. He began by reminding his audience that, with the advent of the wave theory of light and Maxwell's equations, the existence of the aether had, until very recently, seemed practically assured. Now it was obsolete, and even the wave theory was being challenged by certain new phenomena that could be more easily explained by Newton's old emission theory of light as particles. "It is therefore my opinion that the next stage in the development of theoretical physics will bring us a theory of light that may be conceived of as a sort of fusion of the wave and of the emission theory of light."[33]

Then he reviewed the logic that had led him to relativity, dwelling on the resultant equivalence of energy and mass. Objects gained and lost mass by the absorption and emission of light. "The theory of relativity," he pointed out, "has thus changed our views on the nature of light insofar as it does not conceive of light as a sequence of states of hypothetical medium, but rather as something having an independent existence, just like matter." Moreover, this "stuff" transported mass when it was emitted from one place and absorbed somewhere else. In short, relativity was already halfway consistent with a more particulate view of light.

Moreover, the wave theory could not explain such phenomena as the photoelectric effect, various photochemical reactions, or why cathode rays had so much energy. Albert produced all these inconsistencies as if he were opening a bag of twentieth-century curiosities and dumping it out onstage. At the root of these problems, he maintained, was a simple concept. According to Maxwell's classical theory an oscillating ion or electron produced an outgoing sphere of radiation. But the inverse, of course, never happened: Spheres of radiation did not converge from all directions on ions or electrons and make them oscillate. In fact, light comes from a single direction and is absorbed. As usual for Albert, such an asymmetry in the theory suggested a fundamental misunderstanding, and rang alarm bells. "Here, I believe, our wave theory is off the mark." Rather, he argued, *"The elementary process of radiation seems to be directed."* When it is emitted from an atom, he suggested, the light does not go off in every direction, like a spherical wave, but only in a single particular direction, as did Stark's X rays. Like bullets. Left unsaid for the time being was the question of how the light decided in which direction to go.

The biggest blow to the classical wave theory of light, of course, came from blackbody radiation. To achieve his famous formula describing this radiation correctly, Planck had been forced to his infamous quantum hypothesis that only certain energies were allowed for the electromagnetic oscillations in matter. "In my opinion," Albert said, "to accept Planck's theory means plainly to reject the foundations of our radiation theory."

Planck's theory led naturally to a quantum theory of light. "On the basis of this hypothesis, the hypothesis of light-quanta, one can answer the questions raised above regarding the absorption and emission of radiation." Everything seemed to fit.

But did it?

Was it possible, Albert asked rhetorically, that Planck's formula could be derived in some other way without the horrendous quantum assumption, or at least limiting the damage to classical radiation theory? Suppose, he went on, that Planck's formula was correct. Forget about where it came from. Forget the assumptions that had gone into it. Was it possible to use the formula itself to learn something about the constitution of radiation?

Albert proceeded to outline an ingenious *Gedankenexperiment* in which he imagined that there was a tiny mirror suspended in a cavity of blackbody radiation. Drawing on all of his considerable and hard-won skill as a statistical mechanic, he was able to calculate the fluctuations in radiation pressure on the imaginary mirror. That light waves carried momentum and exerted pressure when they were reflected or absorbed was a commonly accepted tenet of standard electromagnetic theory, but the resulting mathematical expression that Albert derived from Planck's formula turned out to have two terms. The pressure on the mirror was subject, it seems, to two kinds of fluctuations. One term was easily attributed to classical electromagnetic waves; interference between the light waves caused the electromagnetic intensity on the mirror to vary slightly.

The other term, Albert pointed out, took a form that would be expected if radiation were composed of discrete chunks of energy raining randomly on the mirror. In short, he argued, both waves *and* particles appeared to be contributing to the statistical properties of the radiation pressure—a result that was sure to please no one.

Though there were simply too few clues to develop a formal theory of light that would explain such behavior, pictures came easily to Albert, and so he offered up his own vision of a wave-particle fusion, in which electromagnetic fields consisted of closely packed points of energy, each surrounded by a force field. From a distance or at large scales the points would blur together, into the standard undulatory field, but up close, the field would desolve into quanta.

"I am sure it need not be particularly emphasized," Albert concluded, "that no importance should be attached to such a picture as long as it has not led to an exact theory." The point was that the wave and particle pictures of light were not necessarily incompatible after all.

Planck was the first to rise, tall, austere, yet avuncular. After thanking Einstein for his talk, which stimulated the audience "to further reflection even where perhaps opposition may have emerged," he said he would restrict his own comments to areas of disagreement. He acknowledged that quanta were necessary. The main problem, he said, was where to look for these quanta. Whereas Herr Einstein wanted to make them part of light itself, Planck believed, "This seems to me a step which in my opinion is not yet necessary."

Planck pointed out that the subjects of emission and absorption, on which Albert's analysis rested, were the weakest and least understood aspects of the classic theory. Perhaps by incorporating quanta into these processes physicists could yet construct a satisfactory radiation theory that preserved Maxwell's equations. "I think that first of all," he declared perhaps a little too stoutly, "one should attempt to transfer the whole problem of the quantum theory to the area of *interaction* between matter and radiation energy."

Stark then rose to take issue with Planck, and they embarked on a debate about whether or not Einstein's quanta were compatible with the phenom-enon of interference. No conclusion was reached.

The Deutsche Physikalische Gesellschaft had gotten its money's worth, and then some. It's not every day that someone stands up and predicts the future history of science. It would be more than two decades before Albert's vision of a wave-particle radiation theory became manifest, and then in a form he detested. If relativity represented a return to classical purity in physics after the complicated jumble of aethers and electron theories, a tri-umph of law and generality over confusion, it was also opening the door, it now seemed, to something wilder and more dangerous, pointing past its own limits to dark, ominous shapes out in the deep water. The wildness was just beginning.

13

IN THE COMPANY OF MICROBES

BY THE END OF THE SUMMER, RELATIONS HAD IMPROVED, AT LEAST ON the surface, between Albert and Mileva. They had spent the week before the Salzburg meeting hiking in the Engadine, a river valley running through the southern Alps that is one of the most beautiful and romantic regions in Switzerland. The River En flows along a spiky fenceline of 3,500-meter peaks below ski resorts and small villages with sweet names like Susch and Sucol, basking in a warm and dry Mediterranean climate. The slopes and ridges beyond are laced with hiking trails for the hardy and cable cars for the less adventurous.

As a place for the couple to rest and perhaps repair their nerves if not their marriage in anticipation of their pending relocation, the Engadine was a poignant choice. The surrounding territory was redolent with associations for the Einstein family. Majola Pass, at the head of the valley, was only a few kilometers east of Splügen, where eight years earlier Mileva and Albert had cuddled beneath a mound of furs in the back of a sledge while the world turned white. Just over the ridge were the headwaters of the Rhine, where they had stomped down through the snow creating miniature avalanches. Somewhere in these folds leading down to milky Como Albert had surrendered his "yes" to the world, and Lieserl had been conceived.

Although Albert and Mileva did not conceive another child on this particular return trip through the Alps, they might have begun negotiations that would lead them shortly to agree to start trying. In the past, Albert had been able to tame Mileva's insecurities with cozy visions of their bohemian future and the work they would share—"our theory on relative motion." By now, except for the sporadic advice she offered about the ever-lumbering *Maschinchen,* there was no prospect of a meaningful scientific collaboration between them. The cozy future had arrived; the avalanches had all crashed down.

Mileva's spirits lifted when they moved on schedule that October back to the familiar setting of Zurich. By the end of the month she was indeed pregnant. Whether by plan or by accident is unclear, but Mileva was definitely enjoying a greater degree of attention from her husband.

Inwardly, though, Albert was still seething over Mileva's embarrassing outburst about Anneli. Three months afterward he was still complaining to Besso that Mileva had destroyed his peace of mind, which he hadn't been able to regain. In the meantime his concentration was destroyed, and his progress suddenly nil. "Brooded only a little and unsuccessfully about light quanta," was a typically dark summation of his activities during this period.[1]

THE EINSTEINS' NEW apartment was a second-floor flat on Moussonstrasse, directly upstairs, it turned out, from the Adlers. The quarters featured gas lighting, a step up from the oil lamps they had to use in Bern. Another feature of the house, often remarked upon in memoirs from visitors of the time, was a leaky coal stove. There was also an extra room, which Mileva, who seems to have been in charge of the family finances, rented out to student boarders, mostly Serbs. The daughter of one of those students, Svetozar Varicak, later told Mileva's biographer Trbuhovic-Gjuric that her father remembered Albert's helping Mileva with the housework. Allegedly he felt sorry for her because she had to stay up past midnight doing mathematical calculations for him.

Aside from Adler, one of Albert's earliest and most frequent visitors was a doctor and professor of forensic medicine at the University of Zurich named Heinrich Zangger, whom Albert had met several years earlier in Bern. Five years Albert's senior, Zangger was a man with wide-ranging interests whose excitable nature showed in his letters, which usually consisted of wild scrawls and half-finished sentences as he dashed from one thought to the next. Zangger had become an international hero in 1905 when he guided the rescue of over one hundred French coal miners trapped in an underground explosion in Courriers. The government had been about to abandon the search for survivors when Zangger, after inspecting the mine shafts, concluded that the miners could still be alive in air pockets inside.[2]

Zannger also followed physics, particularly atomic theory. An interest in Brownian motion and the colloidal properties of milk had led to an introduction to Einstein. Zangger and Einstein had hit if off immediately. Albert was impressed with Zangger's seemingly unlimited curiosity and his knack for making insightful judgments on the basis of limited information. He liked to say that Zangger was one of the most interesting men he had ever met.[3] Zangger, in turn, went back to Zurich awed by Einstein's intellectual prowess and allegedly played a behind-the-scenes role in encouraging Kleiner to recruit him for the physics faculty.[4]

Zangger's experience in rescue medicine came in handy at least once. Dropping by Moussonstrasse one day, Heinrich found the door open and Albert asleep on the sofa in front of the smoking stove. Recognizing symptoms of carbon monoxide poisoning, Heinrich opened the windows and dragged Albert out of the apartment, thereby saving his life.

IN MANY RESPECTS Albert's new existence was not so different from his old one as a student, except that a professor did not have the option of skipping classes. In his first term Albert was assigned a heavy teaching load: four hours a week of introductory mechanics and two hours of thermodynamics, as well as responsibility for conducting the weekly physics seminar.

Like many a novice teacher, he was unpleasantly surprised by the time and effort it took to compose a decent and comprehensible set of lectures. Mindful of his own boredom as a lecturee, he toiled conscientiously over his talks, combining information from a variety of textbooks as well as recent papers, laboring into the night at home, often with Mileva looking on. In the midst of one set now preserved in the Einstein archives is a marginal inscription in her handwriting: "Here give a loving kiss to your poor [————]."[5]

But however much work was required of him, Albert repeated over and over again throughout his first year teaching—to friends, correspondents, and his mother—how much he enjoyed his students. The feelings were not immediately mutual. When Albert first walked into the classroom, in shabby clothes and too-short trousers, sporting a cheap iron watch chain and a couple of notecards, his pupils were not surprisingly rather dubious. There was also the matter of his lecturing style, which was to work out the details as he went along, or not: "There must be a stupid mathematical transformation here, which at the moment I do not see. Did anyone here notice?" Silence. "Leave a quarter page empty, we don't want to lose time, the result reads as follows. . . ." Ten minutes later he would go back and fill in the missing transformation.[6]

After a few encounters, though, the class was hooked. According to a memoir by David Reichinstein, a young physical chemist who became a member of Einstein's circle, Albert treated them as collaborators, his peers, encouraging them to interrupt and ask questions whenever they wanted, as if they were all members of the Olympia Academy. After classes the students and assistants followed after him like a pack of puppies to the Cafe Terrasse, where Albert pursued his usual practice of talking them to death.[7]

The conversations did not always end even at closing time. One of his students, Hans Tanner, recalled being dragged back to Einstein's house in the middle of the night to critique a new paper by Planck that Albert was sure had a mistake in it. After having the error pointed out, Tanner suggested they write to Planck to correct him. "A good idea. But we won't write and tell him he's made a mistake," Albert went on in a hitherto uncharacteristic display of tact. "The result is correct, but the proof is faulty. We'll simply write and tell him how the real proof should run. The main thing is the content, not the mathematics. With mathematics one can prove anything."[8]

These sessions were frequently quite personal. "Einstein often spoke about quite intimate subjects," Reichinstein recalled, "and this not only when one of us was alone with him, but on walks when four or five people were present. I shall never forget his conversation on a certain mountain tour which took a whole day. It is interesting that as soon as a colleague of his own rank—a Zurich University professor, for instance, Professor Zermelo—was present, Einstein became more reserved. Of course we also poked fun and laughed, but the conversation did not overstep a certain limit."[9]

Reichinstein believed he could see Einstein's soul taking flight during that first year in Zurich, as the insecurity born of years of striving and rejection slowly faded. In some respects Albert reminded him of the Buddha, who in turn had once compared himself to a pregnant woman who is justified in avoiding unnecessary risk because she is carrying something valuable. In Albert's case, the load he was bearing was relativity. A scientist nursing a great idea to maturity, Reichinstein thought, had to be ruthless in protecting the precious cargo in his soul; he had to remain unencumbered by cares or conflicts, and to be ready to bear any humiliation from his opponents in the service of his truth. He wondered if Albert was tough enough to withstand the attacks he would attract from smaller minds.

He related Einstein's answer to these concerns as a kind of parable. While they were hiking one day, Reichinstein began to expound in his usual way on the malevolence and envy of the mediocre.

At that moment, as if in response, Albert dropped his cigar, a bad-smelling little stump that he could barely hold on to with two fingers, into the dirt. He picked it up, brushed it off, and stuck it back in his mouth. "But Professor, what are you doing?" Reichinstein protested.

"What about it?" Albert asked.

"And the microbes which are sticking to the cigar?"

Albert glanced at him and growled, "I don't care a straw for the microbes."[10]

Such confidence did have its downside. Albert could be a child one minute, especially with respect to women, and a cynic the next. The targets of his barbs didn't know whether to laugh or to cry. "Here and there," Reichinstein wrote, "one of us would get into Einstein's bad graces. This was manifested by Einstein's keeping away from the person in question, sometimes casting a surprising glance at his victim, a glance which lashed out more violently and was able to inflict greater spiritual punishment than the legendary flash from the eyes of the irate Indian god."[11]

One of the major amenities of Albert's new position was a research assistant. Albert had selected for the position Ludwig Hopf, a good-natured young Bavarian, whom he had first met at the Salzburg conference, and who had just received his Ph.D. from Albert's friend Sommerfeld. In addition to being a fine mathematician and a good piano player, Hopf was an ardent fan of psychoanalysis. He had studied Freud and once in Zurich had attached himself to Freud's ex-disciple Carl Jung, who was in the midst of developing a more expansive theory that would be come to be known as the "collective unconscious." Albert accompanied Hopf to a lecture by Jung, and afterwards the pair went to Jung's house for dinner. Although Albert had not been impressed by Jung's lecture, he returned to Jung's house several times over the years. Albert tried without much success to explain relativity and other aspects of modern physics to the psychiatrists. Jung later reported that he had understood just enough to be hopelessly impressed.[12]

The psychoanalysts had even less luck explaining their work to Albert. It might have been the enthusiastic Hopf who was the subject of another psychoanalytic anecdote. According to Reichinstein's memoir, one of Albert's colleagues, a "Dr. Y," arranged to give a lecture on psychoanalysis, which Albert was cajoled into attending. The audience was full of young men and women hoping to become enlightened about the question of love. Afterwards, Albert went to dinner with a group that included Hopf ("Dr. Y") as well as two young Slavic women, sisters, who had come to Zurich to study science. The older one was married, but the younger, of whom Reichinstein said, "it was her eyes which made the study of stern science a difficult thing for many young men," was still in school. Hopf was badgering Albert for his opinion and arguing that psychoanalysis was indeed science. Albert was bored; his gaze drifted to the young Slavic beauty and stayed there while Hopf railed on.

Finally Hopf noticed that Einstein was not paying a bit of attention. He slammed a book on the table and exclaimed, "Why, Professor, if you were in love you would believe that more important than your quantum theories."

Albert looked up abashed. "No, ladies and gentlemen, my quantum theories are really of great importance to me."[13]

BY THE END of the year, Mileva had slipped back under the cloud of neglect. Between her husband's lectures and his scientific socializing, she barely saw him at all except at musical gatherings and occasional evenings when they met at the theater. On those nights she would bring sandwiches in her bag for his dinner. Albert complained about the lack of plot and chatted with his friends with his mouth full, crumbs scattering, while Mileva watched, alarmed.[14] At home Albert and Fritz quickly developed the habit of escaping to the attic to smoke and to argue about physics and philosophy. Albert was disturbed and disapproving that Fritz would let political ideology get in the way of his physics career. "How an intelligent man can subscribe to a party I find a complete mystery."[15]

When Mileva complained again to Helene about feeling squeezed out of Albert's life, Helene teased her about being jealous of science. Mileva did not appreciate the irony of the suggestion, and even wondered if her old friend was being malicious. But what could she do? The pearls of scientific talent were given to one, not to the other.

"I often ask myself," she went on, "whether I am not rather a person who feels a great deal and passionately, fights a great deal and also suffers because of that, and out of pride or perhaps shyness puts on a haughty and superior air until [I myself] believe it to be genuine." If that was true and her soul "stood less proudly," would Helene still love her?

"You see, I am very starved for love and would be so overjoyed to hear a yes, that I almost believe wicked science is guilty and I gladly accept the laughter over it."[16]

MEANWHILE, genius was continuing to get Albert nowhere with the quantum, even while the quantum continued to invade the rest of physics.

In March of 1910 a physical chemist from Göttingen by the name of Walther Nernst stopped in Zurich to see Einstein with important news about the light-quantum. In the German physics community Nernst, at the age of forty-six, was known as much for his egocentricity as for his brilliance. In 1905 he had enunciated a hypothesis that would become known in a more refined form as the third law of thermodynamics, regarding the entropy of substances as their temperatures approached absolute zero. One of the implications of his idea was that the specific heats of substances would all decline to a uniform low value. (A later version of the hypothesis by Planck said the specific heats should all decline to zero.)

Nernst and his colleagues had set out to test this idea by measuring the specific heats and other thermodynamic quantities of a number of substances at low temperatures. By early 1910 they had their answer: Specific heats not only decreased with lowering temperature, as Nernst had suggested, but it looked as if they were heading for zero. That, of course, is exactly what Einstein had predicted three years earlier on the basis of his quantum theory, a theory that Nernst regarded as "essentially a computational rule, one may well say a rule with most curious, indeed grotesque properties."[17]

He decided to have a look at this Einstein, and so dropped by Zurich on his way to Lausanne. "My predictions regarding the specific heats are apparently being brilliantly confirmed," a delighted Albert told Laub a few days later. "Nernst, who has just been here to see me, and Rubens are busily engaged in the experimental verification, so that we will soon know where we stand."[18]

Nernst's data were the first hard evidence in favor of the quantum hypothesis since the discovery of the blackbody radiation spectrum. It sent him into a paroxysm of enthusiasm for the quantum and for Einstein. "I believe that so far as the development of physics is concerned, we can be very happy to have found such an original young thinker, a 'Boltzmann revividus,'" Nernst wrote to a friend. "Einstein's 'quantum hypothesis' is probably one of the most remarkable ever devised; . . . if it is false, well, then it will remain for all time 'a beautiful memory'!"[19]

Nernst, who was professionally well connected, returned home and began to plan a scientific congress, a summit conference of the brightest minds in the business to see if collectively they could begin to make some sense out of this mysterious quantum, which was beginning to look as if it was here to stay. Among those he spoke with was Ernest Solvay, a chemist, philanthropist, and physics buff in Brussels, who had made a fortune from baking soda and had also been one of Albert's fellow honorees at the University of Geneva ceremony.

Another whom Nernst contacted was a Berlin chemistry professor named Emil Fischer, who would write to Albert six months later on behalf of an anonymous rich industrialist enthused about radiation. "He was very pleased that German researchers like you, Mr. Planck, and Mr. Nernst have taken over the leadership in this fundamental area, and he believes it is the duty of the well-to-do people in Germany to promote these splendid endeavors a bit by providing financial support."[20] To wit, Albert was made a grant of 15,000 marks, to be paid over three years. It had been seventeen years since Albert had renounced his citizenship in Württemberg and scampered through the Alps to Milan. The home country was now reaching out a long crook, which Albert grabbed hold of gratefully.

IN MAY OF 1910 it was the macrocosm, rather than the microcosm, that was holding the rest of the world in sway. Night after night the shivery fairy glow of Halley's comet hung suspended halfway across the night sky, energizing astronomers and scaring the wits out of many less enlightened. In Oklahoma, sheriff's deputies had to subdue a mob intent on sacrificing a virgin to the famous "hairy star." Around the world people sealed the windows and doors of their houses on the night of May 18 to prevent poison gas from entering as the earth passed through the comet's vaporous tail.

If anything should have represented a triumph of science, it was Halley's comet, once more arriving as it did precisely on the seventy-six-year timetable first plotted by Edmond Halley back in 1705. Halley's calculations had been the first demonstrations of Newton's famous law of gravitation, and the comet's clocklike returns had never failed to excite the popular imagination. The comet was the harbinger of cosmic order, of the great celestial clockwork that God had made of the universe, rolling down the grooves in space and time proclaimed by differential equations.

But nowhere in the vast volumes of correspondence and papers accumulating in the Einstein Papers Project is there any mention of the comet. Albert was in Switzerland during its appearance. At the beginning of the month he gave a talk on light-quanta in Neuchâtel, and by the seventeenth, when Halley was pumping quanta around the hemisphere, he was complaining to his friend Lucian Chavan that some Basel firm had just sent him the wrong tea. It is difficult to know quite what to make of Einstein's seeming lack of interest in what turned out to be the most spectacular apparition of Halley's in its recorded history. Perhaps, Switzerland being a cloudy country, Albert never saw the comet at all.

But comets are traditionally omens of change and disaster, and Halley's arrival, unseen or not, coincided with fresh tremors in Albert and Mileva's world. On the heels of Nernst's visit in March, after a mere six months at the University of Zurich, Albert had been quietly contacted by the German university in Prague and asked if he would be interested in a professorship that would soon be opening up there. This was a seductive offer for many reasons. The German university was part of the University of Prague, which, founded in 1348, was the oldest university in Central Europe, and an illustrious center of science and culture since the sixteenth century when an aging Tycho Brahe and a young Johannes Kepler had collaborated on recharting the motions of the planets. Brahe's data and Kepler's ideas had prepared the way for Copernicus. It was in Prague, too, that the great Ernst Mach had spent twenty-eight years teaching and promulgating his own recharting of the philosophy of physics. The university also had an immensely good library. Albert would be a full professor instead of associate, and would get a hefty raise. And intellectually, at least, he would be that much closer to Berlin.

He confided the offer almost immediately to Adler, who he hoped would succeed him at Zurich. Subsequent letters from Adler and his wife, Katya, suggest that Albert was eager for the job,[21] though in Michelmore's account, presumably based on family lore, Albert was ambivalent about the offer, vacillating between accepting it and staying in Zurich, depending upon whom he had last spoken to about it.[22] In all these versions Mileva is portrayed as a brave soldier encouraging Albert to do what was best for his career and to simply make up his mind.

Stoicism was the lot of all wives of physicists and soldiers, but there was no reason for Mileva to be happy with it—and every reason to be unhappy. As much as she was suffering, or thought she was suffering, from neglect and loneliness, she still loved Zurich, their friends, and familiar haunts. Moreover, Zurich was a liberal bastion, friendlier to women and Serbs than Prague, with its population of Czechs and Germans, was likely to be. There was also little Hans Albert's imminent schooling to worry about, and on top of everything else, she was five months pregnant.

Mileva's despair may have been the source of a bad mood Albert's mother noticed in her son later in April, for which he had felt moved to apologize: "The bad mood that you noticed in me had nothing to do with you. This must not concern you; to hammer away with others on the things that depress or anger us does not help in overcoming them. One must knock them down alone." He added, almost as a footnote, that he would soon become a big professor at a university in Prague, but he was not allowed to say exactly where.[23]

Aside from his sister, Maja, Pauline and the rest of the family seem not to have played a significant role in Einstein's life during the preceding ten years. Since Hermann's death Albert's mother had been living in Hechingen in southern Germany with her sister Fanny's family, which included Albert's cousins and old playmates, Elsa and Paula. Pauline had never reconciled herself to Mileva, and on her rare visits to see her son and grandchildren the air, as Albert remarked, was "full of dynamite."[24] As a result, Hans Albert barely knew his grandmother. Albert's appointment to Zurich, however, had brought the Einsteins back into his sphere.

Surprisingly, Albert found that his relatives weren't as bad as he remembered them—Uncle Adolf, one of the banes of Albert's childhood, wasn't quite so nasty, and Aunt Ida had aged well. His mellowing toward his family was doubtless a development that Mileva found even more ominous than moving to Prague.

Word of Albert's possible departure caused a stir on campus. The physics students, all fifteen of them, sent a petition to the Department of Education, urging it to keep Einstein in Zurich.[25] In response the director of the department, Heinrich Ernst, proposed to raise Einstein's salary by 1,000 francs to 5,500 francs a year. At the same time he suggested that Albert's teaching load could be cut back some, since it was heavier than average. The raise was approved, and in July Albert announced at a faculty meeting that he would be staying. Albert had learned meanwhile, in June, that although he was the first choice of the professors up in Prague, the Austrian minister of education down in Vienna had chosen somebody else because Albert was a foreigner. He wasn't getting the job after all.[26]

Mileva perhaps breathed a quiet sigh of relief to herself. Her own parents came for a visit in the summer of 1910, perhaps to help out with the impending birth; the last time they had come to Switzerland was in 1904 under similar circumstances. Back in Novi Sad, Mileva's mother gave a glowing account of her daughter's life. "I didn't know that my Miza was so treasured in the world," she said. "When we were there the greatest and smartest people in the world came to her home and never wanted to start talking until Miza was there. She usually sat off to the side and merely listened, but as soon as she began to speak, everyone turned to her and noted with great attention everything that she said."[27]

In the midst of this turmoil, on July 28, Mileva gave birth to a son, Eduard, and Albert trudged nonchalantly up the Zürichbergstrasse carrying bedding for a cradle on his back.[28] Tete, as they called the boy, turned out to be a delicate baby. Mileva, who had already been tired during her pregnancy, had been done in by the labor. "My wife and the little one feel unwell all the time, so that we must suspend our musical evenings for the time being," Albert wrote to a colleague in August.[29] Trbuhovic-Gjuric reports that a doctor was called in and concluded that Mileva was overworked. To relieve the pressure on her, he insisted, Albert had to make more money. This advice caused an almost hysterical outburst from his patient, who would have none of it: "Isn't it clear to anyone that my husband works himself half dead? His thirst for knowledge, his search for ever newer understanding, his work, this is his genius."[30]

BY THE END of the summer, the possibility of Albert's Prague professorship had been revived. As in the case of the Zurich post the year before, the Prague position had become snarled in politics.[31] On one side of the debate was Anton Lampa, professor of experimental physics in Prague and lead member of the faculty committee in charge of recommending recruits. On the other side was Count Karl Graf von Stürgkh, minister of education for the Austrian half of the Austro-Hungarian empire. Since Prague was—next to Vienna and Budapest—the crown jewel of the beleaguered empire, von Stürgkh and his patron, Emperor Franz Joseph, had the final word on who would teach in his nation's universities.

Lampa, a former student of Mach's, had been determined to find a candidate who would carry on in the tradition of the Great Skeptic himself, as well as reflecting scientific glory on the empire.[32] Einstein, who was true to the spirit, if not the letter, of Mach's teachings, topped Lampa's list. In second place was Gustav Jaumann from the University of Brno, an accomplished older physicist, a strict Machian, and a man who rejected the existence of atoms. Lampa recommended Einstein, but the Austrian educational bureaucracy apparently preferred one of their own, and von Stürgkh had thus offered the job instead to Jaumann.

Jaumann, unfortunately, proceeded to disgrace himself in the ensuing contract negotiations, demanding more money than the ministry was willing to pay. Moreover, he was angry because Einstein had been ranked above him. Perhaps fearing the inevitable, Jaumann finally turned down the job, writing that "I will have nothing to do with a university that chases after modernity and does not appreciate true merit."[33]

In September, then, Albert was duly summoned to Vienna for a meeting with Lampa, Count von Stürgkh, and, of course, the emperor. By that time, Adler had convinced Albert that he had been rejected originally not because of his nationality but because of his religion. Albert officially had none, according to the form he had filled out for the Zurich police department and which the Austrians had undoubtedly seen. The professorship required Austrian citizenship, which required an oath of allegiance to the emperor, which in turn required, as far as the emperor was concerned, a religious affiliation. Any religion would do.

Apprised of this policy, Albert declared matter-of-factly that he was now a Jew. His examiner had no choice but to cross out "unaffiliated" on the form and write in "Mosaic," the official term for Jewish. With that modification to his record, the Prague deal was basically concluded. At the time, Albert regarded this accession to the identity of his ancestors as nothing more than a bureaucratic formality.

VIENNA WAS also the home of Ernst Mach, and Albert could not leave the city without finally meeting this man who was by now less a physicist than a force of nature. Mach was seventy-three, and lived on the outskirts of Vienna, deaf, half paralyzed, and a semirecluse, but age and infirmities had scarcely tamed his famous temperament or softened his views.[34]

In fact, Mach had gotten even more opinionated, and Albert when he arrived found himself in the middle of a war between two of his idols. In

1908 a dispute had broken out between Mach and Max Planck. Ostensibly it concerned the existence of atoms, which Mach still stubbornly spurned, but it had since escalated into a vitriolic debate about the very nature and meaning of science.

Planck started the fight by declaring in a speech that Mach had become a dangerous influence because he was misleading young physicists into not giving credence to the theory of atoms. Planck spoke from painful experience, for as a younger man under Mach's influence, he himself had resisted atoms until forced by thermodynamic arguments to revise his views.

More generally and damningly, Planck also pointed out that Mach's precepts had nothing to do with how physics actually worked. According to Mach's "positivistic" philosophy, physics was about the results of experiments; the scientist's job was to describe the world and organize facts as economically and logically as possible. But none of the great scientists of the past, Planck argued, had engaged in so simple and bloodless an ordering of facts: "Economical considerations were the very last thing that steeled these men in their fight against received conceptions and high-and-mighty authorities." Rather, he explained, they were driven by a passion for ideas. Absent this ideological, yea metaphysical, drive, Mach's followers would produce a physics barren of interest or discovery. "By their fruits ye shall know them," Planck said, echoing Lenin, who had also used Matthew's scriptural phrase to attack Mach.[35]

Planck, who repeated the attack in a lecture at Columbia the next year, thought of himself as an artist, not a stamp collector. The goal of science, he said, was to make contact with the universal truths, valid for "all places, all times, all peoples, and all cultures," disguised behind the appearance of the world.[36] For him, Mach's facts existed only to be boiled away. "If any thought strengthens and elevates us during the patient and often modest detailed work that demands our minds and bodies," Planck said a few months later, "it is that in physics we labor not for the day, not for momentary success, but, as it were, for eternity."[37]

Mach finally struck back in the summer of 1910 with a long article in the *Physikalische Zeitschrift*. To him the notion of scientific truth independent of its own discoverers was pure folly. Without observers, there were no observations; every datum was hostage to the circumstances of its collection. "Concern for a physics valid for all times and all people including Martians seems to me very premature, and almost comic, while many everyday physical questions press upon us."

As for atoms: "If belief in the reality of atoms is so essential to you," Mach went on, "I will have nothing more to do with physical thinking, I will not be a proper physicist. I renounce all scientific reputation, in a word, no, thank you, to the community of believers. Freedom of thought is more precious to me."[38]

Planck responded by attacking Mach's scientific competence and challenging his grasp of mechanics and thermodynamics.

Albert had stayed in the shadows while his mentors fought it out, hoping like a child of divorcing parents that their differences could be settled. He was particularly uncomfortable with Planck's virulence and had tried to soothe Mach's feelings. While sending some papers, he told Mach, "You have had such an influence on the epistemological views of the younger generation of physicists that even your current opponents, such as, e.g., Mr. Planck, would undoubtedly have been declared to be 'Machists' by the kind of physics that prevailed a few decades ago."[39] But this was the kind of debate in which the only resolution occurs at the funerals of philosophers.

Albert was still too young himself to have acquired the philosophical gravitas to pronounce on the meaning of science, not to mention being too busy struggling with the quantum to care much about the niceties of philosophical consistency. In the matter of atoms Albert was, as Adler had remarked, in the school of Boltzmann. Epistemologically, however, Albert seems to have thought of himself as a Machian, although in his heart he agreed with Planck that science and the building of theories were profoundly creative endeavors, not just an economic description of facts.[40] Mach's independence and incorrigible skepticism toward concepts like space and time had been important influences on the road to relativity and Mach himself had been cordial and friendly.[41]

That was where things stood in September of 1910 when Albert made his pilgrimage to the suburbs of Vienna. Mach, an unkempt man with a stiff gray beard, looked like a Slavic peasant. His face displayed a disarming mixture of cunning and charm. He liked to tell his visitors, "Please speak loudly to me. In addition to my other unpleasant characteristics I am also almost stone-deaf."[42]

Albert was determined to wring some concession from Mach regarding atoms and the rules of evidence. Suppose, Albert said to him, that by assuming the existence of atoms it was possible to make a new prediction about the behavior of a gas, and that this new property was then observed and confirmed. If the atomic hypothesis was the only way to tie all these observations together, even though long arduous calculations were involved, would Mach accept it?

Yes, Mach answered, under those circumstances he would consider atoms an "economical" hypothesis. It didn't matter how hard the mathematical computations were. What mattered was the "logical economy" of the hypothesis.

Albert was thrilled; he and Mach were not so far apart on their criteria for a proper scientific theory after all. The glow of his victory was still apparent more than forty years later when he recounted the discussion to the historian I. Bernard Cohen, who wrote in *Scientific American,* "With a serious expression on his face, he told me the story all over again to be sure that I understood it fully."[43] Not only had he beaten Mach philosophically, but he had wrestled the old lion into admitting that there might be some good in the atomic notion after all.

Einstein and Mach parted on good terms, and throughout the following years Albert often referred to Planck's "unjustified criticism" of his adversary. At one point he wrote Mach, "To this day, I still cannot understand how Planck, whom I have otherwise learned to prize like no one else, could show so little understanding for your endeavors."[44]

ALBERT RETURNED home to await the call to Prague, and meanwhile plunged into quantum theory with renewed vigor, still searching for a way out of the destruction of classical electromagnetism. During the summer he and Hopf had investigated in detail one possibility, which involved fiddling with one of the main principles of statistical mechanics that Boltzmann himself had invented—perhaps a measure of Einstein's desperation. This was known as the equipartition theorem, which assured that energy was spread out evenly. But when he and Hopf recalculated the blackbody formula under the new assumptions, they still obtained the same old wrong answer.[45] Albert was not off the quantum hook yet.

In November he came up with an even more radical idea. This one apparently involved modifying the law of the conservation of energy, a pillar of modern thermodynamics and mechanics. Albert was behaving increasingly like someone who was willing to destroy physics in order to save it. "At the moment I am very hopeful that I will solve the radiation problem," he wrote to Laub, "and that I will do so without light quanta. I am awfully curious how the thing will turn out."[46]

By now Laub, von Laue, Stark, Besso, Zangger, Sommerfeld, and the rest of Albert's friends had been receiving letters of this sort for five years. It was probably not much of a surprise when Albert announced a week later, "The solution of the radiation problem has again come to naught. The devil played a dirty trick on me."[47]

BY THE END of 1910 both Mileva and young Eduard were doing better. Tete was developing a striking resemblance to little Hans Albert. The family gradually began to resume a social life in the form of Sunday afternoon concerts at the home of Adolf Hurwitz, a Polytechnic professor with whom Albert had played music during his student days. Adolf had a daughter, Lisbeth, who played the violin. A photograph shows Albert and Lisbeth sitting side by side on the Hurwitz balcony, bows at the ready, Zurich in the background, while Adolf stands over them in a big handlebar moustache, wielding a baton.

At their first meeting Albert and Lisbeth played a couple of Bach sonatas, and Lisbeth found him quite musical. "He is a cheerful, modest, and childlike person with a new view of the notion of time," she wrote in a diary that forms an important source for the Trbuhovic-Gjuric biography.[48]

The Einsteins would arrive faithfully at the Hurwitzes' at five in a bustle of coats, violin cases, and baby gear. They played Bach, Mozart, and Handel. For Mileva the visit was the high point of the week, the music one fragile thread to her past and to Albert.

Another was dance. Perhaps remembering her own love of twirling and hopping, the "little wounded bird" of her youth, Mileva enrolled Hans Albert in a dance school. Already growing into the classic Einstein short-but-steady shape, Albertli was the smallest boy in the school.

But the high point of Albert and Mileva's social life that winter came in February when the great Lorentz invited Albert to come to lecture and stay with him in Leiden. He and Mileva slipped off for a few days on a rare trip together to Holland, presumably with Tete but apparently without Hans Albert. "If the house burns down or some other nice thing like that happens," they wired the Adlers from Basel, a wire should be sent to Lorentz's house.[49]

For Albert the attraction of the trip was not the lecture or a weekend away with Mileva, but the chance to meet Lorentz and discuss the radiation problem with him. "I wish to assure you already in advance," Albert said, "that I am not the orthodox light-quantizer for whom you take me."[50]

The Lorentzes rolled out their legendary red carpet. "Now I sit here again in a monastic cell, brimming with the fondest memories of the wonderful days I had the privilege to spend near you," Albert wrote later in a typically profuse thank-you note.[51] On Friday he lectured. The next evening they dined with other physicists at the Lorentz house.

Albert and Hendrik did not solve the quantum problem during his weekend stay, but they had fun trying, staying up late on Saturday night in Lorentz's magnificent study after everybody else had left.

In fact, Albert was running out of ideas. By then he had spent the greater part of a decade pondering the riddle of why—in apparent transendence, if not outright contradiction, of all the fundamental laws of physics—nature seemed to require light to be parceled out in discrete bits. There was scarcely one of those fundamental laws that he had not considered modifying in an attempt to understand why this quantum, this insult to classical physics, was necessary. He had jiggled the laws of statistics. He had dallied with the idea of waiving the law of the conservation of energy. He had filled countless and ultimately fruitless sheets of paper with modified versions of Maxwell's holy equations. All of them had been consigned to the wastebasket.

No amount of violin playing or walking or sailing and scribbling brought forth any illuminating *Gedanken* or mental pictures like the kind that had spurred him to relativity. Along the way Albert had done important science, elucidating the properties of blackbody radiation and solving the theory of specific heats, but all his digging had only left physics deeper in its quantum hole.

The devil tricked me again.

Albert finally gave up. "I no longer ask whether these quanta really exist," he wrote Besso in the spring of 1911. "Nor do I try to construct them any longer, for I now know that my brain cannot get through in this way." From now on, he told his friend, he would confine himself merely to understanding the consequences of this strange property of nature.[52]

As it happened, Albert surrendered just as the rest of his colleagues were joining the chase. Nernst had done his job of expanding the constituency of the quantum well. A month after Albert sent his note of defeat to Besso, he received an invitation from the industrialist Ernest Solvay to attend an international "Scientific Congress" to be held in Brussels the following fall. There were only twenty-five names on the list of guests from six countries being summoned to discuss "current questions concerning the molecular and kinetic theories," that is to say, the quantum.

The final call to Prague had come just before their trip to Leiden. Albert was to start on the first of April 1911, less than three months hence. Including an expense account his salary was to be 8,672 crowns, effectively nearly twice what he was receiving in Zurich. No amount of added income, however, could make up for Mileva's feeling that both she and her children were being shuttled about. They didn't know anyone in Prague; how would they adapt?

Mileva worried, it seems, that this move might signal the end of too much good luck, or maybe she just worried that she worried too much. "I . . . believe that we women cling much longer to the memory of that remarkable period called youth," she wrote to Helene, the one person to whom she felt safe confiding the Slavic gloom in her soul, "and involuntarily would like things always to remain that way. Don't you find that to be so; men always accommodate themselves better to the present moment.

"Things are going well for mine; he works very hard, gives his courses that are very well liked and attended, as well as many lectures, which I never fail to attend. Since there are rather many musical occasions in our house, we really have very little time that we can pass together in privacy and tranquillity."[53]

On their return home from Leiden, Mileva bent herself to the task of packing once again. She summoned her mother from Novi Sad to help her with the move. Marija is said to have found her daughter obsessed with gloomy anticipation, but Mileva didn't want to talk about it.[54]

14

THE MAN WHO ABHORRED BATHS

PRAGUE, TOO, HAD A FAMOUS OLD CLOCK, ANOTHER WHIRLING, CLANKING, bonging, approximately accurate reminder of mortality. This one featured a dancing skeleton and an ominous Turk as well as an astrological calendar that recreated the motions of the planets. It was perched above the Old Town Hall, a few steps off Staromestske Namesti, the center of the town that Charles IV, Holy Roman Emperor, king of Bohemia and Germany, had raised to splendor in the fourteenth century. Just across the square was the Tyn Church, where Tycho Brahe was buried. According to local lore, the designer of the clock, completed in 1490, had been blinded to prevent him from ever building a copy, a fitting legend for a city of intrigue and occupation in which defenestration had long since been perfected as a political act. Prague and the rest of Bohemia had fallen to the Austrians in the seventeenth century during the Thirty Years War, and it had remained in their heavy, baroque hands ever since, becoming an artistic and intellectual center second only to Vienna.

Prague has been called the city of a thousand gilded spires, but scarcely one of them can be seen from street level. The city is a labyrinth of crooked cobblestone passages hemmed in by stone facades blackened by centuries of industrial smoke. Snatches of haze and light and architectural masterpieces like the old Estates Theater, where *Don Giovanni* was first performed, appear and disappear unpredictably. Navigation by blind reckoning is out of the question for a tourist here. The best vista is from the Moldau River, with its views of Prague Castle and St. Vitus's Cathedral, sprawled across the hilltop of Mala Strana, the lesser quarter, across the water from the Old Town. When the Einsteins arrived, a young clerk named Franz Kafka was living in the warren of streets below the hill, writing strange, mostly unpublishable tales. Kafka spent his whole short life (he died in 1924 at the age of forty-one) in Prague because, one is tempted to say, he couldn't find his way out.

The Einstein entourage of Albert, Mileva, Albertli, Tete, and Marija arrived in this gnarled rococo metropolis at the end of March and checked into a hotel. Their stay there may have been shortened by complaints about Eduard's crying, and after two days they moved into a newly renovated flat down by the river in the district of Smichov, on the castle side of town. Although their only furniture was a baby carriage and a trunk, the flat did have electric lights—a first for the Einsteins—and an extra room for Mileva's mother (and later a maid).

Mileva was miserable almost from the start. The air was full of soot. Because brown water ran from the taps and left a disgusting sludge in the sink, they bought bottled water to drink, and Albert had to haul cooking water from a fountain up the street and then boil it anyway. Bathing was awkward and unpleasant. Bedbugs and fleas were everywhere. Once the mattress in the maid's room caught on fire.[1] Albert doused it with a bucket of brown water only to then find himself covered with fleas. Panicked, he jumped into a bath, for there had been outbreaks of bubonic plague and typhoid in the city only the year before. Besides the disgusting water, Albert had never really been fond of baths.[2]

Mileva was also socially uncomfortable in Prague. Three hundred years of imperial Austrian rule had created a rigid and intricate caste society, with Germans on the top and Czechs on the bottom. Until recently German had been the only official language, despite the fact that only five percent of the population spoke it. Except to sneer and tell jokes about them, the Germans would have nothing to do with the Czechs. Typically, some of the most vehement defenders of this apartheid were those who wanted to forget from whence they had come. Albert's colleague Anton Lampa was in fact the son of a Czech janitor; he made a show of refusing even to buy postcards that had both Czech and German writing on them. One consequence of this social schism was the splitting of Charles University into German and Czech divisions, which functioned as autonomous institutions with no contact between either faculty or students. As a Slav, Mileva had little reason to like Germans and she found it uncomfortable to be confined to the stuffy society of her people's conquerors.

Social life in Prague was complicated further by the fact that half of the Germans in town were Jews. In order to keep the Czechs down, the German gentiles had to make common cause with their Jewish countrymen in schools, shops, theaters, and cafes. At the same time, in order to keep their standing as good German nationalists, particularly in the eyes of southern Germans, where ancient ethnic tensions were rising again, the Prague Germans couldn't really afford to be too friendly to the city's Jews. Life for the

Jews was no less schizophrenic, for to the Czechs they were simply Germans, and thus oppressors.

Albert got to know the city in his first days there by walking around to pay courtesy calls on his fellow faculty members, one of many finely honed imperial traditions that quickly bored him. He charmed the people he did meet, cutting an exotic figure that seemed more Italian than Swiss, but alienating the rest when he never finished making the rounds. "Prague is marvelous, so beautiful that it alone would make a big trip worthwhile," he told Besso, hoping to lure him for a visit.[3] Its people, however, were strange, unfeeling, and cold, "a peculiar mixture of class-based condescension and servility, without any kind of goodwill toward their fellow men. Ostentatious luxury side by side with creeping misery." What he would later remember as well was the city's music. The air was thick with it—solemn pipe organs, joyous chorales, sorrowful ethnic laments, sturdy folk songs, modern concertos and operas, emerging from the churches and concert halls and following him down the street, a medley of human emotion, belief, and harmony.

History was waiting to ambush him on those strolls. It was on such an expedition, according to a biography by his future son-in-law, Dimitri Marianoff, that Albert may have begun to remember that he was indeed a Jew. Down a street in the Old Town and around a corner was the old Jewish Quarter. The ghetto, where 35,000 souls had once lived, had been razed only a decade earlier for hygienic reasons, leaving just a scattering of aging synagogues and the Jewish Town Hall. At the area's center was the high-walled cemetery, where Jews had been buried five deep for centuries under thin tombstones leaning like piles of toppling dominoes. Notable among the graves is that of Jehuda ben Bezalel, otherwise known as Rabbi Löw, the sixteenth-century inventor of the Golem myth, and a friend of Tycho Brahe's.

"The story of his race for a thousand years was told before him on the tombstones," wrote Marianoff, perhaps a bit dramatically. "On them were inscriptions in Hebrew with symbolic records of a tribe or a name. A fish for Fisher, a stag for Hirsch, two hands for the tribe of Aaron."[4]

Perhaps Albert's blood stirred. Perhaps he remembered with an embarrassed flush singing hymns to himself on the way to school.

FROM SMICHOV it was only a scenic twenty-minute walk across the river to Albert's office. The Institute of Theoretical Physics, as it was officially known, occupied the third floor of the university's Science Building, which stood next door to an insane asylum on Vinicna Ulice, a leafy street in the new town. George Pick, chairman of the mathematics department and a former student of Mach's, was conveniently one floor below, and Albert spent a lot of time trundling up and down the stairs asking him for mathematical advice.[5] Although he soon pronounced himself happy with the facilities and the surroundings,[6] particularly the library, Albert never grew accustomed to the doorman bowing and scraping and calling himself "your most obedient servant" every morning.[7] Likewise, the paperwork was also worthy of a three-hundred-year-old bureaucracy. Albert had to write to the Bohemian viceroy for permission to spend 55 crowns to get his offices cleaned.[8]

About a dozen students enrolled in each of Albert's spring lecture classes, and seven (six of them women) in his weekly seminar. Lampa warned him that they were academically weak, but Albert was convinced they would respond to good teaching, as they had in Zurich. "I shall always be able to receive you [in my office]," he told them at the start of the term. "You will never disturb me, since I can interrupt my own work at any moment and resume it immediately as soon as the interruption is past."[9]

Albert's office window looked out over a beautifully landscaped garden. As he sat there he couldn't help noticing a funny pattern. In the morning it was all women who milled about the shady park, in the afternoon all men. Finally he asked Pick about it and found out that the garden was part of an insane asylum. Thereafter Albert loved to tell the story. Those were the crazy people, he liked to say, who were *not* working on the quantum theory.

It was about then that Albert admitted defeat on the quantum: For better or for worse, he was ready to put that craziness behind him. More confused than ever about the foundations of the universe, and perhaps a bit sobered and humbled, he returned at last to the subject he had raised in his big relativity paper back in 1907 and had hardly written a word about since: the strange and ubiquitous phenomenon of gravity.

UNLIKE THE SUBJECT of electrodynamics, in which the brightest physicists in Europe were offering competing theories of the electron when Albert stepped in, gravitation was a quiet field. Newton had written the book on it two and a half centuries earlier, and with the exception of the almost trivially minute discrepancies in the orbit of Mercury there had been nothing to argue about ever since—until relativity arrived on the scene to complicate things.

In 1907 Albert's happy realization that a falling patent clerk does not feel his own weight had led him to suspect that he could use the deep connection between gravitation and inertia—the realization that inertial mass, the mysterious property of bodies that resists acceleration, and gravitational mass, which measures a body's attraction to other bodies, were identical—to investigate gravity. To Albert, this connection, which he called the *principle of equivalence,* meant that in some ways a gravitational field could be thought of as just accelerated motion, and vice versa. More important than that, however, it meant that gravity might be the key to a more generalized relativity theory, one in which the laws of physics would appear to be the same whatever the state of one's motion, whether sitting on a train or plunging down an elevator shaft.

Characteristically, one of the reasons that Albert had been drawn into gravity in 1907 was to try to resolve a logical anomaly that relativity seemed to have left in the picture of the world: Constant motion was relative, but inconstant motion—acceleration—seemed absolute. We can't tell how fast we are moving (or if we are moving at all) on a train as long as it runs smoothly, but if we are jerked back or forward in our seats we know for sure that the train has changed speed and has accelerated or decelerated.

Newton had used this difference in relativities between velocity and acceleration as an argument for absolute space in his famous discussion of the rotating water bucket. Take a bucket of water initially at rest and set it spinning. Due to friction with the sides of the bucket, the water within it will eventually begin to swirl as well, and centrifugal force will deform its surface into a concave shape. If the bucket is then stopped, the water will continue to swirl and maintain its concave surface. Why? Not because of the bucket, which is no longer moving, Newton concluded. Rather, he said, the water feels a centrifugal force because it is rotating with respect to absolute space.

But in the world as Einstein preferred to see it, there should be no such thing as absolute space or absolute motion. Gravity was important, then, not just for gravitation's sake but possibly for the connection it might provide to a more fully relativistic world. And as Albert set out to study gravity, in the years to come, he would have a double agenda—both to explain why apples fall and planets revolve and to generalize relativity to eliminate the last vestiges of Newtonian absolutism. It was a heavy load.

At the tag end of his 1907 relativity review, recall, Albert had exploited his new principle as a heuristic device to discover some novel features that gravity ought to have in a relativistic world. Clocks, he argued, should run slower or faster depending on gravity, and light beams would bend in gravitational fields. When he did the numbers, however, he had found that the predicted bending was minuscule even for a gravitational body as massive as the earth. He had thrown up his hands and disappeared into the quantum swamp for three years.

Now, returning to the subject, Albert realized that he had done his light-bending calculations for the wrong body. Though the earth was too small to exert any meaningful effect, the sun might indeed bend a light beam from, say, a distant star, by as much as a second of arc—enough of an angle to be astronomically noticeable during a solar eclipse.[10]

The prospect that these rather adventurous ideas might be testable after all inspired Albert to embark on a more thorough analysis of them, so that he could write a paper that astronomers could understand. In the process he was forced to convince them of a new and somewhat disturbing conclusion that he had only touched upon in 1907: Not only would gravity bend light, it would also *slow* it. The speed of light, that pillar of relativity, it seemed, was no longer constant, presenting what he called "serious difficulties" in a letter to his friend Laub.[11]

Albert's new arguments were based on a new *Gedankenexperiment* with an apparently transparent logic. He began by imagining a physicist in a windowless elevator.[12] In one instance the elevator is standing on the ground on earth. In the second instance the elevator is being pulled upward through empty, gravity-free space at a rate matching the acceleration with which objects fall downward on earth. Albert's proposition, based on the presumed equivalence of inertial and gravitational mass, was that the physicist inside could not distinguish between the two possibilities by any experiment he might conduct inside the elevator. In either case, for example, the physicist would feel the pressure of his feet against the floor. If he dropped a wrench, it would in each case clang on the floor—in one case because gravity was pulling it down, in the other because the floor was rushing upward toward it.

But what happened if the physicist dropped a light beam instead of a wrench? Albert first considered the case of a lamp shining down from the ceiling of the elevator that was being pulled upward. The light waves would take a small but certain amount of time to make their journey to the floor. During that time the elevator would be accelerating upward, so that by the time the light waves reached the floor, it would be moving faster than it was when the light left the ceiling. One consequence of this motion, according to relativity, was that the arriving light rays would appear to have more energy than they had when they left the ceiling. This extra energy, Albert showed, was just the amount that the light rays would have gained, according to relativity, if they had *fallen* under a gravitational field from the ceiling to the floor.

A different way of thinking about it made use of a phenomenon with which every gymnasium schoolboy was familiar. Motion affected the properties of waves. The well-known Doppler shift that made an approaching train whistle sound higher in tone and a receding one sound lower would make the frequency of the waves raining down on the moving elevator floor look higher (and the wavelengths shorter) than those produced when the lamp was at rest. According to Albert's analogy, gravity should have the same effect, and light from the lamp overhead should have a higher frequency and thus look bluer than that produced by a lamp standing on the floor. The increase in frequency, according to quantum theory, was equivalent to an increase in energy. Conversely, light shining up from below (from a lamp placed down the elevator shaft, for example) would look redder and less energetic than normal.

It was by following this train of thought to its end that Albert came to the conclusion that gravity would slow time. Light was presumably produced by atomic oscillations; the color or frequency of the oscillations depended on the kind of atoms—sodium, nitrogen, mercury—nothing else. Each atom was in effect a clock, and the speed of the clock therefore depended on its position within the local gravitational field—what physicists called its gravitational potential, which was just a number representing the amount of energy it could acquire by falling freely to the bottom, or source, of the field. Looking up, the physicist was always receiving signals from a clock running faster than his own, and so the light appeared bluer than his own; looking down, he was always getting light from a clock running slower than his own, and thus the light looked redder.

The most important sequence of this gravitational time warping was its effect on the speed of light. In accordance with ordinary relativity, each light wave travels at the regulation speed of light, c, with respect to its immediate surroundings and its own local clock. But if that clock is speeded up, Albert realized, that light will appear to be traveling faster than c when its speed is measured from elsewhere in the gravitational field where clocks run slower. Suppose, for example, that our physicist wants to measure the speed of light in the vicinity of his ceiling. In principle he could scribble a series of marks on the ceiling, set off a strobe flash and then click a stopwatch when the flash had passed some specified mark. Suppose it took a tenth of a second, according to a clock mounted on the ceiling, for light to reach that mark. Down on the floor, however, the physicist's stopwatch will read *less* than a tenth of a second, and therefore he will conclude that the light is traveling across the ceiling at faster than the regulation speed. The opposite will happen if he tries to measure the speed of light at any point below him in the gravitational field, when he will see light apparently slowed down.

From this hypothesis, deducing the bending of light was easy. When a wave front traveling through space hit a gravitational field, the part of the wave closer to the center of the field would slow down relative to the parts farther out. The light wave would curve around the center of the gravitational field like ocean waves around a sandbar. Albert calculated that a light ray from a distant star that just grazed the edge of the sun would be deflected by an angle of .83 second of arc—about one two-thousandths of the width of a full moon.[13] That was a small quantity, but astronomers were accustomed to measuring the positions of stars to a tenth of an arc second.

"Since the fixed stars in the portions of the sky that are adjacent to the sun become visible during total solar eclipses, it is possible to compare this consequence of the theory with experience," Albert concluded his paper. "It is greatly to be desired that astronomers take up the question broached here, even if the considerations here presented may appear insufficiently substantiated or even adventurous."

Another way to test the theory, Albert pointed out, was to measure the degree of reddening in light that had risen through a gravitational field. Here again, the sun was a perfect laboratory. The spectrum of sunlight was notched by dark bands at characteristic wavelengths where the atoms of various elements in the sun absorbed light. The slowing of time deep in a gravitational field of the sun should manifest itself by the shifting of these so-called absorption lines to appear at slightly longer and redder wavelengths (slower frequencies)—a so-called gravitational redshift—than they did in laboratories on earth.

Albert finished the paper and submitted it to the *Annalen* in June, and then he began looking for ways to enlist astronomers in his cause. He didn't have to look far. One of Albert's Prague colleagues, an astronomer named Leo Pollak, who was taking a course from him, was enthused about Einstein's new theory and took it upon himself to write to an astronomer he knew in Berlin, pressing upon him the need to pursue astronomical confirmation. His words fell on fertile ground. The astronomer who got Pollak's letter was one Erwin Finlay Freundlich, then twenty-six, ambitious, and bored with performing mundane practical astronomical chores for Karl Hermann Struve, the director of the Royal Observatory in Berlin. Freundlich had studied mathematics and astrophysics at Göttingen before being shipped off to the Berlin Observatory, and in Einstein's theories he sensed a career opportunity. Freundlich did some research and soon discovered that Hamburg Observatory had a collection of photographic plates taken during an eclipse in Algeria in 1905. He wrote to Albert announcing his intention to analyze the plates. Albert was gratified but justifiably skeptical about the chances for success, owing in part to the difficulties for correcting for refraction in the haze of electrical particles surrounding the sun.

"If only we had a truly larger planet than Jupiter. But Nature did not deem it her business to make the discovery of her laws easy for us."[14]

Nature, as it happened, was not making it easy for Freundlich, who obtained no conclusive results from the plates. But it was the beginning of what would be a long relationship between him and Albert.

THAT SUMMER was a hot and sultry one in Prague, and things were not going well for Albert. He had hoped to construct an entire theory of gravity based on the variability of the speed of light, but he was getting off to a slow start. "The formidable heat must have softened my brains," Albert told Zangger. In August his stomach flared up again—the old "poetic ailment" that he had complained about to Mileva back in 1899.[15] One of Zangger's medical students happened to visit and treated him for gastroenteritis, but Albert was still sick when Zangger himself came to see him a month later.[16]

Meanwhile, a familiar melodrama was playing itself out in the Einstein household. Albert's new salary afforded the family the luxury of a live-in maid, and they hired a woman named Fanni after Mileva's mother left town. Fanni turned out to be pregnant, with no husband, or even father to the child, in sight. Albert, who had once deferred to Mileva's father in a similar situation as an experienced man who knew the world better than "your overworked, impractical Johnnie," had to take charge.[17] After several fruitless attempts to find a place for the child either by adoption or in an orphanage, Albert and Mileva took it temporarily into their own home. "After some time it will go to its grandmother," Albert told Besso.[18]

That little reminder of their own younger days probably didn't have a very positive effect on Mileva's spirits, already dampened by her isolation and the strain of caring for her sickly baby as well as a sturdily healthy young boy. Mileva's characteristic taciturnity had only increased on their move to Prague. By all accounts she seemed to be slipping into depression, and with her limp and a growing disregard for her appearance, she projected a rather gloomy and withdrawn image. Philipp Frank, a physicist who succeeded Einstein in Prague and later became his biographer, met Mileva around this time and concluded that she was possibly schizophrenic.[19]

Albert was slowly coming to the same conclusion. The anger that had been festering in him ever since Mileva had exploded over the incident with Anneli had been transferred into a clinical distance and recognition that "She was thoroughly not mean-spirited, but distrustful, short on words and depressive, that is, gloomy," as Albert later told Carl Seelig.[20] "That is doubtless traceable to a schizophrenic genetic disposition coming from her mother's family. Also, a certain bodily deformation, which is obviously due to childhood tuberculosis, added to her fundamental psychic disposition."

The irony, of course, is that Albert had rejected or failed to carry relationships through with sunnier women—Marie, Anneli—in favor of Mileva. It was Mileva who had understood the black side of his own soul, as he once wrote. It was as if a certain edgy gloom was the coin of the realm of the unphilistine, of the intellectual life he wanted to live. Of course she had once helped him understand kinetic theory and electrodynamics as well. But that had been long ago.

Mileva was intelligent and was capable of a certain amount of insight, but was, as he once put it, "without ease at conceptions."[21] And, of course, Albert's work was becoming nothing if not conceptual. Now he had what he later called his *Rechnenpferde* (literally, "calculation-horses"), like Hopf, to do the heavy mathematical lifting.

In late September the annual convention of German scientists took place in Karlsruhe, which began several weeks of travel for Albert. After the meeting he went to Zurich, where he had agreed to give a series of lectures to Swiss middle-school teachers. He and Mileva had discussed her joining him in their old town, but it apparently never came to pass. "It must have been very interesting in Karlsruhe; I would have loved only too well to have listened a little, and to have seen all those fine people," she wrote him halfway through the trip. "It's been an eternity since we have seen each other, will you still recognize me?"[22]

With few friends and not much chance to get out, Mileva collapsed her life even more tightly around the children. Hans Albert had grown into a sturdy boy, with his mother's eyes and mouth. In photographs of the time, he seems like a dour child, his pouty lips usually downcast, with a bearing that implied graveness and responsibility. The birth of Tete seemed to have already inspired a fatherly attitude in him toward his little brother.[23] Almost alone among the Einsteins, Albertli seemed to be having a good time in Prague's smoke-scarred surroundings. Years later when he was a distinguished hydraulic engineer and authority on silt at the University of California, he recalled spending hours staring fascinated at the swirls and eddies in the Moldau River, often becoming so engrossed that he was late for dinner.

Like his parents, he loved music and had begun to play the piano. Another trait he might have inherited was his father's independence. "Father, we are quite alone, nobody can see or hear us," he said one day when he was about eight. "Now you can tell me frankly—is this relativity story all bunk?"[24]

Albert beamed whenever he repeated that story in the cafes along the Moldau, where he spent his time drinking white coffee and holding forth.[25] One of his steadiest companions was George Pick, the veteran Machist, who loved to reminisce about the old man, and especially to expound upon the threads he perceived between Machian philosophy and relativity—one of Albert's favorite subjects as well. Albert charmed his way into Pick's musical circle and began playing in a regular duo with the sister-in-law of a Sanskrit professor.[26]

By the end of the summer Albert's assistant Hopf had had enough of Prague and left for a job at the *technische Hochschule* in Aachen. He was replaced by Emil Nohel, a farmer's son from rural Czechoslovakia, who was studying mathematics because Lampa had convinced him that all the great problems in physics were already solved. Nohel spent hours at home with Albert, presumably helping him with calculations, although they never published anything jointly. Perhaps because they were both peasants at heart, Nohel became fond of Mileva. Because he was also Jewish he took it upon himself to educate Einstein on the predicament of the peasant Jews in Bohemia.[27]

On Tuesday evenings the action was downtown, only a few steps across the square from the Old Town Hall clock. There, under the hospitality of Berta and Otto Fanta, owners of the landmark Unicorn Pharmacy, a group of young Jewish intellectuals gathered weekly to discuss philosophy. This salon was an offshoot of a larger group of artists, thinkers, and musicians who frequented the Cafe Louvre to talk about art and literature. Their leader was a philosophy student named Hugo Bergmann, who would one day be a professor at the University of Jerusalem. Bergmann was an early Zionist of a secular sort, and he and his friends were interested in carving out a distinct Jewish culture that was neither German nor Czechoslovak. Among those who made appearances at the Cafe Louvre at various times were Franz Kafka; Max Brod, a novelist and later Kafka's biographer; the sculptor Frantisek Bilek; and the composer Leos Janacek.[28] Albert may have been drawn into the circle by Bergmann, who worked at the library and often walked home across the river with him. Although Albert had no interest in anything as parochial as Zionism, he enjoyed fraternizing with the eclectic mix of personalities.

The Tuesday night crowd at the Fanta house was a more scientifically and philosophically inclined gathering, resembling to Albert, no doubt, his beloved Olympia Academy. The core members included Brod, Bergmann, a philosopher named Baron Ehrenfels, and a sprinkling of mathematicians and scientists like Hopf. Kafka occasionally showed up to sit quietly in the corner. Brod referred to the meetings as *"Kant Abende,"* because Kant's work was one of the main continuing subjects of discussion. Albert joined up as the group was finishing a long encounter with Hegel and embarking on a two-year study of Kant's *Prolegomena* and the *Critique of Pure Reason,* which Albert had read and admired as a boy.[29] From time to time the menu was enlivened by presentations on scientific topics, including, apparently, relativity.

Albert cut a swath through other parts of society, as well. He took part in a series of public debates on relativity, "circus performances," as he called them, with a grim law philosopher named Oscar Kraus.[30] As far as the spectators were concerned, the high point of these evenings was usually when Albert played the violin.

Albert made a strong impression on Brod, so much so that the author later reportedly used him as the basis for the character of the great astronomer Johannes Kepler in his novel *The Redemption of Tycho Brahe.*[31] Kepler, who worked at the Prague Observatory around 1600, had struggled upward from abject poverty and abuse. Luckless in love, he was a mathematical mystic and Platonist who sought God in the geometrical underpinnings of the the world. Brod described him as a cool genius with a doglike horror of baths, who wore his innocence like a shield and seemed to take a perverse joy in the perplexities of nature. "The tranquillity with which he applied himself to his labors and entirely ignored the warblings of flatterers was to Tycho almost superhuman. There was something incomprehensible in its absence of emotion, like a breath from a distant region of ice. . . . He had no heart and therefore had nothing to fear from the world. He was not capable of emotion or love."[32]

"No, I am not happy, and I have never been happy," Kepler says at one point, with a dull obstinacy. Then he adds quite gently, "And I don't wish to be happy."

Brahe concludes that Kepler is so absorbed in his science and so innocent of intrigue that he does not even realize that he is happy. The world outside of science is a kind of dream to him.

"You are this man Kepler," Walther Nernst, the egocentric chemist in Berlin, is reported to have told Albert when he read the novel.[33]

As for intrigue, it just seemed to transpire around Albert, like water swirling around a rock in a stream. By the time he came back from Karlsruhe and Zurich, he was involved in maneuvers and negotiations for a new job.

Earlier in the summer, Albert had been offered a job at Utrecht University, in Holland, but had refused it, having just moved to Prague. Apparently his initial refusal of the Utrecht position had not been convincing enough, and his name had been brought up again.

Utrecht, next door to Leiden, where Lorentz worked, would seem not a bad place to do physics, but Albert had his eye on another prize. Zangger had hopes of securing Einstein a position at the Swiss Polytechnic, and had made Albert promise to inform him in advance if he received any offers to leave Prague. Albert quickly wrote to Zangger and told him, "I am now giving really serious thought to the call to Utrecht."[34]

Albert wrote back to Willem Julius, his contact at Utrecht, teasing him with questions about pensions and moving expenses, while explaining that he had promised the Polytechnic the first chance at making him an offer. Whether there actually was such an arrangement or it was just a fiction to enable Zangger to see if he could produce a counteroffer is unclear. Although he was neither a physicist nor on the faculty of the Polytechnic, Zangger's high standing in the medical community gave him easy access to the Swiss authorities in charge of the institution.

In fact, Zangger had already taken steps earlier that summer after sensing, perhaps from Grossmann, that Einstein was uncomfortable in Prague.[35] But Robert Gnehm, the president of the university, thought that Albert was an unnecessary luxury, and not a very good teacher either. Undaunted, Zangger went over his head to Ludwig Forrer, the member of the Swiss Federal Council, the Bundesrat, in charge of education, who had the idea of obtaining Einstein as a "gift" to the faculty. He had sent Zangger to Prague to discuss it with Einstein in August when Albert was ill.

In October, while Albert was lecturing in Zurich, Zangger bolstered his case to the council with a memo, pointing out that, as a theorist, Einstein needed neither a laboratory nor an assistant, and that contrary to prevailing opinion he enjoyed teaching.[36] "He is not a good teacher for intellectually lazy gentlemen who only want to fill up a notebook and learn it by heart for the examination; he is not a fine talker, but whoever wants to learn how to develop his ideas about physics in an honest way, from deep within, how to check all premises with circumspection . . . he will find in Einstein a first-class teacher." Zangger followed up his defense by taking Einstein to Bern to meet Forrer in person.

Albert returned to Prague to find Julius and Utrecht clamoring for an answer, and he begged for a few more weeks to make his decision, assuring Julius that he did not think the Polytechnic would ultimately come through with an offer. A few days later, however, Zangger forwarded an encouraging telegram that he had received from Forrer. Albert handed it to Mileva without opening it. "To enjoy the friendly attitude of and the trust of a man like Forrer is a great pleasure for me, even if I must admit to myself that his trust is not based on anything real," Albert replied to Zangger.[37] The next day, however, Gnehm protested Albert's appointment to the Bundesrat, and back in Zurich the professors were also stirring, angry at having been left out of the decision making.

That is where matters stood—Einstein playing Julius off against Zangger playing off Forrer playing off the Polytechnic—in late October when Albert boarded a train for Cologne and thence to Brussels, Belgium, and Nernst and Solvay's long-awaited "scientific congress" about the quantum problem. Preparing for this congress had recently taken up the rest of his time that had not already been consumed with lecturing, traveling, and conniving with Zangger. He had almost nothing to show for the last three months on his theory of gravitation. The quantum, it seemed, would not leave him alone.

"Don't be angry because of my long silence. This time I do not feel guilty, because I really did not have a single free moment," Albert told Besso, running down his list of recent travels and lectures. "Add to this the many shoptalks and personal obligations! But now—once the witches' sabbath in Brussels is over—I will be my own master again, except for my courses."[38]

Halfway to Cologne, in the middle of the night, Albert found some ham and a few apples that Mileva had packed for him as a midnight lunch. A wave of tenderness and affection swept over him. He ate ferociously.

15

THE WITCHES' SABBATH

AT SOME POINT DURING THE WEEK OF OCTOBER 29, 1911, TWENTY-THREE men and one woman paused in their deliberations to pose on one side of their giant conference table in the Hotel Metropole for a historic photograph. Their ringleader, Walther Nernst, perched smugly on a chair at one end of the assembly with his feet extended, bald and looking hard-boiled in his pince-nez and drooping moustache. Planck stands in the back row with the air of a mild clerk, while Lorentz, bushy-bearded, and looking a little hollow-eyed, stares straight ahead over the table, which is littered with papers and books. The tall New Zealander Ernest Rutherford is grinning open-mouthed into the camera next to his English colleague, the dapper-looking James Jeans, the only clean-shaven man there. Marie Curie slumps over the table, head propped up by an elbow, huddled over a piece of paper with Poincaré, who is gray and slightly disheveled, his face in the shadows. Albert stands behind them looking callow and distant, thumb and forefinger clenched as if making a silent point to himself. Fully a third of the subjects of that photograph either were or would be Nobel laureates. They had been paid 1,000 francs apiece to appear in Brussels to try collectively to shed some light on the quantum problem. The world had never seen a gathering as intellectually elite as the First Solvay Congress, nor had physics ever seemed so tenuously poised between deliciousness and disaster.

By then, the scientific community (with the exception of Mach) was approaching a consensus on at least one aspect of the nature of microscopic reality—namely, the existence of atoms. Albert had been one of the heroes in that effort. In his 1905 paper on Brownian motion he had argued that the jostling of atoms or molecules would cause small particles suspended in a fluid to dance around. If these motions could be accurately measured, it would be possible to calculate Avogodro's number, which told how many atoms there actually were in a gram of material.

In the years following, a group of Parisian physicists had taken up the work. Paul Langevin, a young theorist who taught at the Ecole de Physique et Chimie, had redone Albert's calculations, getting the same result more gracefully. His friend Jean Perrin set about measuring the Brownian motion and calculating Avogodro's number, getting an answer that differed from Einstein's by only about 15 percent.

Perrin's work, along with continuing investigations of radioactivity and the electron, had pretty much ended the atomic wars. Even Ostwald, who had vehemently debated Boltzmann on the topic, conceded in 1908, admitting that the results "entitle even the most cautious scientist to speak of an experimental proof for the atomistic constitution of space-filled matter."[1]

Atomism may have triumphed as a theory of matter, but did it apply to radiation and energy as well? The Solvay Congress was the first scientific meeting to be arranged around one particular subject, namely the threat and promise of quantum theory.

Nernst had asked a select group of Solvay invitees to prepare reports on different topics, ranging from the status of atomic theory to derivations of the Planck formula, which would serve as a de facto agenda for the week's discussion. Lorentz, naturally, presided as chairman, guiding the discussions with his usual tact and scientific acumen, smoothly switching from Dutch to German to French. "He is a living work of art!" Albert gushed to Zangger.[2]

Albert had arrived in Brussels with a sense of scientific obligation and curiosity about the characters with whom he was about to spend a week, but with no illusions of real hope of progress. He himself, as he kept saying, had already given up trying to understand the basis of the quantum, and he was anxious to get back to his new fledgling theory of gravitation.

As the week wore on, Albert began to feel that he had heard all the themes before. There was Planck, for example, stubborn as he was honest, still trying to maintain that the quantum condition only applied to the emission and absorption of light, not the light waves themselves. Albert had been assigned to report on his quantum theory of specific heats. When he finally spoke, however, he expanded his talk into a general survey of paradoxes and mysteries, discussing and dismissing the now-familiar arguments by which Planck and others sought to dodge the quantum bullet. "These discontinuities, which we find so off-putting in Planck's theory," he insisted, "seem really to exist in nature."[3]

As he later said, it was like listening to the lamentations on the ruins of Jerusalem.[4]

Meanwhile, in the corridors and during coffee breaks the talk swirled around relativity. One might suppose that Albert, the inventor of relativity, and Poincaré, who had almost invented it, would have a lot to say to each other. Albert was astonished and dismayed, however, to find that Poincaré didn't seem to understand the concept, but had continued to pursue his own garbled version of Lorentzian mechanics.[5] "Poincaré was simply negative in general," Albert reported to Zangger, "and, all his acumen notwithstanding, he showed little grasp of the situation."[6]

Albert found himself having more in common with the younger members of the French contingent, Langevin, Perrin, and Curie, who had been so instrumental in verifying his atomic theories, and who were all good friends—too good, as it would turn out. He told Zangger that he had become "quite enchanted with these people,"[7] and was particularly impressed by Langevin, a tall, slender man with a handlebar moustache, who had been one of the first to respond to relativity and who had published one of the earliest popular expositions on the subject in 1911, *L'Evolution de l'espace et du temps.* Albert later paid him the ultimate compliment by saying that if he, Albert, had not invented relativity, then Langevin would have made the breakthrough.[8] While the conference was going on, however, Langevin achieved a very different and unwanted kind of fame. He and Mme. Curie were embroiled in a scandal that threatened to overshadow the more serious side of the meeting, at least as far as the public was concerned.

Marie Curie was already famous, the role model for every female scientist in Europe, as well as controversial, before she arrived in Brussels. The year before, she had been rejected by the all-male French Academy of Sciences amid an impassioned debate about the sacredness of tradition and French womanhood. Curie had been castigated in some circles as immodest and unfeminine for even attempting to enter the Academy. These criticisms were mild compared to what happened when the world found out she and Langevin were having a secret affair.

Marie had been widowed in 1906 when her husband, Pierre, with whom she had shared the Nobel Prize, was run over and killed by a carriage in the street. Langevin, Pierre's former student and protégé, had been unhappily married for some time. His wife, Jeanne, beat him and stole his personal correspondence, but was also the mother of his four children. He had once turned up at the laboratory covered with bruises and admitted to Perrin that his wife and mother-in-law had attacked him with an iron chair. Drawn together by mutual grief after Pierre's death, Marie and Paul began to spend time together and eventually became intimate, keeping a pied-à-terre near the Sorbonne for trysts. When Langevin's wife learned the truth she threatened several times to kill Curie, but settled for dispatching a bagman to burgle the lovers' apartment and steal a number of incriminating love letters.

It was just as the scientists were gathering for Solvay's congress that

Langevin's mother-in-law began showing those letters to a reporter from *Le Journal*. "The fires of radium which beam so mysteriously . . . have just lit a fire in the heart of one of the scientists who studies their action so devotedly; and the wife and children of this scientist are in tears," said the resulting story, which accused Langevin and Curie of having eloped to Brussels. The press was soon in full cry. In defiance of accepted practice, one newspaper even published the love letters themselves. Langevin challenged its editor to a duel, but neither of them could bring himself to fire his pistol at the other.[9]

This hue and cry obscured that fact that Langevin and Curie had already stopped seeing each other in the wake of the death threats and the possibility of public exposure. From his ringside seat in Brussels, Albert had, he claimed, detected no signs of a special relationship between them. The "horror story" in the newspapers, he insisted at first, was nonsense. Even if they were in love, which he doubted, Mme. Curie and Langevin could get together in Paris anytime they wanted—they didn't have to come to Brussels. Whatever the true case, his dispatches lacked any indignation on behalf of the allegedly betrayed wife and children.

To him, Marie seemed scarcely capable of being a seducer. "She is an unpretentious, honest person with more than her fill of responsibilities and burdens," he told Zangger. "She has a sparkling intelligence, but despite her passionate nature she is not attractive enough to represent a danger to anyone."[10]

Later that very same week, Curie learned that she had won the Nobel Prize for chemistry. The Swedish Academy suggested to her that it might be best if she were to decline to attend the ceremonies in Stockholm and to delay accepting the award until the Langevins' upcoming divorce trial cleared her name. Curie announced that she was going to Stockholm anyway, on the logical grounds that the prize was for her professional achievements, not her personal ones. This only fanned the flames of vilification in the press, which did not hesitate to note that many of Curie's supporters had also been Dreyfusards only a few years earlier.

Mileva might have for once thanked her lucky stars that she wasn't a famous woman physicist after all.

Back in Prague, Albert was again outraged on Marie's behalf. "I am convinced, however, that you hold the rabble in contempt, whether they feign reverence or seek to satisfy their lust for excitement through you," he wrote her.[11]

It is intriguing to consider what effect the public hanging of Curie and Langevin might have had on Albert and his feelings about his own strained marriage. Albert clearly identified with the suffering Langevin, who was

only a few years older than himself, even to the extent of imagining him as
the inventor of relativity. Paul and Marie could have been Albert and Mileva twelve years earlier, embarking on their great bohemian adventure. But
if Albert felt he too had martyred himself for love, it had long since become
clear that Mileva was no Marie Curie. Marriage, Albert once remarked sar-
donically, was the attempt to make something lasting out of an accident.
"In ten years . . . she will no longer satisfy you; you will find the marriage
an unbearable fetter; she will become insanely jealous." So he would write
to his own son a decade hence to attempt to dissuade him from marrying a
certain older woman.[12]

In the wake of Albert's return from Brussels, one other event resonated
with echoes of his past. That same November Marie Winteler married Albert Müller, the manager of a watch factory in Büren, outside Bern. Nine
months later they had a baby boy; Albert remarked to Besso that he was
therefore an uncle, of sorts. The guilt and the strange vulnerability he had
for so long felt toward her finally began to fade. "I frankly welcome Marie's
marriage," Albert said. "With it a dark point in my life dwindles. It is all
now as it should be."[13]

WHAT CAME TO BE known as the First Solvay Congress thus cast long shadows personally and stylistically, as well as scientifically. It was followed by a
series of biennial summit meetings that ran into the 1930s. These congresses
helped inaugurate a new style of international celebrity science: conferences jointly sponsored by corporations, government agencies, and universities at resort hotels, organized around a particular problem or field,
drawing experts from around the world. Although today these scientists can
be seen scribbling their talks on airplanes as they jet from enclave to enclave,
the world in 1911 was much smaller, and they rode trains from Berlin to
Leiden to Göttingen to Zurich to Paris to Vienna in upholstered comfort,
sipping coffee, changing currency and language at each frontier as casually
as a man trading his scarf for another one. Whereas Albert once had been a
man without a country, he was now a member of a society almost above
country, with a passport stamped by God—or at least Lorentz—admitting
him to the company of saints, sinners, feuders, egomaniacs, neurasthenics,
and worshipers and granting him all the privileges accruing thereto, namely
the right to make pronouncements about the universe and then to fall flat
on his face.

Of all the pronouncements in Albert's young career, none had been
greeted with more incredulity and had survived against higher odds than his
suggestion that light was composed of quanta. In the decade leading up to
the Brussels meeting, no one had done more to breathe fire into the strange

embers that Planck had discovered. As much as the Solvay Congress marked Albert's debut on the international scientific stage, it served as the formal debut of the quantum, enshrining it as the central scientific mystery of the modern age. The Solvay conferees did not solve the mystery of the quantum, but in the aftermath of the meeting a series of advances showed just how strange and deep the quantum mystery was, and how prescient Albert had been to worry about it.

It fell to the grinning New Zealander in the group portrait, Ernest Rutherford, to take the next step inward toward the quantum. Although atoms were now acknowledged to exist, their constitution was still a mystery. According to the most favored theory, sometimes known as the Plum Pudding model, put forth by J. J. Thomson, an atom consisted of a large, amorphous ball of positive electrical charge studded with thousands of tiny negatively charged electrons. By the time of the Solvay Congress, Rutherford had already been getting indications that this prototype was incorrect, but it was too early in his research to publish any formal report. At his laboratory in Manchester Rutherford had been experimenting with a form of radioactivity called alpha rays, which were actually positively charged helium atoms that came shooting like bullets out of the Curies' radium samples. When he in turn aimed these "bullets" at a sheet of gold foil most of them passed through it as if nothing stood in their way. About one in eight thousand, however, bounced straight back.

Rutherford concluded and announced early in 1912 that the atom—and thus all matter—was in fact almost entirely empty. His atom was structured like a solar system. Squatting at the center of a vast amphitheater of inner space was a tiny, dense, positively charged kernel that he dubbed the nucleus, which contained almost all of the atom's mass, and was orbited by buzzing electrons.

Nobody took Rutherford's model very seriously, because it violated the laws of physics: Electrons whirling around under the intense electrical forces inside an atom would radiate away all their energy and crash down into the nucleus within a fraction of a second. The existence of our own bones, let alone the Pyramids, suggested that matter is stable and hence could not be subject to such destruction. Rutherford's theory was not even discussed at the next Solvay meeting in 1913. By then, however, it was being investigated by a shy young Dane by the name of Niels Bohr.

The son of a physiology professor at the University of Copenhagen, Niels had grown up in the shadow of his younger brother Harald, a brilliant mathematician and halfback on the silver-medal-winning Danish Olympic soccer team. Niels had the unfortunate habit of only being able to formulate his thoughts by speaking them aloud, which led him to expound in long, rambling, complex sentences that hardly anybody could follow. Moreover, he mumbled. About the time of the First Solvay Congress, Bohr, at the age of twenty-five, had gone to Cambridge on a fellowship sponsored by the Carlsbad Brewery to spend a year with J. J. Thomson. Unable with his stumbling English to communicate with Thomson, who had a brusque, impatient manner, Bohr gravitated to Manchester and the gregarious Rutherford, who was by then announcing his results about the structure of the atom.

Even Rutherford thought the model implied by his experimental results was beyond sensible explanation and warned the young Bohr away from trying to delve into it. Bohr was an adventurous thinker, however, and wondered if the idea of the quantum that Einstein had been advancing could be applied to the structure of matter as well as to light, and could thus keep the atom from collapsing. In a series of papers published in 1913 he suggested that the electrons in an atom were only allowed to occupy certain particular orbits. It was as if, in the solar system, planets could orbit the sun at the distance of the earth or of Mars, but not at any place in between. These orbits were like rungs on an energy ladder; each rung was a multiple of a quantum of energy. An electron could jump from a lower orbit to a higher one by absorbing a light-quantum of exactly the right energy, or it could hop down an orbit by releasing a light-quantum. This scheme explained why atoms couldn't collapse—the orbital rungs stopped just short of the nucleus.

More important, Bohr's scheme also explained why atoms absorbed or emitted light of only a few characteristic wavelengths, the so-called spectral lines, which also served as the fingerprints of the different elements in light from the stars. The orbital structure of each atom, and thus the characteristic frequencies it could absorb or emit, depended on the mass and charge of its nucleus. Applying these ideas to the hydrogen atom, Bohr was able to derive a famous formula that described the observed spectrum of hydrogen.

Albert learned of Bohr's work from a friend of Rutherford's during a meeting in Vienna in 1913. It reminded him of some ideas that he himself had entertained but had never dared publish, and he was properly encouraging, if skeptical at first. But as he began to appreciate the details, doubt turned to wonder. "Then it is one of the greatest discoveries," he finally announced.[14] The atom, the spectrum, the photoelectric effect, X rays, the blackbody radiation formula, and the quantum were here all rolled together in an intuitive master stroke, the kind of imaginative leap for which he himself was becoming recognized. He later referred to Bohr's atom in his autobiography as "the highest form of musicality in the sphere of thought."[15]

However genuinely laudatory Albert was about Bohr's work on the quantum atom, he must have felt a little like a lover hearing that a woman he had long ago courted unsuccessfully, and perhaps too delicately, has gotten married to another suitor. Ever since he had read Planck's blackbody paper in 1900 and felt the foundations sliding out from under physics, Albert's main goal—the problem that he cared about more than any other and kept coming back to again and again—had been to solve the quantum and save physics. His quantum treatment of blackbody radiation had been the one paper from his 1905 efforts that he had called "revolutionary." He told Otto Stern, his new assistant in Prague, "On quantum theory I use up more brain grease than on relativity."[16]

Perhaps if he had been more ruthless he might have cracked the puzzle, but he had hung back out of respect for the old ways. All that expended brain grease had prepared him to comprehend instantly the magnitude of Bohr's achievement and the brilliant tragedy that it was for physics. For what Bohr's theory could *not* do was say where the electron was while it was jumping *between* orbits, predict when an electron might decide to make such a jump, or even provide a framework in which such questions could be sensibly posed. None of the laws of classical electrodynamics or mechanics applied to the inside of an atom. As he once told Besso, "The more success the quantum theory has, the sillier it seems."

"I could probably have arrived at something like this myself," Albert told Bohr later when they finally met, "but if all this is true then it means the end of physics."[17]

IN THE WAKE of the Solvay Congress, the manuevering for Albert's academic services became even more intense and convoluted than it had already been. For his part, Albert was weary of Prague, and his enthusiasm for his students had long since departed.

When not attending to his official duties at the Solvay meeting, Lorentz in a bumbling way had tried to lobby Albert to accept Utrecht's offer, indicating "he would be pleased" if Albert accepted the appointment.[18] Out of propriety, though—he was not, after all, a member of the Utrecht faculty—Lorentz had not pressed the issue too strongly, which suited Albert, who didn't want to be in the position of refusing the man whom he professed to admire most in physics.

Shortly after returning from Brussels, Albert and Mileva made a quick and quiet trip to Zurich to sound out the possibilities of an appointment to the Polytechnic. What they learned must have sounded promising, because on the next day Albert wrote to Julius finally declining the Utrecht offer.

However, a snag developed almost immediately in Zurich, where the federal authorities had decided, in the face of Gnehm's continuing opposition to Einstein, not to intercede further in the appointment process.

Zangger was furious. Albert advised his friend simply to remove himself from the matter. "I just wrote a letter of refusal to Utrecht," he complained, "and the dear Zurich folks, you excepted, can kiss my . . ."[19]

Zangger, however, persevered, shuttling between Bern and Zurich with letters and arguments. He solicited recommendations from Curie and Poincaré, who praised Einstein's imagination while implying that he really hadn't achieved anything yet.[20] "The future will show, more and more, the worth of Einstein, and the university which is able to capture this young master is certain of gaining much honor from the operation."

Angry at having been left out of the decision-making, some members of the Polytechnic faculty resisted being so honored. "Great rage of many Poly professors about me," reads Zangger's account of the battle. "Attack of the professors of the Poly. The students are coming to my aid. A poster was made: A thin man with a moustache [that is, Zangger] squats behind the Poly building. He pulls on a rope that reaches to the railroad station. To this rope is attached "One Stone [*ein Stein*] a big stone, he pulls it to the Poly. The humor makes everything well."[21]

Hoping to keep the pressure on the Polytechnic, Albert wrote to Julius asking him to delay the announcement of Utrecht's second choice, Peter Debye, lest Zurich lose its fervor and keep him "forever in suspense."

In the meantime he sought to make his peace with Lorentz. "I write this letter with a heavy heart, like a person who has done some kind of injustice to his father," Albert told the Dutchman. "You have probably sensed that I revere you beyond measure. If I had known that you wanted me to come to Utrecht, I would have gone there."[22] Lorentz wrote back in a similarly conciliatory spirit. Fate this time had not been kind to them; if only he had written to Einstein sooner. "As I already said, I will console myself with the thought that you can achieve great things in Zurich, too."[23]

But two days later Lorentz wrote again, demanding to know how the negotiations with Zurich were going, asking for an immediate reply. This threw Albert into a panic. Lorentz clearly was about to lean on him to accept the Utrecht post after all. He wrote hurriedly to Grossmann asking him to get the negotiations speeded up *"prestissimo."*

Noting "with great joy that your friendly attitude toward me has remained unchanged," Albert nevertheless assured Lorentz that arrangements with Zurich were moving right along. Moreover, he added, he had a personal obligation, namely, Zangger, to see it through with the Polytechnic.[24]

So why was Albert conniving so hard to avoid going to Utrecht where he could work with Lorentz? Everybody loved Lorentz, which was perhaps the problem. Albert liked and respected Lorentz a little too much. When Lorentz spoke, everybody listened, including Albert. But what Albert needed was to listen to the small voice and look at the pictures in his own head. Friend after friend has testified to the sense of aloneness that Einstein projected. Albert was the eternal outsider. He knew what it was like to be smothered. His whole life had been a flight out from under the influence of domineering women like his mother, drill-instructor teachers, and over-bearing mentors. On the whole, he had thrived as an outsider. In his heart Albert knew that he was a man without defenses—helpless to the blandish-ments of sex, of a groaning table, or of a melody, of an elegant idea, or of suggestions and demands—easily led by the nose, he feared. As he said in his autobiography, by virtue of possessing a belly a man was condemned to partake of that opera. His moments of surrender had only strengthened his resolve to guard against the honeyed touch.

In rejecting the call to Utrecht, Albert told Lorentz that he felt as if he had turned down his father. But he was more comfortable dealing with Lorentz and the other lions of physics at a distance; the *idea* of a mentor was more palatable than the mentor himself. Only in some psychic recess out of the wind, it seems, could he remain himself.

For Albert, every season was the season of the witch.

JUST BEFORE Christmas Albert slipped back down to Zurich and finally closed the deal with Gnehm. "This is really a good piece of news!" said Paul Habicht, who had just successfully demonstrated the *Maschinchen* to the Berlin Physics Society. "If you come to Zurich, then there will again be a mensch in Europe."[25]

Shortly thereafter Zangger announced that a telegram from the Federal School Council was on its way to Einstein. Albert was to be appointed pro-fessor at the Polytechnic for 11,000 Swiss francs a year, a substantial raise over his old Zurich salary. "Although my life is not going so smoothly," said Zangger, "I still feel like standing on my head—dancing with joy."[26]

So, for that matter, did Mileva and the "two little bears."

Three days later Albert submitted his resignation to Count von Stürgkh and the German University. It was at this point that Lorentz decided to re-veal that he was planning on retiring from his chair at Leiden, and that the faculty there could conceive of nothing finer for the university or for physics than if Albert would take up the position as Lorentz's successor. For Lorentz personally, it would be a dream come true to have young Einstein as a colleague and be able to hear all about his ideas and new work. "I know, of course, that there may be circumstances that could make all of this just a beautiful dream."[27]

Luckily for Albert, those circumstances were now just about ironclad. Thus safely prevented from actually considering Lorentz's proposition, Albert could afford to express honestly his feelings—a little. "And now the most admired and the dearest man of our times offers me a place close to him, in that he holds out to me the prospect of a friendly personal relationship. I can think of nothing more beautiful. . . . However, to occupy your chair would be something inexpressibly oppressive for me. I cannot analyze this in greater detail, but I always felt sorry for our colleague Hasenöhrl for having to occupy Boltzmann's chair."[28]

Lorentz accepted Albert's rejection gracefully, and within a year ceded his chair to Paul Ehrenfest, the young Austrian theorist who had taken up relativity. Ironically, the appointment actually strengthened the ties between Albert and the Leiden circle, because by then Ehrenfest and Einstein were close friends.

Ehrenfest, a voluble man with a ragged beard, was an effusive writer who kept voluminous journals and penned long letters to *"Lieber lieber Einstein."* Born in Vienna, he had been awarded his doctorate under Ludwig Boltzmann in 1904 and was a master of statistical and atomic theory. His Russian wife, Tatiana, had studied mathematics at Göttingen, and after their marriage they had eked out a living as independent scholars, producing a series of major papers on statistical mechanics while they shuffled from Vienna to Göttingen to St. Petersburg, depending on their inheritances for income. Ehrenfest's search for a regular job was complicated by his religious status. A secular Jew, he had been forced by Austro-Hungarian law to renounce his faith when he married Tatiana, who was raised Christian. Unlike Einstein, he now stubbornly refused to reenroll in Judaism simply to get work. By 1911, however, he was getting desperate.

Ehrenfest and Einstein had already corresponded and crossed swords in the literature regarding relativity. Ehrenfest was interested in the seemingly paradoxical effects that relativistic contractions could have on so-called rigid bodies when they moved. In 1909 he had asked: What happens to a spinning disk? Shouldn't its circumference contract while its radius stays the same, thus violating the laws of geometry? It was a question that would have a profound influence on Albert's desire to generalize relativity.

When Ehrenfest decided to travel through Europe in search of a position, he wrote asking if he could visit Einstein en route. Albert wrote back inviting him to stay with them on the way back from his trip. It was on this journey that Max von Laue warned Ehrenfest not to let Einstein talk him to death.

Albert and Paul became fast friends almost instantaneously. When Ehrenfest arrived in Prague late in February, Albert and Mileva met him at the train station and took him directly to a coffeehouse, where the three of them engaged in small talk.[29]

When Mileva left for home the conversation immediately turned to science.

PART FOUR

Gravity's Rainbow

Why shouldn't one let the servants live
happily in the madhouse?

Albert Einstein to Heinrich Zangger, 1915

16

THE JOY OF FALLING

AS 1912 DAWNED, ALBERT EINSTEIN, AT LEAST IN PROFESSIONAL SCIENTIFIC circles, was certifiably a rising star. Wilhelm Ostwald, the Austrian chemist who hadn't believed in atoms, had already twice nominated him for the Nobel Prize, for the discovery of relativity. His newfound fame also meant that Albert's commitment to moving to the Polytechnic was continually being tested. The Austrians, for example, had let it be known "under the table" that 20,000 crowns a year were waiting for him if he came to the University of Vienna, and there were recent murmurs from Berlin.

The Prague newspapers made the usual fuss when it was announced that Albert was intending to leave the German university to return to Zurich. Eventually he had to publish a statement in the local newspaper denying that he was leaving because of religious or ethnic discrimination or that a new laboratory had been built for him in Zurich. Rather, he explained, it was simply that, as a "paterfamilias," he felt that Zurich with its lake and mountains was a more congenial place to raise his family. And, he added, he would consider offers to return to Prague if he should ever leave Zurich.[1]

Technically, as an Austrian civil servant Albert had to petition the emperor to be released from service. In the meantime he was assigned to the committee to search for his successor. Albert's new friend Ehrenfest ranked high on the subsequent list of recommendations, but highest of all was a young Viennese physicist and philosopher by the name of Philipp Frank. Born in Vienna in 1884, Frank had been one of Boltzmann's last students and as a *Privatdozent* was a regular in Vienna coffeehouses with other up-and-coming young intellectuals talking science and philosophy in what would become the famous Vienna Circle.

Frank had made a name for himself with a 1907 paper about cause and effect in which he argued that causality was a convention rather than a scientific conclusion, and could therefore never be experimentally proved or refuted. His argument drew the attention of Vladimir Lenin, among others, who attacked Frank as a Kantian idealist in a 1909 essay.[2] Albert, who also read the piece and liked it, sent a letter (which apparently no longer exists) back to Frank recasting the argument in more pragmatic terms. Disproving causality would always be problematic, he said, because you could never be sure that you had properly accounted for all the effects in any given situation.[3]

In 1910 Ernst Mach decided that he lacked the mathematical background necessary to understand relativity and made it known that he wanted someone to explain to him the theory that he had allegedly grandfathered. Because Frank had gone on to write several important papers clarifying relativity, he was nominated and eventually wrote a little essay for Mach that managed to avoid the more abstruse aspects of Minkowski's version of the theory. In the process Frank became fond of the old man and a disciple of Machian thinking. In the 1920s he commuted from Prague to Vienna to take part in meetings of the Vienna Circle with philosophers like Rudolf Carnap, Moritz Schlick, Otto Neurath, and others who sought to carry on in Mach's spirit the belief that science must consist of empirically verifiable statements—to which they added the requirement of stringent logical scrutiny. Whole forests have been consigned to the printing press in the quest to explicate those principles, and an apogee of sorts was reached with Karl Popper's conclusion that a scientific statement is one that is falsifiable, that is, a prediction that can be tested and found wrong. If you ask the average physicist today what science is about and how it works, the answer will most likely be some variation on Popper's theme.[4]

At about the same time that Albert wrote to the faculty in Prague recommending Frank as his successor, an open letter espousing some of these thoughts was published in the *Physikalische Zeitschrift,* signed by thirty-two scientific luminaries, including Freud, Mach, Einstein, and Hilbert. Headlined *"Aufruf"* (Appeal), it called on scientists and philosophers "whose expectation is to reach by themselves valid knowledge only through the penetrating study of the facts of experience, to join a Society for Positivistic Philosophy." Although Frank was not one of the signers, he was clearly active and known in these circles, and was probably still too junior a scientist to be listed by name.

Frank got the job and, once in Prague, began to appreciate the power of Einstein's aura. His students told him anecdotes of a professor who was always available, no matter when they came to see him. Frank had his own experience with that characteristic. Once, planning a visit the two of them would make to the Potsdam Observatory, Albert suggested that Frank meet him on a particular bridge. Frank warned him that he didn't know the city very well and might be late, but Albert replied that he didn't mind waiting. "Oh, no," Einstein explained, "the kind of work I do can be done anywhere. Why should I be less capable of reflecting about my problems on the Potsdam bridge than at home?"[5]

What he was reflecting about, of course—from the time he left Brussels and through his remaining months in Prague—was gravity. Perched above the madman's garden, with Prague's new electric lights beating against the gloom of approaching winter, Albert could not help being struck by the contrast between his new, vaguely formed theories and the material he was of necessity reviewing for his winter mechanics course. "I am just reading the foundations of the poor dead mechanics, which are so beautiful," he wrote to Zangger. "What will their successor look like? I torment myself with this endlessly."[6]

Already the quest for a more universal form for the laws of physics had led him to postulate the equivalence of gravity and acceleration. That in turn had led to the slowing of time, and the bending and reddening of light. What next? Unlike the case of electrodynamics, there was little body of speculation or experiment on gravity, however contradictory or wrong, to serve as a signpost. And few other scientists were willing or able to follow his lead. His old friend Max von Laue wrote to say that, after reading and even lecturing on Albert's light-bending paper, he didn't believe it could be correct.[7]

The next step, Albert decided, should be to stop concentrating on special effects like light bending and to try instead to formulate a real, mathematically rigorous theory of a simple gravitational field. The simplest example he could think of was that of a static field around a lump of matter—a world, in short, with no moving parts. This was the gravitational equivalent of calculating the electrostatic field around an electrical charge.

In classical Newtonian physics as it had evolved over the previous two hundred years, the way to define a gravitational field of, say, the earth was to determine the potential energy of every point in space around it. The potential energy (or potential, for short) was a measure of the amount of energy an object would acquire if it was dropped from a given point. The farther, or "higher," this point was from the source of gravity, the higher its potential. Let loose in a gravitational field, an object (or a light beam) would go "downhill" in the direction of lower potential. Albert's problem, then, was to discover or invent an equation that related the gravitational potential to the distribution of mass, while at the same time incorporating the presumed equivalence of gravity and acceleration. Then he had to solve that equation.

Luckily, there was an obvious place to start.

From his earlier thought experiments with the man in the elevator, Albert knew that there was a simple relation between the gravitational potential and the speed of light. Albert proceeded to attempt to simply substitute the speed of light in a well-known classical equation for the potential. "I work like a *horse,*" he wrote Hopf at one point, "even though the cart does not always move very far from the spot."[8]

At first the results were gratifying. Simple and elegant features emerged automatically from the calculations. The speed of light was variable, as he had figured it must be. As a result, light rays were bent and clocks were slowed by gravity. In the absence of any gravitating masses, Albert's formulas looked like those of regular relativity, with a constant light velocity. By the end of February he was readying a paper for the *Annalen*. In his reports to friends, words like "staggering," "stunning," and "beautiful" were cropping up. To Ehrenfest, he said, "I really believe that I discovered a piece of the truth."[9]

In the meantime, however, he learned he had competition in the study of gravitation. Max Abraham had read Albert's light-bending paper and had been similarly inspired to publish his own theory. Like Albert, Abraham sought to define gravitation as the slowing of the speed of light. But while Albert sought to map the slowing throughout ordinary three-dimensional space, Abraham's equations were developed within the context of four-dimensional Minkowskian space-time. When Albert first read Abraham's paper, even as he was working out his own theory, he later told Besso, he had been bowled over by the clarity of his approach.[10]

After more scrutiny, however, Albert changed his mind. Abraham's theory seemed to have been created out of thin air on the basis of nothing but mathematical beauty. "Abraham's theory is completely wrong," he told his former assistant Ludwig Hopf. "We're probably going to have a big battle in the journals."[11]

At the end of February Albert finally sent his own paper off to Wilhelm Wien, the editor of the *Annalen*. He admitted that he was opening a Pandora's box by allowing the speed of light to vary and asking that the laws of physics be the same for observers in any state of motion. The simplicity and elegance of the original relativity theory was sure to be lost, and the notions of time and space themselves might become problematical.[12]

Two weeks later he had second thoughts; a number of his equations were turning out to be inconsistent with some of the standard truths of physics. In Einstein's new universe, it seemed, a pair of weights hooked together could spontaneously burst into motion on their own in violation of Newton's old law of action and reaction. In short, Einstein's theory was not in accordance with the sacred principles of the conservation of energy and momentum. He wrote Wien asking him to ignore the paper, but a few hours later wrote again and told Wien to go ahead and publish it. Parts of it might be wrong, he explained, but readers might find it interesting to see where the equations had come from.

Albert followed those letters with another paper explaining why the first one didn't ultimately work. The net result was that he had to add a term to his original so-called field equation. The nature of this adjustment was an early indication of the kind of looking-glass maze the marriage of gravitation and relativity was going to produce. One of the conclusions of ordinary relativity had been that mass and energy were interchangeable: A moving object gained mass, a radioactive one lost it. Even an aethereal light beam—one of Albert's *Lichtquanten*—by virtue of being pure energy, represented an equivalent mass and thereby could be bent by a gravitational field. What Albert had forgotten, or at least failed to account for in his first try at a field equation, was that the gravitational field *itself* was energy and therefore also represented some mass, which of course generated gravity of its own. Which, of course, added yet more energy and mass to the field, making it pull even harder, and so forth. Not even gravity could escape its own clutches.

Albert absorbed the lesson well. To avoid the faux pas of publishing a theory that violated the conservation laws, and thus common sense, Albert made it part of his mental checklist to make sure that future theories obeyed this constraint before loosing them on his colleagues.

It turned out, however, that he had traded one problem for another. Mathematically, the new adjustment to his theory had an unexpected and puzzling effect. Now it seemed that the equivalence of gravity and acceleration only held for infinitely small systems. This was mysterious, to say the least, since it was that very equivalence that had been the original basis of the theory. "It has not yet become clear to me why the equivalence principle fails for *finite* fields," Albert confessed to Ehrenfest.[13]

Max Abraham pounced on this development, responding to Albert's criticism of his theory with a vehement note in the *Annalen*. He argued that Albert's recent modification of his own theory had undermined the validity of relativity itself. "This most recent theory of Einstein's thus also rests on shaky ground."[14]

Albert quickly replied to the *Annalen* with assurances that relativity was in no danger of being subverted. His inability to make equivalence work outside of small systems was no reason to renounce the principle altogether. "No one can deny that this principle is a natural extrapolation of one of the most general empirical laws of physics. On the other hand, the equivalence principle opens up for us the interesting perspective according to which the equations of a relativity theory that would also include gravitation may also be invariant with respect to acceleration (and rotation). . . . I would like to ask all of my colleagues to have a try at this important problem!"[15]

Albert, after all, had his eyes on something grander than just a theory of gravity; he hoped to complete the Machian view of the universe that he had begun with relativity. It had been Mach's genius to insist that science eschew the abstract and metaphysical in favor of the concrete. The universe, space, and time, he argued, consisted simply of objects and their relations. In such a universe there should be no privileged observers or conditions; the laws of physics should be the same regardless of how fast you were going or whether you were falling off a mountain, sitting in an armchair, or spinning around in a carousel down by the Zürichsee. It was the quest to expand relativity into such a universal form of physical law that had led Albert to equate gravitation and acceleration in the first place. Gravitation was somehow the key to this relativistic view of the universe. And vice versa.

One of Mach's ideas in particular seemed pertinent to Albert. In his famous refutation of Newton's rotating bucket argument, Mach had suggested that rotation itself was relative. Remember the experiment: A bucket of water is set to spinning, say, on a turntable. Why does the water in the bucket form a concave surface and continue spinning if the motion of the bucket itself stops? Newton argued that it was due to centrifugal forces caused by the water's rotation with respect to absolute space. Mach, in turn, said the water was simply rotating with respect to the fixed stars, and suggested that the experiment might come out differently if the walls of the bucket were expanded to be several leagues thick. His implication was that the inertia of the water was somehow determined by some spooky interaction with everything else in the universe; if the walls were massive enough, the water might take its cue from them, instead, and remain flat. (Of course, speculating about the nature of this spooky interaction or about an unperformable experiment would itself be distinctly un-Machian.)

Although Mach never spelled it out any more clearly, the idea that the inertia of an object depended on everything around it became one of Albert's polestars in the quest for a sensible theory. Electrons, atoms, planets, the nearly massless test particles that starred in numerous thought experiments being dropped from imaginary towers or skittered through electromagnetic fields—they were all social creatures, citizens and creations of the universe of which they partook. This idea, which Albert later referred to as the relativity of inertia, and later called "Mach's principle," would one day play an important part in his attempts to construct a theory of the whole universe.

Early indications were that Albert was on the right track. In late March, after correcting his field equation for a static mass, Albert tackled the infinitely more difficult problem of the gravitational fields of objects in motion. He began by considering a point mass—a lonely atom, say—suspended inside a massive spherical shell. What happened to the atom when the shell was moved? Albert's calculations showed that the particle inside tended to be dragged along, although it had experienced no direct force—as if its inertial field were controlled by the surrounding mass. You could think of it as the gravitational analog of electrical induction, in which a charge on a conducting plate induces the opposite charge on another nearby plate. Or, alternatively, it could be viewed as a provisional confirmation of Mach's spooky ideas, as Albert pointed out in a footnote to his paper.[16]

On the other hand, the theory didn't work for things that were rotating. Einstein and Ehrenfest spent several letters discussing one simple example, of light bouncing off a rotating ring, but the answer, they agreed, kept coming out wrong.

By the time summer approached, Albert had made little progress, and Abraham was *still* on the attack, continuing to propose new versions of his own theory while challenging Einstein's. In September the following brief article appeared in the *Annalen:*[17]

COMMENT ON ABRAHAM'S PRECEDING DISCUSSION
"ONCE AGAIN, RELATIVITY AND GRAVITATION"

by A. Einstein

Since each of us has presented his point of view with the necessary degree of detail, I do not find it necessary to respond again to Abraham's present note. For the present, I would only like to ask the reader not to interpret my silence as agreement.

Zurich, August 1912

Albert loved a good fight as well, and surprisingly, this combat had not dimmed his respect for Abraham the physicist. In April, when the University of Zurich sought his advice on filling a vacancy (Einstein's old post) in their physics department, Albert recommended Ehrenfest and Laue, and then wrote again to say that he should have mentioned Abraham as well: "It is incomprehensible that this really important man is being avoided like the plague because of a few cocky sarcasms he indulged in a few years ago. Could you not consider bringing him to Zurich all the same? Of course I do not know whether he would accept."[18]

IT WAS A couple of weeks later, during the spring holidays in 1912, that Albert took a long-planned trip to Berlin. By now he had many friends in the city and a good deal of business to attend to there, scientific and otherwise. Emil Warburg, a physicist and president of the Physikalisch-Technische Reichsanstalt (where blackbody radiation had first been accurately measured), who had been corresponding with Albert about quantum photochemistry, wanted to talk to him about coming to work at his institute, and offered to put Albert up at his apartment for the week. Walther Nernst, the motive force behind Solvay, wanted to discuss his new results about specific heat, work that confirmed for Albert just how silly quantum theory really was. "How nonphysicists would scoff if they were able to follow the odd course of developments," he confided to Zangger.[19] Fritz Haber, head of the newly formed Kaiser Wilhelm Institute for Physical Chemistry and Electrochemistry, whom Albert had met earlier at a meeting in Karlsruhe, was eager to speak about quantum chemistry. There is no record of whether he also got over to the observatory to talk to Erwin Freundlich, who was then combing astronomical records looking for evidence of light bending.[20]

In the course of a week's stay, Albert had time to make all these stops and some personal ones as well. Berlin had been a nexus for the extended Einstein family. His mother had settled there briefly a few years earlier with his aunt Fanny and uncle Rudolf before moving back to Heilbronn as a housekeeper. Fanny and Rudolf were living in Schöneberg, an old middle-class residential area on the southern outskirts of the city.[21] Albert paid them the obligatory visit, during which he was reacquainted with his cousin Elsa.

Elsa, three years older than Albert, was about as close as family could get. Her and her younger sister Paula's frequent visits were part of the rosy image Albert still carried of his youth in Munich.[22] With blue eyes, curly cornsilk hair, and a *gemütlich* manner, Elsa was as light as Mileva was dark. Upon meeting her for the first time, Dimitri Marianoff, a frequent visitor to the Haberlandstrasse household who later married Elsa's daughter Margot, was struck by "her clear blue eyes, which throughout her life had the faculty of appraising people with an almost uncanny exactness."[23] (Then there is the testimony of Evelyn Einstein, Albert's adopted granddaughter, who typically refers to Elsa as "that social-climbing bitch.")[24]

In 1896, at the age of twenty, Elsa had married Max Löwenthal, a Swabian cotton merchant who lived in Berlin, even though his business was farther south, in Hechingen. They divorced in 1908, and since then Elsa had been living with her parents, along with her two daughters: Ilse, now thirteen, developing into a quietly headstrong, swan-necked beauty; and Margot, eleven, artistic and so notoriously shy that she would dive under a table when a stranger came into the room.

Elsa had become a force on the Berlin social scene, with wide acquaintances in theatrical, literary, business, political, and even scientific circles. On occasion she gave dramatic readings of German poetry in the theater. She was also locally famous for her vanity, refusing to be seen in glasses even though she had terrible eyesight. An oft-told tale had it that at one dinner party she had started eating a flower arrangement, mistaking it for the salad.

Albert and Elsa took a day trip together to the nearby Wannsee, one of many lakes and waterways that weave in and out of Greater Berlin. The result seems to have been instant romance. No details remain; upon her orders, Albert burned Elsa's subsequent letters to him as fast as they arrived.

We can only imagine Albert's state of mind at the time. He might have felt that it had been ten years since a woman had smiled at him and seemed happy to be with him—not needful, or disappointed, complaining, depressed, or demanding, merely happy. What weary frost-ravaged traveler fleeing his own particular wasteland has not stepped off the gangplank or train platform and scented the soft amaretto wind of some paradise and felt a new self stirring, has not thought to himself that he could be a better person, a nobler, wiser, healthier, warmer, sweeter man, a happier and richer, more just soul lingering here, basking, embraced . . . here, in the sunshine of your smile?

Albert was an easy catch. When it came to women he could be like a child. It was, Albert knew very well, one of his charms, a yin to complement the yang of his aloof aura, that distance that was like an electrical charge. You might never find out what was on his mind, but you could please him with a cookie, anytime. Striving and smoldering desire were not his style; it was not in his makeup to plot and scheme to get a woman, nor to record his conquests afterwards in little black books. He took what nature offered without resistance. It's a sure bet that there weren't any cookies being handed his direction in Prague. The way Albert saw it, Mileva had abandoned him. "I suffer very much because I'm not allowed to love truly," he told Elsa.[25]

As for Elsa, she had undoubtedly heard through her mother the saga of Pauline's antipathy to Mileva and the resulting tension in her aunt's side of the family. Albert strode back into her living room and her life at the age of thirty-three with a cocky squint and the knowledge that he already had something that the world couldn't take away from him, a far cry from the honest Biedermeier of her adolescence.

"Like coins struck off for some special occasions, faces like his arrive out of the centuries," Marianoff later wrote. "The first thing you see is that rare and powerful head. The power lies in the colossal splendor of the benevolence of expression which wraps the whole countenance in a kind of radiant mildness. The nose is not Hebraic, but wide and fleshy. The tenement of bone structure that holds the mold eases at the mouth, and from there to the chin it becomes decidedly feminine in its sensitiveness. But the eyes are liquid, racial, and unmistakable eyes of Jewry."[26]

As her married cousin, however, Albert was an unpromising match for Elsa. She was in a tight spot, dogged by her adolescent daughters on one hand and her parents on the other. Still, she found herself writing to him a week later, suggesting a stratagem by which they could keep in touch and avoid their respective household mail police.

Albert was already despairing of seeing her again. "How happy I was then today when I saw from your letter that you found a way that will allow us to stay in touch with each other," he exclaimed. "I am in seventh heaven when I think of our trip to Wannsee. What I wouldn't give to repeat it."[27] In fact, he was considering coming to see her again at the end of the semester, when everybody else would be away.

Apparently Elsa's letter was not entirely romantic; Albert's family and his relationship with his mother and with Mileva had also been ruthlessly dissected. She accused him of being henpecked, which Albert vehemently denied. "But I acknowledge that the sum total of what I do out of pity for her and for . . . myself in her presence creates such an impression. . . . Let me categorically assure you that I consider myself a full-fledged male."

As for his mother, Albert conceded that she was a sorry witch and unlikely to change at this age. "I used to suffer tremendously because of my inability to really love her. When I think of the bad relationship between my wife and Maja or my mother, then I must admit to myself, sadly, that I find all three of them quite unlikable, unfortunately! But I have somebody to love, otherwise life is miserable. And this someone is you; you cannot do anything about it, since I'm not asking you for permission. I am the absolute ruler in the netherworld of my imagination, or at least that is what I choose to think."

Hans Albert later recalled that it was when he was about eight, in the spring of 1912, that he first noticed discord and tension between his parents.[28] He was not alone. According to Michelmore's biography, Albert and Mileva's friends began to worry that Mileva's dark moods were both lasting longer and growing more frequent. She was far too introverted, and never talked about herself. Even in the bosom of the family, according to Hans Albert, she rarely had anything to say.

Whether this behavior was the cause or effect of Albert's being drawn to Berlin is a question best left to the reader. That Albert would spend the holidays in Berlin, a place full of Einsteins, instead of home with her and the boys could only have been wounding. That he came back whistling only exacerbated the situation.

Within a few weeks and the next exchange of letters, a more sober assessment of their situation was already beginning to settle in for both Elsa and Albert. Her next epistle saddened him. "We are both poor devils each chained to his ruthless duties," he answered.[29] Only confusion and unhappiness would result if they were to give in to their feelings for each other. "I would be very happy just to walk a few steps at your side or have pleasure some other way in your presence." His suffering was worse than hers, he averred, because she only suffered from what she did not have. Albert had to contend with Mileva. "I am as tortured as you and must always take a new approach so that I do not fall victim to bitter voices."

Two weeks later, in late May, only a month after their Berlin idyll, Albert wrote to break off the romance. "I have the feeling that it will not be good for the two of us as well as for the others any good if we form a closer attachment. So, I am writing to you today for the last time and submitting myself again to the inevitable, and you must do the same. You know it is not hardness of heart or lack of feeling that makes me talk like this, because you know that, like you, I bear my cross without hope."

He said he hoped that she would still think kindly of him; he promised to cherish always the memory of her warmth. "If you ever have a hard time or otherwise feel the need to confide in somebody, then remember you have a cousin who will feel for you no matter what the issue might be."[30]

17

KING OF THE HILL

ALBERT, MILEVA, AND THEIR BROOD MOVED BACK TO ZURICH IN August 1912, and took up residence at 116 Hofstrasse, a sweaty hike or a swerving trolley ride halfway to the clouds up the steep Zürichhof from the Polytechnic. As if reflecting Albert's steady rise in the world, each of his and Mileva's Zurich abodes was farther and farther above their old student-day quarters. Their new apartment was in a five-story stucco building, to-day mustard-colored with a matching wall and iron gate, with bay windows at its corners and a red-tiled mansard roof. It perched like a small castle across from a church on the brow of the hill. From its roof the view in-cluded the lake, the city, and the surrounding hills, off to the distant Alps. The family celebrated its homecoming with a day in the mountains. Mileva and Eduard picked mushrooms while Albert and Albertli explored glaciers. On the other side of that Alpine bulwark, however, the world was increas-ingly troubled.

In the years since Mileva had left her homeland to seek her education and career in Switzerland, the map of southeastern Europe had been re-drawn almost annually as the Ottoman and Austro-Hungarian empires continued to splinter and the Serbs pursued their vision of liberating and joining these splinters into an independent nation of the Southern Slavs. A snapshot of the situation circa the summer that Albert and Mileva returned to Zurich shows a crazy quilt of states, territories, and jurisdictions sewn haphazardly together by invisible threads of loyalty, treachery, nationalism, ambition, and whatever other qualities enrich the lives of nations. Greece had been independent since 1829. Serbia, landlocked directly in the heart of the Balkan Peninsula, had been granted independence at the Congress of Berlin after the Russo-Turkish War in 1878, as had Rumania and tiny Montenegro on the Adriatic coast. The Turks still owned a swath across the middle of the peninsula from Constantinople to the Adriatic that included Macedonia and Albania. Bulgaria, technically part of the Ottoman Empire, had been divided into three zones of different autonomy. The Vojvodina region, Mileva's old home, was still part of Austria-Hungary, as was Croatia. Bosnia had been occupied by Austria-Hungary but had technically remained part of Turkey until 1908, when the Austrians annexed it outright.

Most of the nations of Europe had long since staked some portion of their fortunes on the fate of these little patches of real estate. It began with Bismarck, who, in the late nineteenth century, seeking protection for the nascent German Empire, had entered into an alliance with Austria-Hungary, pitting Germany more or less squarely against the Slavic aspirations in the south. Italy, when it became a nation, had subsequently joined what would come to be known as the Triple Alliance. Concerned about Germany's growing influence, France had formed a loose alliance with Russia, which had its eye on gaining some sort of permanent access to the Mediterranean through the Bosporus; as a result of this unlikely union, dancers like the immortal Nijinsky from the legendary Ballets Russes, and Russian musicians like Stravinsky were soon running through the streets of Paris creating artistic and cultural havoc. Russia, as a Slavic nation, had ethnic ties to Serbia, as well as a common geopolitical interest in driving out the Hungarians. Finally in 1907, Britain, which was engaged in a competitive, expensive naval buildup with Germany, had agreed to set aside colonial rivalries and moderate a tradition of remaining aloof from the Continent to join France and Russia in the so-called Entente Cordiale—which of course only intensified Germans' paranoia that their aspirations were being held captive in an "iron ring."

The year 1912 brought these tensions to a crisis. Italy had attacked Turkey the year before and won several more pieces of the Italian peninsula to call its own. Emboldened by Turkey's weakness, Serbia, Greece, and Bulgaria had then launched their own offensive against the Turks. By the end of the fighting, Serbia had conquered a chunk of Albania, to which almost everybody had some sort of claim. If Serbia kept it, the Serbs would have access to the Adriatic, a prospect that Austria so particularly dreaded that it mobilized its forces to attack the Serbs, to whose defense Russia quickly came. A peace conference was called for December 17, 1912, at which Russia backed down, abandoning the Serbs for the second time in two years. Albania became another free state.

In Zurich these events were anxiously followed by Albert and Mileva. "If the Austrians will only remain calm; a conflict with Austria would be bad for the Serbs, even in victory," Albert remarked to their "Serbian heroine," Helene, back in Belgrade. "But I believe that the saber rattling does not mean much."[1]

Albert and Mileva resumed their previous routines, including the Sunday afternoon concerts at the Hurwitz house. "Here comes the whole Einstein henhouse," Albert would announce on the steps.[2] By now these gatherings had acquired considerable cachet in social and academic circles; important scientists and local celebrities began dropping by or seeking invitations to sit in with the professor and his friends. Hans Albert was old enough now to join in sometimes on the piano.

It didn't take long, however, for Lisbeth Hurwitz to notice that Mileva's heart was no longer in the occasions. At times she simply didn't show up, or if she did, she sat glumly through the festivities. "She is somewhat different since their time in Prague, but it will pass," Hurwitz wrote. "It's just remaining sadness from the unhappy time in Prague."[3] The Hurwitz diaries tell of repeated attempts to bring Mileva out of her shell by scheduling all-Schumann concerts every now and then. Sometimes the ploy worked, while at others the announcement was greeted by silence.

At home Albert retreated into his violin. When the weather turned cold, Mileva began to get pains in her legs, which she attributed to rheumatism. She had trouble walking and was terrified on the icy, hilly paths around town. Hurwitz recalled Mileva arriving for a concert one afternoon clinging desperately to Albert's arm.

Professionally, Zurich could hardly have been cozier for Albert. Upon his recommendation, his old friend von Laue had been hired for the vacant post at the University of Zurich. (Ehrenfest, meanwhile, had been appointed to Lorentz's old chair in Leiden, which prompted Albert to remark, "You now live in an incomparable scientific environment, which will also, of course, make terrible demands on you. When Lorentz offered me a job, it really gave me the creeps.")[4] There were frequent guest lectures at the Polytechnic, which became occasions to watch the young lions needling each other about relativity and the quantum affairs of atoms. "It is strange," Ehrenfest remarked from the audience after one of Albert's lectures, "that Einstein always gets the results he wants from his work."[5]

Albert had also brought his new young assistant, Otto Stern, with him from Prague. The son of a mill owner, Stern had received a Ph.D. in physical chemistry from the University of Breslau, with a stop in Munich and an apprenticeship with the lordly Sommerfeld along the way (time which, he admitted, had been devoted mostly to drinking beer). In Prague he had been welcomed like a brother by Albert, who seemed to have nobody else to talk to about physics there. When he was a grand old man in Berkeley fifty years later, Stern reportedly wept recalling "the beautiful days with Einstein in Prague."[6]

In Zurich those beautiful days continued. According to at least one report they included not just carousing and arguing about quantum theory, but occasional trips to brothels, which, outside of moving to Paris and getting a mistress, were probably the most socially acceptable way for a man to escape the frustrations of bachelorhood or an unhappy marriage.[7]

Stern, von Laue, and Ehrenfest were all future Nobel Prize winners, but arguably the most valuable person in Zurich, as far as Albert was concerned, was his old friend Marcel Grossmann, whose lecture notes he had borrowed as a young student, and who was now the chairman of the Polytechnic math department. Grossmann had only a tenuous grasp of physics. He once recalled that he had always felt uneasy when he sat down on a seat that was still warm from its previous occupant, until Albert eased his squeamishness by explaining that the heat was just molecules in motion—nothing personal. But Grossmann would prove to have what it took, at least mathematically, in relativity.

In Albert's last work on that subject in Prague, he had been floundering in mathematical mysteries and unexpected consequences of his determination to define gravitation as the speed of light. By the time he resettled in Zurich, he had abandoned that simple approach for a more powerful but more complicated conception. His new idea was that gravity was geometry.

It was a concept whose ramifications would prove to be literally cosmic, but Albert never explained in detail how he had made the leap. On the basis of Einstein's later correspondence and popular talks, however, John Stachel, the original editor of the Einstein Papers Project, has made a strong argument that this "missing link" in Albert's thinking arose from pondering a relativistic paradox known as the case of the rotating disk.[8] It hardly sounded like the kind of paradigmatic problem that turns the direction of physics; it was not a contemplation on the origin of the Milky Way, or even as poetic a conception as imagining riding a light beam. The problem was in fact rather humdrum and pedestrian, but Albert's genius was to be able to extract the miraculous from thinking—*really* thinking—about the ordinary.

The rotating disk was known as Ehrenfest's paradox, one of several conundrums that had been popping up in letters and in the hallways at places like the Solvay Congress ever since 1909, when the scientific community had begun taking relativity seriously.[9] The question was simple on its face: What is the ratio of the circumference of a spinning disk to its diameter? Any schoolchild would know the answer is *pi,* the sacred, irrational, and eternally incalculable number roughly 3.14159265 . . . , prescribed by the holy laws of Euclidean geometry. Pi was a truth as Platonic and transcendental as they came, except for the fact that Albert and the other young relativity theorists claimed that, from the standpoint of the inhabitants of the disk, spinning around like elves on a record turntable, that timeless truth wasn't true any longer. If a spinning elf measured the circumference and diameter of his disk, he would find that the ratio between them was greater than the prescribed 3.14159.

It was a subtle argument, and Albert spent the rest of his life answering letters from people who got it wrong the first time they tried to think about it. Imagine, Einstein explained on one such occasion, that you take a photograph of a spinning disk face-on.[10] In the picture, the edge of the disk will be a simple circle, to which the normal geometrical axioms apply. If we measure the circle, and its diameter is, say, 10 inches, the circumference will be about 31 inches. Because the disk is spinning, however, according to relativity, it must have undergone a contraction in the direction of its motion, that is to say, around its circumference. So its circumference, as measured in the photograph, must be smaller than the circumference that would be measured by an elf riding around on the edge of the disk. Meanwhile, the diameter, being perpendicular to the spinning motion, would not undergo any relativistic change. As a result, the elf would find a different, larger, non-Euclidean, value for pi.

What was going on? In one of his earlier papers on the static theory, Albert had noted in passing that the normal laws of geometry probably did not hold for a uniformly rotating system. There was one other special thing about being an elf spinning around on a disk, of course: The elf had to endure tremendous centrifugal force pushing him outward toward the edge of the disk. According to the equivalence principle however, he was entitled to imagine this force as a gravitational field with the edge of the disk being "down." So Albert was left with a gravitational field, and anomalous geometry.

Sometime that summer, Albert put these two conditions together. The rotating disk convinced him that "in a gravitational field Euclidean geometry could not hold," as he later said. Gravity would produce a distortion in the normal space-time geometry, or distorted geometry would produce the effect known as gravity: They went hand in hand. Albert knew from his college days and Olympia Academy reading that there was a large body of work on non-Euclidean geometries dating back to the 1820s, when Lobachevsky and Gauss had invented "imaginary" geometries by suspending or modifying Euclid's controversial "fifth axiom" about parallel lines. Each of these geometrical variants could be thought of as describing the properties of some uniquely curved surface, and strange shapes had bloomed like exotic flowers in the nineteenth-century mathematicians' imaginations: the pseudosphere, which looked like a trumpet lip; the saddle-shaped hyperboloid; the torus; the Möbious strip; and the Klein bottle.

One non-Euclidean shape that was already familiar to mathematicians and surveyors was the spherical surface of the earth. Gauss was also a surveyor. He developed a series of formulas by which a surveyor could determine what kind of surface he was standing on—the earth, say—by measuring the sides and areas of triangles laid out on the land.

The so-called Gauss curvature showed up as a deviation from the predictions of ordinary plane geometry. There was "positive" curvature, as on a sphere, in which the area of the triangle was less than expected from Euclidean laws, and "negative" curvature, as in the hyperboloid, in which the area amounted to more than the expected value. The formulas also yielded a "radius," *R*, for the curved surface—the smaller the radius, the more highly curved was the surface.

Albert recognized that his solution to the motion of a falling particle bore a striking resemblance to the mathematics that Gauss had developed to describe curved surfaces. Gravity caused objects to move along curves in space-time, like marbles in a bowl. The bending of light beams was another clue to the ultimate geometrical nature of gravity. In the Minkowskian version of relativity, a light beam was the very definition of a straight line, a so-called geodesic. If it curved, then space itself must be curving. The conclusion was inescapable. Gravity was simply geometry, but geometry *bent,* non-Euclidean geometry, space-time warped like a sagging mattress by the presence of matter and energy, those heavy cosmic sleepers.

Reaching that conclusion and expressing it mathematically, however, were two different things. That was the problem Albert brought to Grossman in the summer of 1912, pleading, "You must help me, or I'll go crazy."[11] Grossman agreed to go to the library and do a little research on Albert's behalf.

The name he came back with was Georg Friedrich Bernhard Riemann, by any measure one of the greatest mathematicians in history. Riemann had succeeded Gauss at Göttingen, even to the extent of living in Gauss's old quarters in the observatory. A brooder and a perfectionist who kept polishing and refining his work, rarely sharing it, his entire life's published output amounted to a stack of paper only an inch high. But in a series of lectures at Göttingen in 1854, Riemann had exploded with enough ideas to keep astronomers and mathematicians busy for the next hundred years.

Riemann went Gauss a number of dimensions better. He extended the notion of Gaussian curvature beyond two-dimensional surfaces to three- and four-dimensional volumes. In the process, the notion of curvature became a little less intuitive and a lot more abstract. Technically, Riemannian curvature was defined as the degree and kind of deviation of a space's geometry from the Euclidean ideal. It's easy to visualize the curving two-dimensional surface of, say, an apple, because the apple itself exists in three dimensions and we can see and feel its roundness. We can't appreciate the curvature or non-Euclidean nature of three-dimensional space in the same way, because there is no fourth dimension in our imaginations, or concretely in the world, within which to view it. That lack doesn't mean we can't employ Gaussian techniques to measure the shape of the universe, Riemann pointed out. We should measure space with no preconception, he argued, to find out exactly what kind of geometry we actually live in: "The properties which distinguish space from other conceivable triply-extended magnitudes are only to be deduced from experience."[12]

Riemann even invented his own version of a possible space-time geometry, modeled on the sphere, but in four dimensions. It was called the hypersphere, and was impossible to picture except in analogy with its lower-dimensional cousin. Technically it was a world with constant positive curvature; poetically it was a world without end or edge because, as on a sphere, a line or light ray extending in any direction would eventually curve back on itself (through the mysterious fourth dimension) and come back to its beginning. A person with a powerful enough telescope and a lot of time to sit still could look out in any direction and eventually see the back of his own head.

There were various experiments by which one could verify that the geometry of space-time was indeed a hypersphere, Riemann said. On a normal sphere, for example, if you stand at the North Pole and draw a series of circles with larger and larger radii, a curious pattern will emerge. The circles will get bigger and bigger until you reach the equator, and then they will get smaller and smaller, even as the so-called radius increases. Finally, when the radius reaches all the way to the South Pole, the circle shrinks back to a point. In Riemann's geometry, the analogy to a circle would be a sphere. The same behavior would be observed if you drew a sphere around yourself at larger and larger radii. The volume of the sphere would get bigger and bigger, reach a maximum, and then start to diminish. At that point the backside of the universe would have come into view, and the sphere would eventually shrink down to nothing when the radius finally reached around the universe to the point opposite your own, the *antipodal* point.[13]

In practice, of course, the universe need not be so regular. Both Gauss and Riemann allowed for the possibility that the curvature of space could vary from point to point in their mathematics. That was just the kind of feature Albert needed to be able to exploit to model gravity in a lumpy universe.

Riemann's brilliance was extinguished by tuberculosis in 1866, when he died at the age of forty. At the time of his death he had been trying to formulate a theory to unify the laws of electricity, magnetism, and gravity, an effort toward which Maxwell would take his own first steps seven years later.

Over the next half century Riemannian geometry became a small but sophisticated industry. Following in Riemann's footsteps, the Italian mathematician Gregorio Ricci and his student Tullio Levi-Civita, among others, had evolved a new branch of mathematics, known as absolute differential calculus, to explore the properties of these multidimensional spaces (called manifolds), and life inside them. This was exactly the language Albert needed in which to express his new theory.

Grossman agreed to collaborate as long as he didn't have to be responsible for any of the physics. As Einstein put it, "Grossmann at once caught fire."[14]

Within a week Albert had made major progress. "The work on gravitation is going splendidly," he told Hopf in the middle of August. "Unless I am completely wrong, I have now found the most general equations."[15] Part of the beauty of the work for him was that the equations involved were almost pure geometry. Everything proceeded from fundamental principles; there was no place to "patch up" the equations with fudge factors, which meant that the theory would be either right or wrong.

Which is not to say that beautiful was easy. By October, Albert was still struggling. "But one thing is certain," he told Sommerfeld by way of a progress report. "Never before in my life have I troubled myself over anything so much, and I have gained enormous respect for mathematics, whose more subtle parts I considered until now, in my ignorance, as pure luxury! Compared to this problem, the original theory of relativity is child's play."[16]

In Riemannian geometry, a simple number like a Gaussian radius no longer sufficed to define the curvature of a space at some point. Instead, such information was represented by more complex mathematical entities known as tensors. A tensor was a sort of grown-up version of a vector, something whose essential mathematical properties do not depend on how you look at it. To say, for example, that Boston is two hundred miles northwest of New York is to describe a vector, an arrow with a direction and a magnitude that don't depend on how they are named. We can relabel the earth so that south is north, and convert miles to kilometers, but that won't affect how sore our feet get slogging from New York to Boston, or what landmarks we use to get out of town. In a somewhat analogous but more abstract way, tensors encoded information about, for example, the properties of a surface, in a way that was independent of whatever way the points on that surface were labeled. The manipulation and transformation of tensors was what the new calculus was all about.

The geometry of a four-dimensional space-time of the kind that Albert was working with was represented by the so-called metric tensor. It was an array of sixteen numbers, six of which repeated, in the following form:

a	e	f	g
e	b	h	i
f	h	c	j
g	i	j	d

The metric tensor was a kind of machine for calculating the geometrical properties of a space. If you punched in the coordinates of a pair of nearby points, for example, it would find the spatial distance along the geodesic that described the shortest line between those two points. There was another tensor, closely related to the metric tensor, called the Riemann tensor, which represented the curvature of a space. Albert was a long way from trying to picture running alongside a light beam in his head, or even shining flashlights around in an elevator. There were, in fact, no pictures that would help at all.

That there were ten different numbers in the metrical tensor meant that there were ten different gravitational potentials describing each point of space-time, not just one, as he had naively and hopefully assumed the year before—ten different solutions of ten different equations for every point in space-time. Each of them constituted part of the answer to the question of the gravity of space-time.

The grand "field equation" that would describe the relationship between gravity and geometry took shape in Albert's mind, therefore, as a relationship between tensors. On the left-hand side of the equation was an array of numbers or symbols that represented the distribution of matter and energy responsible for the bending of space. It was called the stress-energy tensor. On the other side of the equation was another tensor, some variant of the metric tensor, presumably, that represented geometry, the bending of space itself. The two quantities, matter and space-time, had a kind of yin-yang relationship aptly expressed in an aphorism coined by John Wheeler much later: Matter bends space, and space gives matter its marching orders. Matter was both the lord and prisoner of space-time, warping geometry and then encaged by the possibilities thus arranged. Gravity itself had gravity. Geometry was destiny.

The stage was thus seemingly set, by the winter of 1912, for Albert to achieve his holy grail: a "generalized" theory that not only explained gravity, but also extended the principle of relativity to all observers, no matter how they were moving, falling, spinning, or bouncing around with respect to one another. The laws of physics, Albert believed, would then be truly universal and would have the same form for all observers when he could write them in this tensor form that was independent of the coordinate system used by any particular observer—a property that was technically known as *covariance*. Covariance was the goal that Albert had proclaimed as far back as 1907, when he first invented the equivalence principle. Indeed, it seemed built into the notion equating gravity and acceleration.

So it came as a shock nine months later when Albert and Marcel announced that they had failed. Their admission came without warning halfway through their report, published as a pamphlet called *Outline of a Generalized Theory of Relativity and of a Theory of Gravitation,* and usually referred to as the *Entwurf,* after the German word for "outline."[17] "We have not found a method for the solution of this problem [referring to gravitation]," Albert wrote. The best they could do was formulate a theory that was consistent with the rules of the old, ordinary ungeneralized relativity, which was at least *some* improvement over Newtonian gravity. They couldn't prove that a more general solution did not exist, but they couldn't prove that it *did* exist, either.

In short, the universe didn't seem to be going along with Albert's dreams of universal relativity. There was what he called an "ugly dark spot" on his theory.[18]

This was a catastrophe, but not for the reason Albert thought. It would turn out that Albert was wrong in his acknowledgment of failure. Not only *was* there a solution that would fulfill his grand expectations, but sometime during those nine months of calculation, Albert had actually written it down. The answer had been in his grasp, but he abandoned it, and three years would pass before he found it again.

THE QUESTION OF exactly how and why Albert could have gone so wrong has provoked debate and a flurry of dense papers by historians and general relativists during recent years. Did he and Grossmann commit a simple mathematical blunder? Or, in the fog of physics, did a deeper, subtler issue lead Albert astray?

The answers to these questions seem to lie in a series of scratchy unlabeled calculations interspersed with lecture notes and unrelated quantum equations in a blue-bound notebook titled "Lecture Notes 1909–1915" that was left behind when Einstein died in Princeton in 1955. It had sat neglected in the Einstein library for years before John Stachel realized that it had seen double duty. On the back cover Albert had written the word *"Relativität,"* and started working inward, back to front, doing mathematical calculations that appeared to be related to gravitation and the *Entwurf.* Over the last few years, an international team of physicists, historians, and philosophers[19] has been scrutinizing this notebook in order to retrace Albert's steps through the mathematical and physical maze that ultimately led to general relativity. Although many details and subtleties are still being debated (and probably will be forever), a rough consensus has recently emerged on the main outlines of the story.

In general, Albert's method for finding his way through the fog of this multidimensional physics was to use Newton as a lantern. In what Jürgen Renn has dubbed a "double strategy," he looked first for equations that had the mathematical properties he aspired to for his theory, that is covariance. Then he would check by a trial calculation whether the answer for some simple straightforward situation that he thought he understood looked if it was coming out to agree with the Newtonian solution. By going back and forth between physics and mathematics—or between intuition and elegance—he could home in on a final theory. But in this case his intuition wound up outsmarting him.

The logical place to begin constructing his so-called field equation was with what is now called the Ricci tensor, an offshoot of the Riemann curvature tensor, which was in turn a relative of the basic metric tensor. Albert proceeded to write the equation with that tensor. At that point he was, as John Norton, a philosopher and historian of science at the University of Pittsburgh, says, on the "royal road" to the generally covariant solution he sought.[20] The final answer was only a few relatively easy mathematical steps away, but first Albert had to resort to the physics side of his strategy in order to make sure he was on the right track. He tried to solve his equation for a very simple case, that of a very weak and unchanging gravitational field.

In this case, however, Albert took the Newtonian analogy too seriously. He assumed (wrongly as it would later turn out) that space would be "flat" in a static, weak field. It was a reasonable assumption. Space was flat without gravity; space should still be flat if you added just an unchanging smidgen of a gravitational field. In that case, his ten differential gravitational equations should reduce to one single familiar equation describing the old relativity, Minkowski's space-time.

When Albert didn't get the simple answer he was expecting, he concluded that he was using the wrong tensor after all. Through a series of mathematical manipulations, he broke apart the Ricci tensor to derive yet another candidate for the gravitational tensor and solved it. Once again he got what he thought was the wrong answer when he checked it against Newtonian physics.

Frustrated, Albert changed his approach. This time around, he didn't worry about the mathematical properties of his tensor; instead he first looked for one that satisfied the basic requirements of physics: that energy and momentum must be conserved (he was not going to repeat the debacle of the previous spring) and that a weak gravitational field should result in a flat space. There was a third physics requirement that Albert had in mind, namely that the laws of physics would be unaffected by rotation on the part of the observer. But once Albert found a tensor that satisfied the first two conditions, he neglected to check to see that it worked for rotation, presuming that it somehow did. For once, Albert's intuition had led him astray.

Thus seduced by its apparent successes, Albert anointed this new creature as his gravitational tensor. The only problem was that it was not covariant. The result was that as far as gravitation was concerned, the new Einstein-Grossmann theory would have only the limited invariance characteristic of regular relativity; it was valid, he concluded wistfully, for rotating observers and those in steady motion at any speed, but nobody else. Although he did succeed in finding covariant tensors for other fields, such as electromagnetism, despite his best intentions and most fervent hopes, Albert had failed to write the laws of *all* physics in a form that would be the same for all observers in the universe. But it was a theory of gravity, perhaps the best that could be achieved.

By then it was March of 1913. "In the past half year I've worked harder than I ever have in my whole life," he wrote to Elsa, "and finally a few weeks ago I solved the problem. It is a bold continuation of the theory of relativity in addition to a theory of gravitation. Now I must allow myself some rest, otherwise I'll soon make myself kaput."[21]

ALBERT'S VOW to keep away from Elsa had lasted less than a year. Whatever hopes Mileva (and Albert) might have harbored that the return to Zurich would restore their relationship had melted into the snow by February. Mileva's legs had become so painful that she could hardly bear to leave the house. Albert, she told Helene drily, had disappeared into the Ricci tensor and was ignoring her: "My great Albert has in the meantime become a famous physicist, esteemed and admired in the world of physics. He works at his problems indefatigably. You could truly say that he only lives for them."[22]

The Einsteins' appearances at the Hurwitz Sunday afternoon concerts became more and more erratic.[23] On February 14, Albert arrived alone to play his violin. In order to entice Mileva out of the house, he and Eva planned another Schumann program for the following week. Mileva greeted this news with a silent glare but she showed up, practically immobilized with pain.

The week after, the twenth-eighth, neither Albert nor Mileva showed up. August Piccard, one of Albert's graduate students, was scheduled to play with them, and Albert arranged to have his violin sent over so that Piccard could use it. Albert came alone again on the following Sunday, providing no explanation for his failure to appear the week before.

The following Sunday, March 14, was Albert's birthday, an occasion of which Elsa took advantage to end the nine-month silence between herself and him.

Albert wrote back immediately, offering to explain relativity to her personally. "There is no understandable comprehensive book about relativity for the layperson, but what do you have a relativity cousin for? If your path takes you to Zurich then we'll take a nice walk (without my unfortunately jealous wife)." He would rather come to her, he admitted, but he was too tied up. There was a certain misery associated with his fame.

"If you would like to make me very happy, then arrange to stay here for a few days."[24]

That same birthday afternoon neither Albert nor Mileva appeared at the Hurwitzes', but Albert sent word excusing himself on account of "family concerns." In a note to Clara Stern that evening, he reported that Mileva had a toothache. The next day Lisbeth Hurwitz and her mother went up to Hofstrasse to see Mileva and found her with a very swollen face.[25]

Whatever was going on between Albert and Mileva, by the time of his birthday and Elsa's note, in Albert's mind a line seems to have been crossed. Whatever doubts he had about the wisdom of pursuing his relationship with Elsa had disappeared or seemed irrelevant. "I would give anything if I could spend a few days with you, but without my cross," meaning Mileva, Albert wrote a week later. He offered to visit Elsa in the fall if she wasn't going away. "My colleagues won't be there at that time, and we would be undisturbed."[26]

During Easter, Albert and Mileva went to Paris, where Albert had agreed to lecture (in French!) on photochemistry and thermodynamics to the French Physics Society. It was Albert's first visit to the City of Light, and he was duly bedazzled. "I was so overwhelmed with honors that I thought I would die of shame."[27] The Einsteins stayed with Marie Curie and her family, and Mileva met her role model at last. She and Marie, who was originally Polish, seem to have established a bond; the two families discussed a hiking trip together in the Alps during the coming vacation. In the course of his trip he got to see Jean Perrin's "wonder kitchen," the laboratory where his Brownian motion theory had been tested and confirmed.

The passions of the Langevin affair had long since cooled, and Curie had retreated into a resigned sexless life. If Albert felt any eerie parallel between her situation and his own, he did not remark upon it. Albert, the master of the thank-you note, later wrote, "I know of nothing more inspiring than seeing beings of your quality live together so perfectly. Everything at your house appeared to be so natural, as though in the workings of different parts of a beautiful work of art."[28]

BACK HOME, Albert and Marcel started writing feverishly their long-awaited theory of gravity and relativity. In May Albert broke a long silence with Ehrenfest, characteristically apologizing: "But my excuse is the frankly superhuman effort I have invested in the gravitation problem. I am now deeply convinced that I have gotten the thing right, and also, that a murmur of indignation will spread through the ranks of our colleagues when the paper appears, which will take place in a few weeks."[29]

The report was published a month later. It consisted of two parts, a physics section written by Einstein, and a mathematical section, an impressive contribution in its own right, in which Grossmann consolidated what had been separate strands of tensor analysis and non-Euclidean geometry into a unified body of work.

Albert's part begins confidently enough: "The theory expounded in what follows derives from the conviction that the proportionality between the inertial and the gravitational mass of bodies is an exactly valid law of nature that must already find expression in the very foundation of theoretical physics." This identity had been tested, he noted, by the Hungarian Baron Vasarosnemeny Lorant Eötvös around the turn of the century out on the ice of Lake Balaton. By means of a delicate torsion balance, Eötvös had measured the ratio of the centrifugal force due to the rotation of the earth (inertial mass) to the gravitational force due to gravitational mass of a number of objects of different composition and mass. Eötvös found that the two quantities were identical to within the limits of his accuracy.

Albert reported how he had proceeded to use this as a basis to generalize the theory of relativity and how he had been able to derive covariant expressions for most of the laws of physics *except* gravity, where he had been forced to choose a tensor of less universality.

Despite its ugly spot, Albert was pleased by many features of his new theory. It seemed to incorporate Mach's ideas on inertia, for example. If one accelerated a hollow sphere, a free-floating mass inside would also feel acceleration. Moreover if you set the sphere spinning, a pendulum inside would start also start to rotate, as if its natural "rest frame" of reference was being dragged around with its surroundings—shades of Newton and Mach's rotating bucket argument. The new theory also, of course, predicted the bending of light beams, and Einstein hoped that the warpage of space and time near the massive sun could be shown to account for the long-standing mystery of Mercury's perihelion shift.

In June there was a de facto month-long gravitational summit conference in Zurich. Besso, still working in Glorizia, was recruited by Albert to use the new theory to calculate Mercury's perihelion motion. Besso moved into the apartment on Hofstrasse, soldiering through the horrendously complicated computation despite an inflamed stomach. Halfway through the month he had to return to Glorizia for a business meeting, leaving the manuscript with Albert. Ehrenfest and his family came to stay in a boardinghouse for a month. At some point during the month Albert was also in contact with Freundlich. Although the astronomer was about to give up on finding evidence of light-bending in old eclipse plates, a total solar eclipse was due in the Crimea on August 21, 1914, and Freundlich proposed to go and measure the apparent star positions himself.

Another visitor that month was a Finnish physicist by the name of Gunnar Nördstrom, who had been pursuing his own relativistic theory of gravity for the past year. As of the summer of 1913, notwithstanding Abraham's "stately horse with three missing legs," it was the leading rival to Albert's theory. There was no light-bending in Nördstrom's theory, which made the prospect of Freundlich's eclipse expedition especially alluring. The presence or absence of light-bending would provide a clear way to decide between the theories.

At the end of June, Albert sent a copy of the *Entwurf* to Mach, extolling the Machian triumphs of the theory. "Next year, during the solar eclipse, we shall learn whether light rays are deflected by the sun," he declared.[30]

"If yes—in spite of Planck's unjustified criticism—your brilliant investigations on the foundations of mechanics will have received a splendid confirmation. For it follows of necessity that *inertia* has its origin in some kind of *interaction* of the bodies, exactly in accordance with your argument about Newton's bucket experiment."

As it happened, Mach was engaged at the time in writing the preface to a new book, *The Principles of Physical Optics,* which would not be published until 1921. By then Mach would be dead, and it would be from the grave that his voice rose to mock and condemn relativity:

> I gather from the publications which have reached me, and especially from my correspondence, that I am gradually becoming regarded as the forerunner of relativity. I am able even now to picture approximately what new expositions and interpretations many of the ideas expressed in my book on mechanics will receive in the future from the point of view of relativity.

It was to be expected that philosophers and physicists should carry on a crusade against me, for, as I have repeatedly observed, I was merely an unprejudiced rambler, endowed with original ideas, in varied fields of knowledge. I must, however, as assuredly disclaim to be a forerunner of the relativists as I withhold from the atomistic doctrine of the present day.[31]

He was hardly alone. Einstein's new theory, Max Planck admitted to Wilhelm Wien, "does not appeal to me at all."[32]

THE LAST WALTZ

ON JULY 11, 1913, MAX PLANCK AND WALTHER NERNST BOARDED THE train at the Berlin station and headed for Zurich, on a mission for science and for Germany. Their task was to persuade Albert Einstein to move to Berlin. Once that would have seemed a far-fetched notion indeed; anyone who took the trouble to inquire could have learned that Albert had famously quit Germany and all it stood for at the age of fifteen. To lure him back, however, Planck and Nernst were prepared to offer Einstein the job of a lifetime.

This unlikely-sounding expedition had its origin in a unique confluence of geopolitical, professional, and personal needs. Nernst, who had been following Einstein's work as early as the Solvay Congress, had an entrepreneurial spirit, which was well suited to the times. Mindful of the importance of science and technology to a growing economy, Kaiser Wilhelm had in 1911 approved the creation of the Kaiser Wilhelm Gesellschaft, a collection of research institutes to be financed largely by industry and patriotic individuals in return for the usual honors. One of the first to step forward was a financier named Leopold Koppel, who donated a million marks to establish the first institute, which was devoted to chemistry. It was headed by a recruit from the Karlsruhe Polytechnic named Fritz Haber, who had invented a process for extracting nitrogen from the air to make ammonia, the key ingredient in both fertilizer and explosives. Producing guns and butter at the same time—and literally out of thin air—represented an achievement that was the very definition of "science in the public interest."

Bald and delicate as an underboiled egg, Haber was a brilliant, driven man who hungered for acceptance into the German military-industrial elite. Born Jewish, he had renounced his religion at the age of twenty-four. His first wife had committed suicide, and his second divorced him. The strain of overwork and ambition left him prone to episodes of depression and guilt. It was during a leave to recuperate, at a resort castle perched above the sun-carved valley of the Engadine in January of 1913, that Haber formulated a plan to bring Einstein to Berlin. Although there was no physics institute yet, quantum theory, which Albert had pioneered, was the key to the future of chemistry and technology.

The idea of making a place for Einstein at the new institute had already been raised by Haber's superior at the Ministry of Education, but Haber had been slow to act upon it. "After having turned this idea over in my mind for quite some time, I have become convinced that the realization of this idea would be of the greatest advantage for the Institute, and that, from the personal side, it could probably be attempted with some chance of success," he wrote to his friend Hugo Krüss in the Ministry of Education. "It is a very rare coincidence that such a man is not only available, but that his age (34) and his other traits make me very confident of a beneficial relationship." Haber estimated that 15,000 marks would cover Einstein's salary and expenses; as a theorist, he wouldn't need equipment.[1]

Haber might have been encouraged to write this letter by none other than Elsa Löwenthal, Einstein's comely cousin. Sometime during this period, unbeknownst to Albert, Elsa apparently "popped in" on Haber one day.[2] Exactly when this occurred and what was said are not known; Elsa reported the meeting to Albert only much later, and her letter was destroyed. However, Elsa clearly let Haber, and through him the German scientific establishment, know that Albert might be more available and amenable to a Berlin move than a cursory examination of his past might reveal.

Planck was then recruited to be the go-between. Whatever his own judgment of the new gravitation theory and the light-quantum hypothesis, Planck knew that Einstein had the most original and penetrating mind of his generation. By June, Haber, Planck, Nernst, and their confederates had raised promises of enough money to guarantee Einstein a salary of 12,000 marks a year—half of it from the generous and patriotic Mr. Koppel. In the course of time the original plan had changed: Instead of being appointed to Haber's chemistry institute, Einstein would be named director (and sole member) of his own Kaiser Wilhelm Institute for Physics. He would also be appointed a professor at the University of Berlin, but without any teaching duties. Planck took the first step, in a letter cosigned by Nernst, Heinrich Rubens, and Warburg, by proposing Einstein to fill a recently vacant seat in the Prussian Academy of Sciences.

In sum, it can be said that among the important problems, which are so abundant in modern physics, there is hardly one in which Einstein did not take a position in a remarkable manner. That he might sometimes have overshot the target in his speculations, as for example in his light-quantum hypothesis, should not be counted against him too much. Because without taking a risk from time to time it is impossible, even in the most exact natural science, to introduce real innovations.[3]

Apart from the Nobel Prize, there was no greater honor on the Continent than to be elected to the Academy. For a thirty-four-old-old scientist it was unheard of, but the physics and chemistry section of the Academy quickly seconded the nomination. By then Albert must have been aware that he was being considered, because the Prussian Ministry of Education had been making inquiries all over Europe.

In the second week of July Planck and Nernst took the train to Zurich. Albert met them and took them to the Polytechnic, where they made their offer. As the famous story goes, he was unable to decide immediately, so he sent them off on one of Switzerland's famous cog-rail excursions to the top of the Rigi Mountains. When they returned, he told them, he would meet them at the train station with a red rose if his answer was yes, a white one if no.

The balance sheet that Albert might have tallied mentally as he sat down to ponder the unthinkable could have looked something like this:

He was making good money in Zurich; he would make more money in Berlin. In Berlin he would have Fritz Haber as his local friendly uncle; in Zurich he had Heinrich Zangger. At the University of Berlin he would have no teaching duties, while at the Polytechnic he taught electricity and magnetism, optics, the weekly seminar, and supervised the lab. In Berlin he had Elsa; in Zurich he had Mileva. In Berlin he would be a few minutes from Max Planck; in Zurich he was half an hour from his mother.

When Planck and Nernst came back down from the mountain, Albert was carrying a red rose.[4]

Naturally, the deal was far from done. The entire Academy still had to approve his membership, and then his appointment had to be confirmed by the government itself in the form of a royal decree. Initially, however, Albert hoped it might all be concluded by the fall, so suddenly eager was he to escape northward. In the meantime, Albert arranged with the post office to have his mail forwarded to his office, so as to avoid the chance that Mileva would intercept the renewed stream of mail from Elsa.

"In recent days I had a visit from Planck and Nernst," he whispered to Elsa, cautioning her to keep it confidential.[5] "At the latest next spring I'll be coming to Berlin forever."

Mileva reacted to the news with predictable gloom. She was returning into the bosom of Albert's horrible relatives. She had always feared them, and now, especially, there was a new one to be wary of.[6]

THE JOINT HIKING TRIP with the Curies had been scheduled for early August. At the last minute, however, Eduard became so ill that Mileva had to stay behind and take care of him while Albert went on alone. He met Marie Curie, her two daughters, Eve and Irene, and a governess in Samedan, a major town just downstream of St. Moritz in the Upper Engadine, from where they set off on foot.

The Engadine rises gradually toward Maloja Pass and the Italian border, past a series of long meandering lakes and terraced villages. By the time they reached the pass Mileva and Albertli had joined the expedition. According to Eve's biography of her mother, however, Mileva did not hike.[7] That Mileva, whose legs had been too painful for her to leave the house five months earlier, could even make the trip to the mountains is slightly astonishing. Eve's account of the journey portrays the children gamboling over the ridges and hillsides, scrambling up cliffs to picturesque villages while Albert and Marie strolled in their wake, talking about everything from the geology of the mountains to gravitation and relativity. At one point Marie playfully challenged Albert to prove he was a real mountain man by identifying all the peaks within sight. Albert grabbed her by the wrist once as they negotiated a ledge and said that he wanted to know about the man falling in the elevator. Marie recorded in her notebook that she was having a good time.

In his report to Elsa, Albert painted a slightly different picture of his adventures. "Mme. Curie is very intelligent but she has the soul of a herring, that is, lacking in any kind of joy or pain. Practically her only emotional expression is cursing about things she doesn't like. And she has a daughter who is even worse—like a grenadier."[8]

The tour ended at Lake Como, where Albert and Mileva had begun their fateful sledge ride up to Splügen Pass twelve years before. This time, however, they took the train home.

Determined not to be left out of the action, Elsa began to keep up a steady drumbeat of news and commentaries from Berlin. In between lecturing Albert on the need to bathe more regularly, pestering him to come and visit, and planning her campaign to get on good terms with Albert's mother, she dropped the worrisome news that she had been diagnosed with an enlarged heart.

Albert reassured her that a few months of rest would do the trick. "That such a thing must happen to the most medical of womenfolk!" he wrote.

"I would laugh at you if I were there to make it good through a little kiss or by way of some other sweetness.

"I have firmly decided to bite into the grass with a minimum of medical help when my little hour has come. But until then I plan to sin away as my wicked soul bids me. Diet: smoking like a chimney, working like a horse, eating without consideration or being selective, walking *only* in truly comfortable company, that is, unfortunately, seldom, sleeping irregularly, etc."[9]

ALBERT CONTINUED to fret over his new gravitational theory, puzzling over its curious failure to meet his expectations of covariance. The final equation had two parts, one describing how curved space affected matter, the other describing how matter curved space. The first part consisted of laws of motion and electromagnetism that fulfilled Albert's initial prescription of being the same for any observer in any kind of motion—in short, they were covariant. The tensor that Albert had mistakenly settled on for the second part of the theory, however, did not held true. "The theory thus refutes itself," he complained to Lorentz. Why?[10]

By the middle of August Albert had convinced himself that he knew the answer. His theory's limitations, he explained to Lorentz, were a necessary mathematical consequence of ensuring that it conform to one of the most fundamental laws of physics, namely the conservation of energy and momentum. The year before he had been embarrassed when his first attempt at a gravitational theory had turned out to violate those principles. This time he had built them in from the beginning, in effect by limiting the theory to those reference systems in which the conservation laws appeared to hold. Could there be anything more beautiful than to have the theory in a sense limit itself on the basis of such fundamental pillars of life as the conservation laws?

Albert professed himself at last to be happy and satisfied. "Only now, after this ugly dark spot seems to have been eliminated, does the theory give me pleasure."[11]

Albert, however, wasn't quite telling everything he knew, for the theory was shakier than he was willing to admit. When he and Michele used it to calculate the mysterious motion of Mercury's perihelion the results were disastrous. Their attempts were recorded in a fifty-six-page manuscript that only became known to scholars in 1995, when it was discovered in Europe and hastily incorporated into volume 4 of the *Collected Papers*.[12] They demonstrate just how sly or stubborn Albert really was, and reveal him weaving yet another thread in the net of self-deception that kept him tied to his theory.

By the second decade of the twentieth century, astronomers were aware with exasperating precision that Mercury did not follow a fixed path around the sun. Rather, the ellipse that represented its orbit itself revolved at the rate of once every 2,273,684 years, or about 570 seconds of arc per century. As a result, over the millennia the planet would trace out a rosette pattern around the sun. After using Newtonian gravity to take the influence of other planets into account, there remained a stubborn, minute discrepant change amounting to 41 seconds of arc per century.[13] Astronomers had invoked a hypothetical planet Vulcan and bands of interplanetary dust to explain it, and Lorentz had attempted to calculate the motion with his aether-based theory of gravity a decade before. Albert was banking on space-time curvature.

But when they finally inserted numbers into their painfully wrought formulas, Albert and Michele came up with an answer of 1,821 seconds (more then 30 *minutes*) of arc for the perihelion motion, leaving them off by a factor of 40. It turned out, however, that they had used the wrong mass for the sun. Albert, now carrying on the work alone, put in the correct mass, redid the calculations, and got a new answer of 18 seconds of arc—closer, but now far too small.

This left Einstein in an awkward position. His theory gave the wrong answer, and as there were no fudge factors to adjust in the equations, he would have had perfect grounds for dropping it. At best, he would have had to contend that as-yet-undiscovered dust bands made up the difference in Mercury's motion—a theory he loathed. At worst, in Abraham's hands, these results could have been used to attack the *Entwurf*. But Abraham and the rest of the gravitational community never found out about the results, because Albert never published them.*

Perhaps just a *little* worried about the discrepancy, Albert plunged on with an entirely different kind of calculation, designed to test one of the

---

*By withholding unfavorable evidence, can Einstein be regarded as having failed his duty as an officer of the court on this occasion? That seemed to be the implication of an essay on the Einstein-Besso calculations in the Christie's auction catalog written by Michel Janssen, an affable young Dutch physicist and member of the Einstein Papers team. Originally from the University of Amsterdam, Janssen had joined the project as its general relativity expert after receiving his Ph.D. at the University of Pittsburgh under John Norton, who had been instrumental in unraveling Einstein's notebook.

"I do think Einstein could have been more candid about the eighteen seconds fiasco," Janssen says. "But I don't think airing the eighteen seconds would have hurt the *Entwurf* theory that much. No theory before general relativity got that right." Janssen pointed out that Nördstrom's theory predicted a seven-second effect in the opposite direction, *increasing* the discrepancy. At least the *Entwurf* theory made things better. As for Albert himself, he seems to have decided to ignore the discrepant Mercury calculation.

central tenets of his theory: Mach's notion that even rotation was relative. According to Mach's rotating bucket argument, it made no difference whether the water rotated or the universe rotated around it. The net effect of water spinning was the same. In Einstein's theory, this meant that it should be possible to construct a gravitational field that would mimic the effects of rotation. Albert concluded that this was in fact the case. Once again, however, having gotten the answer he expected and hoped for, Albert failed to check the calculation carefully. It would be two years before he realized he had made a mistake.[14]

Meanwhile, plans for the other test of the theory, light-bending, were proceeding fitfully. Freundlich had obtained permission from his boss, Struve, to go to the Crimea to view the eclipse the following year, but had received no funds. Albert invited him to come to Zurich in September. It happened that Freundlich was getting married then and planned to honeymoon in the Alps, so when he and his bride arrived in Zurich they were met at the station by Albert, wearing a floppy straw hat, and Haber, who happened to be visiting.[15]

Albert immediately led them off to Frauenfeld, a few kilometers away, where the Swiss Natural Science Society was meeting. He bought them lunch and then discovered he had no money. Otto Stern slipped him a hundred-franc note under the table. Afterward Albert gave a talk on gravitation and mortified Freundlich by introducing him to the crowd as "the man who will be testing the theory next year."

On the way back to Zurich, Albert and Erwin talked eclipse while Erwin's new bride, Käthe, getting a good taste of the married scientific life, studied the scenery. The basic plan was straightforward: During totality, Freundlich would photograph the sky with the eclipsed sun and stars visible. That photograph would be compared with another of the same star field sans sun. Stars whose light had been bent would appear displaced slightly outward from the sun compared to their normal positions. The amount was slight indeed—only 0.8 second of arc near the sun's edge, or "limb."

Albert had a few wrinkles in mind, however. He suggested that they might want to use two cameras, mechanically yoked, so that the eclipsed sun could be in the center of one field of vision and the stars of interest in the center of the other, thus avoiding spurious optical effects. He also wrote for advice to George Hale, the director of the Mount Wilson observatory in California and a famous solar astronomer, wondering if it was really necessary to wait for an eclipse. Hale confirmed Einstein's suspicion that it would be impossible to photograph stars near the sun during daylight, and passed Einstein's letter on to W. W. Campbell, the director of Lick Observatory in Northern California and an experienced eclipse astronomer.[16] A correspondence ensued between Freundlich and Campbell, who agreed to take plates of the star field during the Crimean eclipse, if he managed to get there—a big *if*. Freundlich began to make plans to do the experiment himself.

FOLLOWING FRAUENFELD, the Einsteins took another trip, one that was partly professional and partly family fun and for Albert was to culminate in a secluded visit with Elsa—a plan that was, obviously, unknown to Mileva.

Their first stop was Kac, still barely a part of the old and fading Hungarian empire, where they spent a week with Mileva's family at their old country estate, The Spire. The weather was hot, and Zangger had warned him to watch out for cholera, but Albert reported seeing none. They passed the stay, he said, in "bucolic tranquillity," indulging in the innocent jocularity he had enjoyed with the relatives and townspeople during their previous visit eight years before.[17] It was Tete's first trip east to see his grandparents. Mileva's brother Milos, who had been studying medicine in Switzerland, was also there.

Yielding to the pleas of her father and various other grandparents and godmothers, Mileva took the children to town one day and had them baptized into the Greek Orthodox church. Albert did not attend the baptism.[18]

From Kac they went to Vienna, where Albert was scheduled to perform his "trapeze act" before the 85th Congress of Natural Science. By now Albert practically owned the yearly peripatetic congress, the scene of many of his triumphs. This time, however, he learned of one that was not his, when Georg Hevesy whispered to him about Bohr's theory of the atom with electrons making quantum jumps from orbit to orbit. The irony of the moment was undoubtedly not lost on Albert. Here he was, about to proclaim and defend his most radical and far-sweeping theory, a veritable reconstruction of space, time, gravity, and the whole universe, a stunning triumph of logic and beauty, only to learn that there was something inherently, spookily absurd about the atoms that inhabited that universe.

I might have done something like this, but it would have been the end of physics.

Vienna, the city of Mach and Mozart and of Albert's old friend Fritz Adler, was still the glittering capital of Eastern Europe. The lecture hall was packed with his peers in their formidable moustaches and stern beards, eager to see Einstein propound his new theory in person, if not necessarily to accept it unqualifiedly.

Albert began by comparing the development of gravitation theory to the history of electromagnetism. Suppose all we knew about electricity was Coulomb's law of electrostatic attraction and repulsion between charges at rest. "Who would have been able to develop Maxwell's theory of electro-magnetic processes from these data?" Yet that was exactly the situation in gravity: "All we know is the interaction between masses at rest." How, then, could Newton's law be extended into a more comprehensive theory? Without relativity as a guide and illuminating principle, there were no limitations at all on the imaginations of the theorists.

He proceeded to outline the two relativity-based theories in contention. Nördstrom's theory was basically ordinary relativity with gravitation thrown in as an extra force. It did not fulfill the principle of equivalence. The speed of light was constant; there was no light-bending, no space-time curvature, no pretensions to a more generalized relativity. "Only one thing remains unsatisfactory," Albert concluded, "namely, the circumstance that, according to this theory, it appears that the inertia of bodies, though indeed *influenced* by other bodies, is not *caused* by them, because according to this theory, the inertia of a body is greater the farther we remove other bodies from it."

Albert then introduced his own theory by recapitulating the story of the physicist in the box, who had no way of telling whether he was sitting on the earth or being towed through space. "If it is really impossible in principle for the two physicists to decide which of the two points of view is correct, then *acceleration* possesses as little absolute meaning as *velocity*." What the Michelson-Morley experiment was to uniform motion, the Eötvös experiment, which showed that gravitational and inertial masses were identical, was to acceleration. This meant that from a mathematical standpoint the laws of nature must be covariant with respect not just to uniform motion, à la relativity, but to accelerated nonuniform motion. His and Grossmann's not-quite covariant theory was the closest one could come within the framework of sensible physics, Albert maintained.

Toward the end of his address, Albert said, "To avoid misunderstandings, let me repeat here that, just as Mach, I do not think that the relativity of inertia is a logical necessity." But it was, he acknowledged, certainly more beautiful. Nördstrom's theory, he concluded, was simple and consistent but did not fulfill that condition. The Einstein-Grossmann theory did. "It is true that in that case one arrives at equations of considerable complexity; but, in exchange, the equations to be sought follow from the basic premises with the help of surprisingly few hypotheses, and one satisfies the conception of the relativity of inertia.

"Whether the first [Nördstrom's] or the second [Einstein's] path corresponds in essence to nature must be decided by photographs of stars appearing close to the sun during solar eclipses. Let us hope that the solar eclipse of 1914 will already bring about this important decision."[19]

When Einstein concluded his speech, Gustav Mie, a professor at the University of Greifswald, leapt to his feet. "First of all, I would like to round out Mr. Einstein's interesting lecture by adding a few words on the historical development of the theory," he declared. Obviously Mr. Einstein hadn't read Mie's own theory of gravity. "No, no," protested Albert. Wasn't it true, moreover, Mie said, that Abraham had been the first to set up reasonable gravitational equations, and that Nördstrom had taken off from there?

"Psychologically, yes, but logically, no," Albert answered to the former, for, as he explained, the theories were fundamentally different. As for Mie's own theory, it failed to follow through on the principle of equivalence. "It would have been illogical of me to start out from certain postulates and then not to adhere to them."[20]

Mie announced that he would shortly publish a paper showing that the Einstein-Grossmann theory itself did not live up to the equivalence principle. From there the discussion grew heated. At one particularly tense point, somebody pressed the button that moved the blackboard from one side of the room to the other, saying: "Look. The blackboard moves against the lecture hall and not the lecture hall against the blackboard."[21]

It was this fracas that finally propelled Albert and relativity into the popular press. Three days later an article entitled "The Minute in Danger, A Sensation of Mathematical Science" appeared in a Vienna newspaper.[22] For all Einstein's fame on the scientific circuit, few laypeople had ever heard of him. This was the first account of relativity, the leading edge of what in ten years would become a thriving subgenre of journalism.

After Vienna Mileva headed on to Zurich, where she was preoccupied for the next few days showing a visiting countryman around town. Meanwhile Albert went alone on the circuit of relatives, traveling to Heilbronn, Ulm, and, at last, Berlin.

Elsa later fondly recalled cooking mushrooms and goosecrackle for him on the visit, which, apparently, proved to be everything he had been looking forward to for the past year. Perhaps encouraged by Haber to tidy Albert up, Elsa sent him home with a new hairbrush—a "bristly girlfriend," as he referred to it—and a toothbrush, along with admonishments to bathe more regularly.

As soon as he returned he wrote her. "Here I am in Zurich again, but no longer the same as before. I now have someone of whom I can think with unalloyed pleasure and for whom I can live."[23]

AFTER THE cheeriness of Berlin, home seemed more ghastly than ever. Without anything having been said, it seemed that all the parties involved knew that something had changed as a result of Albert's stay in Berlin. An icy silence emanated from Mileva. Albert told Elsa that he had moved into a separate bedroom and avoided being alone with his wife. He gradually stopped all conversation about his relatives, especially Elsa.

Meanwhile, Albert directed his emotional energy northward in an awkward blend of longing and bile, as he counted down the months until he and Elsa would be united, free to meet and have their cozy walks and talks. "Itinerant folk that we both are, chosen by fate from among the crowd of philistines to be tightrope dancers, though—thank God—not on a real tightrope, but only in the headroom of human madness!"[24]

In the middle of October he wrote: "It's half a disgrace that I'm already sitting down ready to write to you when I just received your letter today. But the hours that I spent with you so contentedly have left behind in me such a longing for cheerful conversation and intimate togetherness that I cannot withstand reaching for the miserable paper surrogate of reality. . . .

"How beautiful it could be if we could sometime run a little Gypsy household together. You have no idea how charming such a life with tiny needs and without grandeur can be. Who is to say that we will not realize it one of these days?"[25]

A month and a half later: "You pinch me with your letter but you stroke me with goosecrackle. How good it is . . . Have you realized that half of our period of separation is nearly over? I am thoroughly tired of this sadness beyond loneliness."[26]

As for Mileva, the cascade continued. "She is an unfriendly humorless creature who has nothing from life herself and smothers the joy in life of others through her mere presence (*malocchio!*) [evil eye!]."[27]

IF THESE LETTERS are all too reminiscent of those he wrote to Mileva back in the days of their college and postgraduate courting, when the party standing in the way of Albert's happiness was his mother, Albert's passion for Elsa seemed more centered on the products of her kitchen than on the sweet kisses she might bestow. Overall, his correspondence with Elsa is curiously sexless; gravity, perhaps, was taking its toll.

In early November of 1913 Albert headed for Brussels and the Second Solvay Congress, where he argued with Nernst about thermodynamics and fumbled the job of making a toast to the founder Solvay at the final banquet. "All those who had known me as a quick-witted debater without fear and reproach found it amusing that my mastery of words deserts me so completely when I eat and drink."[28]

A week later Emperor Wilhelm II finally confirmed Einstein's election to the Prussian Academy of Sciences, and Albert formally petitioned the

Federal School Council to be released from his contract. The council responded by offering him a raise, lifetime tenure, freedom from teaching assignments, and even extra research grants as inducements to change his mind and stay, but they were all to no avail. Zangger, who was on sick leave and was spending the semester convalescing on the Italian Riviera, was not available to try to sway his friend. The move was set for April 1, the same date on which Albert had departed Zurich three years earlier for Prague.

With commitment, naturally, came second thoughts. Einstein told Otto Stern that he felt as if he was being collected like a rare stamp. "The Berliners are betting on me as on a prime laying hen. I am not certain I can still lay eggs."[29] He suddenly remembered that he had never really liked Germans. Haber, for example, whose picture was posted everywhere around his institute, seemed to have succumbed to a tasteless personal vanity, a failing he railed about to Elsa. "This lack of refinement is, unfortunately, just the way of the Berliners. When these people are in the company of the English and French, what a difference! How coarse and primitive they appear. Vanity without real self-esteem. Civilization (well-brushed teeth, elegant tie, well-groomed moustache, impeccable suit) but no personal culture (coarseness of speech, movement, voice, feeling)."

Albert had already taken steps to distinguish himself from these vulgarians by abandoning the toothbrush Elsa had given him, on the grounds that it was dangerous. Hog bristles, after all, could bore through diamonds. The more Elsa scolded him, the more perverse pride he seemed to take in his stand. "Therefore a kiss of the hand from a dainty distance from the incorrigible filthy pig shitbird," he ended his tirade.[30]

A PLAN FOR the move to Berlin was already under way, one that reflected the tension between the Einsteins. Albert sent Mileva to Berlin by herself over the Christmas holidays to look for housing, for, as he explained to Elsa, "I don't care about the thing with the apartment. She should go herself and find one that suits her taste. If you take pains you have the devil's thanks for it."[31] She was, Albert emphasized, under no obligation to call on his relatives, and would stay with the Habers, Fritz and Clara. This arrangement enabled Albert and Elsa to plan a New Year's rendezvous, but, unfortunately, Elsa had already committed herself to spending the holidays with relatives in Munich.

"It is a pity," Albert commented, "but both of us are accustomed to looking at pleasurable things as something destined for others only. Do you know Busch's beautiful little verse: 'It's a pleasure to abstain from those things we can't obtain'?"[32]

Elsa, in fact, was far too shrewd a woman to give up pleasurable things simply because she had an admirer (a married man!) who lived in another country. She arranged another poetry recital at the Künstlerhaustheater, even though a previous recital a year earlier had received a tepid review from the *Vossische Zeitung*. Its critic had complained about her lack of dramatic range, but acknowledged that her performance had been received with vigorous applause.

More significant, Elsa was also keeping time with one Georg Nicolai, a brilliant physician and physiologist, pioneer of the electrocardiogram, as well as one of the most notorious womanizers in Berlin. Everything about Nicolai (who will figure later in this book) was larger than life, from his experience as a fanatical dueler and playboy to his scientific adventures, which included studying psychology with Pavlov, to his list of sexual conquests and illegitimate children. Perhaps to provoke Einstein, Elsa asked if she had his permission to continue seeing Nicolai. Albert, however, did not take the bait, and told her that as long as Nicolai was "well behaved," Elsa could keep on enjoying his company.

Elsa also raised the issue of divorce, which brought a very strange reaction indeed from Albert. "Do you think it is so easy to get a divorce if one does not have any proof of the other party's guilt, if the latter is shrewd and—with all due respect—a liar? Actually, I do not even have a proof of such an act—which is the only one the court recognizes as 'adultery'—that would be convincing to myself," he wrote, as if it were Mileva who was suspected of infidelity.

It was not the first time that Albert had imagined someone else's doing unto him what he was in fact doing to her. There was poor Marie Winteler, for example, whom he had accused of leaving him when it was Albert who had gone away to study in Zurich and then called an end to the relationship. And of course, in his mind it was Mileva who had abandoned him these last few years, withdrawing her affection, hiding in a shell of depression and jealousy, not Albert who had abandoned her for Besso, Habicht, von Laue, Planck, Ehrenfest, Anneli, and Elsa. To a man with imagination, and a little experience, the shadows were alive with dangerous intentions.

"On the other hand, I treat my wife like an employee whom I cannot fire," he went on. But because he had his own bedroom, "In this form I can endure the 'living together' quite well. In fact I don't understand why you are upset by that. I am my own master and, if you do not want to join me, also my own—wife."[33]

As the move neared, Mileva's apprehensions grew, and Albert seemed to take a kind of perverse pleasure in imagining the catastrophic conjunction of personalities of the two women. "My wife whines incessantly to me about Berlin and her fear of the relatives," he reported. "She feels persecuted and is afraid that the end of March will see her last peaceful days. Well there is some truth in this. . . . *Ach,* and Miza is the most sour sourpuss that ever existed. I shudder at the thought of seeing her and *you* together.

"She will writhe like a worm if she sees you even from afar."[34]

Things were hardly better between Mileva and the other woman in his life, his mother. Once she realized its potential, Pauline seems to have quietly encouraged Albert's relationship with Elsa, volunteering how nice it would be for them to be reunited in Berlin.

With her usual tact, Pauline sent a Christmas note that year addressed to Albert alone, announcing the imminent arrival of packages. She sent them, however, by way of Maja, in Lucerne, a pillow for Albert and Mileva and toys for the children. Furious at having been left out of the loop, Mileva threw a tantrum and sent the toys back along with a letter announcing that henceforth neither she nor the boys would have anything to do with Pauline. Albert kept his distance from the hostilities, convinced that the two of them deserved the misery in which they had collaborated. "No wonder that the love of science thrives under these circumstances, for it lifts me impersonally, and without railing and wailing, from the vale of tears into peaceful spheres."[35]

Mileva left for Berlin shortly after Christmas. The Habers lived in Dahlem, a suburban section on the southwest side of town preferred by academics, and near the new chemistry institute on Faradayweg. After a few days of searching, Mileva rented an apartment in the same part of town, at 33 Ehrenbergstrasse.

As Albert had predicted, things did not go smoothly in Berlin. Inevitably Mileva went to call on her in-laws; moreover, Elsa had insulted Mileva by offering to help her find housing. Albert acknowledged that he was happy to be home in Zurich sitting on the sidelines for this one, where he could "watch with perfect calm how the wretched folks are coping with each other. . . . Interaction with the family would bring me nothing but incessant annoyance. I am not letting Miza meddle in my private affairs, but I don't meddle in hers, either. I find Miza's absence very pleasant. As you see, I too get pleasure out of my marriage."[36]

EINSTEIN'S DESCRIPTION to the contrary, it was scarcely less fractious in the so-called peaceful spheres of science. Abraham launched a new attack on relativity at the beginning of 1914 with a review article called "The New Mechanics," masterfully dissecting the logical and mathematical inconsistencies of the Einstein-Grossmann theory and dismissing it. He concluded, "The relativist ideas are evidently not broad enough to serve as a frame for a complete image of the world. But there remains a place for the theory of relativity in the history of criticism of the concepts of space and time. . . . This fact guarantees the theory of relativity an honorable burial."[37]

"To be sure, he fulminates against all relativity in *Scienza,* but he does it with great understanding," Albert told Besso in a kind of survey of his colleagues' attitudes toward his work. Langevin and Lorentz were interested; von Laue and Planck were both opposed to it on fundamental grounds. Sommerfeld was leaning away. A free, unprejudiced look was not at all characteristic of Germans, he complained.[38]

Mie's long-promised polemic landed on Albert's desk not too long afterwards in the form of an article in the *Physikalische Zeitschrift*. It was heated but hardly a deathblow, for Mie was mainly confused about the theory—a result, perhaps, of Albert's own muddled state when he wrote the *Entwurf.* Albert spent the month of January grinding out rebuttals to his critics.

"I enjoy controversies," Albert admitted to Zangger. "In the manner of Figaro: 'Would my noble Lord venture a little dance? He should tell me! I will strike up the tune for him.'"[39]

Meanwhile, Freundlich had been trying to hold up his end of the relativity enterprise. In December he had submitted an official proposal to the Prussian Academy of Sciences for an eclipse expedition to the Crimea in August. Albert wrote to Planck, who was now the rector of the University of Berlin as well as secretary of the physics and chemistry section of the Academy, asking if he could help get the funds for the trip approved.

Freundlich's boss, Struve, the director of the Berlin Observatory, was a different matter. "*I shall not write to Struve,*" Albert told Freundlich emphatically, all too aware that the practical-minded director was not likely to be sympathetic to Einstein's theory and already resented the amount of time Freundlich had devoted to it. "If the Academy shies away from it, then we will get that little bit of mammon from private individuals." He promised to enlist Haber to help convince Koppel. If everything else failed, Albert promised to put up his own money—at least up to 2,000 marks. He urged Freundlich not to waste any more time worrying and just go ahead and order the requisite photographic plates.[40]

Planck, though not surprisingly hating the Machian aspects of Albert's theory, went to consult with Karl Schwarzschild, the director of the Potsdam Observatory, twenty-five kilometers outside Berlin. Schwarzschild was an astrophysicist ahead of his time, who had taught with Hilbert and Minkowski at Göttingen and had published his own speculations on the possible curved geometry of the universe back in 1900. He had also done eclipse observations.

Presumably Schwarzschild was supportive, for in spite of everyone's misgivings, Freundlich's proposal was referred to the appropriations committee, which eventually approved it. In January the Academy followed suit, subject to the confirmation of the Ministry of Education. Einstein and Freundlich were overjoyed; everyone had acquitted themselves honorably, particularly the stubborn old man, Planck. "Hats off!" Albert exclaimed to Zangger.[41]

Albert's experience with Mie, however, had convinced him that he could make his theory clearer, and he went back and redid the gravitational calculations to his great satisfaction. So pleased was he by the work, in fact, that he told Besso the eclipse results didn't matter anymore; the logic of the theory was simply too evident to ignore. "Now the harmony of the mutual relationships in the theory is such that I no longer have the slightest doubt about its correctness," he told Zangger. "Nature shows us only the tail of the lion. But there is no doubt in my mind that the lion belongs with it even if he cannot reveal himself to the eye all at once because of his huge dimensions. We see him only the way a louse that sits upon him would."[42]

Elsa sent another package of goosecrackle to Albert, but all that made it through the mails intact was the cover of a box and a few grease stains.[43] Albert feigned "a purely moral pleasure" in the gift, unsullied by vulgar appetites. "Another milestone on the way to Buddha victoriously passed."

MILEVA CAME BACK from Berlin more distrustful of the Einstein relatives than ever. "I have noticed that she sniffed out some kind of danger in you,"[44] Albert told Elsa. From Mileva's point of view the situation was growing even worse. In February Albert's least-favorite aunt, Julie, died, and his mother decided to move in with the newly widowed Uncle Jacob in Berlin. A critical mass of critical in-laws was gathering in the city of her future residence, like crows nesting noisily in the tree outside a bedroom.

Promptly upon Mileva's return, Tete fell sick with whooping cough, the flu, a middle-ear infection, and general exhaustion. Doctors recommended that the boy go south for a lengthy recuperation at some spa once the weather got better and he was healthy enough to travel. Mileva made plans to take him to Locarno, a resort on the banks of the Lago Maggiore, another of the meandering resort lakes on the Swiss-Italian border. This meant that Mileva and the children wouldn't be able to accompany Albert when he made his move to Berlin late in March. He promptly decided to pass up a physics meeting in Paris so that he could get to Berlin faster and spend time with Elsa.

ON MARCH 16, 1914, a Sunday, Lisbeth Hurwitz recorded in her diary that Professor Einstein had come to play his violin with them for the last time.[45] In honor of Mileva they played Schumann as well as Mozart. Mileva sat by herself and didn't say a word the entire evening. Two weeks later she left with Tete and Albertli for Locarno, where they planned to stay until mid-April.

Albert had already left for Antwerp and a visit with his uncle Caesar, and then Leiden. Along the way he sent a card to his family, on which he had drawn a duck and a ship with smoke curling out its stack.

"I hope things are going well with you," he wrote, and added, "Tete should eat a lot. Heartfelt greetings from your father."[46]

He arrived in Berlin a week later, where he was greeted by the gift from Koppel of a grandfather clock.

19

THE LANDSCAPE OF BAD DREAMS

PERPETUALLY BATHED BY FLOODLIGHTS, THE BOMBED-OUT TOWER OF the Kaiser-Wilhelm-Gedächtniskirche rises up from Berlin's Breitscheid-platz at night like a jagged gold tooth, carved and scarred by black holes and shadows. It is a clocktower dedicated not to the passing of time but to its freezing, rooted forever in one apocalyptic moment of the kind that the skeletons and dancing Turks on all those other clocktowers across Europe had been trying to warn against for all those centuries. The Gedächt-niskirche stands at the fulcrum of modern western Berlin, the first thing a visitor stepping out of the Bahnhof Zoo train station sees. To the north stretches the Tiergarten, once Frederick the Great's private hunting pre-serve and now Berlin's central park. Angling away to the southwest is Kur-fürstendamm, today the jewel of western Berlin, glittering with all the neon and noise of twenty-first-century capitalism, though in Albert's time it was a seedy strip of cabarets and bars.

On the eastern edge of the Tiergarten, the Reichstag, the seat of the German parliament, stands stony and gray as a permanent headache, drain-ing the color from the sky, crushing the earth. The grandest of Berlin's old ruined streets, Unter den Linden, runs east from the Brandenburg Gate past the reopened Adlon hotel and the surviving vestiges of Prussian and Fred-erickian splendor, gray neoclassical edifices with gods adorning their roofs, culminating in the massive Berliner Dom, a nineteenth-century cathedral lording it over a cluster of museums. One of them contains an entire Greek temple, retrieved piece by piece from Turkey. The old University of Berlin, a former palace and now renamed Humboldt University, fronts the avenue, gazing across through iron gates to the old Royal Prussian Library, where Lenin once studied, and the square in front of it, filled with cars and buses, where Hitler burned books by Einstein, among others, in 1933.

Albert arrived in Berlin at the end of March 1914 and, finding his new apartment in the midst of repairs, temporarily moved in with his uncle

Jakob in Charlottenburg. Mileva arrived with Hans Albert and Tete, his ear infection healed, toward the end of April.

His new career commenced with Albert splitting his time between a small office in Haber's institute nearby in Dahlem and the Royal Prussian Library, where the Academy had its quarters. For the most part he found his new colleagues quite human except for "a certain peacocklike grandeur in writing," as he put it,[1] but beneath their surface friendliness, there was a coolness that bothered him. "They have no psychological comprehension of others," he confided to Frank. "Everything must be explained to them very explicitly.[2] Even with Planck, he felt, there was a reserve, an inner barrier between them that no amount of friendliness, or scientific camaraderie, could bridge. Only later would he conclude that it was because Planck was not a Jew.[3]

Albert's sponsors in the Academy—Planck, Haber, and Nernst—clearly expected him set up a research program in quantum theory, especially as it pertained to chemistry and specific heats, although no formal mission statement was ever formulated. Albert, however, was mainly interested in finding experiments that could verify the new Einstein-Grossmann theory of gravity. In that endeavor he found a willing spear-carrier in Freundlich. A dinner at the Freundlich house soon after the Einsteins had settled in in Berlin ended with Albert scribbling equations all over Käthe's best tablecloth. (Many years later she would rue the day she had ever washed it.)[4]

Such vestiges of their former life would be rare for the Einsteins, however. It was not long after arriving in Berlin that Mileva's worst fears about the move began to come true. By now used to being able to run free in Berlin, Albert continued to act as if Mileva were not there, disappearing for weeks at a time and not telling her where he had been. Having told Mileva for twenty years that she had the advantage over his family because he knew he could take her anywhere without worrying that she would "chatter stupidities," he now confided to Haber that he felt more comfortable with his relatives than with his wife.

Much of this time, Mileva was pretty sure, Albert was spending with Elsa. The atmosphere at home got increasingly tense, with even Hans Albert accusing his father of becoming nasty since the move to Berlin.

The last straw seems to have come in July, when, without consulting Mileva, Albert began to look for someone to sublet the apartment in order, Mileva believed, to force her out.[5]

There was a huge row. Mileva accused Albert of being led around by the nose. "You must know that people take an interest in how the great man behaves," she warned him. Albert was so stung by these accusations that he was still muttering them, as it were, to himself days later when he scrawled them under the heading of "bad jokes" on the back of a seething note to Mileva.[6] A reference to Clara Haber among these innuendos suggests that Mileva may have complained about him to the closest person she had to a friend in the city. Fritz Haber, of course, was ultimately responsible for having brought Albert to Berlin; in a sense he was Albert's boss.

Outraged and no doubt feeling betrayed, Albert fled the apartment back to his uncle's place. To escape the tension, meanwhile, Mileva yielded to Clara Haber's entreaties to take the boys and move in with her. For the next few days the only communications between Albert and Mileva took place in the form of notes hand-delivered by Fritz Haber.[7]

Missing his children, Albert compiled a list of conditions under which he would be willing to resume life with Mileva. Numbered and subdivided like the clauses of a contract, or the protocols for an experiment, they included demands that his laundry and clothing be kept in an orderly fashion, that Mileva serve him three meals a day in his room, and that she not disturb his desk. She would also have to forgo all personal relations with him and not criticize him in front of the children. He would not travel or appear socially with her except to keep up appearances. His instructions left little to chance: "You explicitly oblige yourself to observe the following points in interaction with me. 1. You expect no tenderness from me nor do you make any accusations of me. 2. When you direct your speech at me you must desist immediately if I request it. 3. You must immediately leave my bedroom or office on my request without opposition."[8]

Mileva sent a letter through Haber indicating that she was willing to abide by these rules, but her seeming acquiescence was not sufficient for Albert, and he wrote back, demanding a more explicit answer. "I am prepared to return to our apartment because I don't want to lose the children and because I don't want them to lose me, but only for that reason. After all that has occurred, a comradelike relationship between us is out of the question." If she couldn't keep things businesslike between them, he warned her, he would pursue a separation.[9]

The next day Haber prevailed on him to mellow his tone slightly, to which he agreed. "It is not out of the question," Albert told Mileva, "that if you behave correctly on your side that I would assume a greater trust in you again. Be that as it may, no one can see the future. In any case I don't find remarks about it to be useful." He asked her once again for a clear answer.[10]

Faced with such implacability, Mileva really had no choice but to give up the marriage—at least for now—if she wanted to keep her sanity. On Friday, July 24, Albert came over to the Habers' for a sit-down, during which Fritz drew up a contract for a formal separation. Albert would give Mileva 5,600 marks a year (or 7,000 Swiss francs). The children were to remain with Mileva, who made it one of the conditions that she would never be asked to give the children over to Albert's relatives. Albert balked only at the request that he increase his life insurance, but then gave in to that too. They made an appointment with a lawyer, but Albert did not show up, sending Michele Besso, who had been summoned all the way from Trieste, in his place.[11]

Two days later, on Sunday, Mileva and Albert had a three-hour meeting, which nailed down the specifics and smoothed the way to a divorce, as he told Elsa. Mileva would take the boys to Zurich. Albert could see them henceforth, but only on "neutral ground," not in Elsa's house. "Now you have proof that I can make a sacrifice for you," he wrote to Elsa, who had left town on a vacation with her daughters. "Such an affair is bit similar to a murder!"[12]

Albert went straight from Haber's house over to Elsa's after the meeting, even though Elsa wasn't there, and told her parents, who received the news with mixed feelings. Then he went to sleep in Elsa's bed, bemused that he found it comforting, "like a tender confidence."[13]

In the days to come he often referred to the meeting with Haber and Mileva as "that day of misfortune (Friday)," because it had cost him his children. "They used to shout with joy when I came in," he told Elsa. Now they were gone forever and their image of their father would be systematically torn down. Nevertheless, he stood firm when Mileva made one last attempt to reconcile.

On July 29 Albert took Mileva, Hans Albert, and Tete to the train station. The parting was harsh. Mileva made it clear once again that his conduct was a crime against her and the children. Albert kissed the boys and bawled like a baby for the rest of the day. Haber had to help him home from the station. The next day Albert packed up most of the furniture and sent it after her.

Although they shouldn't see each other for a while, out of discretion, Albert assured Elsa that she would eventually become his wife, a union of which apparently even Haber approved. Within days, however, he had backed off, after telling Mileva that he was not demanding a divorce—provoking a minor storm, especially with Elsa's parents, who were already annoyed that Albert had been too generous in his settlement with Mileva.

Nevertheless, he reminded Elsa, there was no other woman for him. "It is not a lack of true affection which scares me away again and again from marriage!" he told her. "Is it a fear of the comfortable life of nice furniture, or the odium that I burden myself with, or even of becoming some sort of contented bourgeois? I myself don't know; but you will see that my attachment to you will endure."[14]

While the Einstein household was thus collapsing, so finally was the house of cards that a half century of European diplomacy had built. On June 28 in the fine old historic city of Sarajevo, which was then still part of the Austro-Hungarian Empire, a Serb terrorist assassinated Archduke Francis Ferdinand, the heir to the venerable Hapsburg throne. Ferdinand, in one of the ironies that marks Balkan history, was a moderate, who would have made things better but not good enough for the oppressed Balkans. Backed by Germany, Austria issued an ultimatum to Serbia demanding to participate in bringing the cowardly assassins to justice. Serbia, after consulting with its putative big brother, Russia, which in turn consulted its ally, France, refused.

Austria responded by declaring war on Serbia. Having failed at crucial moments twice before in the previous five years, Russia could not withhold its support, and mobilized its troops along the German and Austrian borders. The Ring of Iron was being tightened and sharpened. Germany protested the mobilization and, receiving no reply, declared war on Russia. Two days later, Germany declared war on France as well. As far as Germany was concerned, its actions were strictly an issue of self-defense. Germans had been reminded for decades that they might have to fight again for the existence of their nation. The German high command had long ago devised a plan for just this sort of emergency; their strategy was to hold off the Russians in the east while concentrating most of their forces in the west to make a quick sweep of France, by way of neutral little Belgium. On August 4, accordingly, the German army invaded Belgium, the land of the Solvay Congress, burning the library at Louvain, and generally shocking the rest of the civilized world. Britain had no choice but to join the war on the side of its treaty partners France and Russia. After being courted by both sides, Italy eventually threw in its lot with the latter three.

But the German plan proved a failure. As it turned out, not even the occupation of tiny Belgium went according to the master schedule. The Germans would be the last to know, however, for their high command lied from its onset about the progress of the war. In September the German offensive bogged down near Paris in the Battle of the Marne. Both sides dug in for what would be a four-year standoff. The Western Front—a swath of trenches, barbed wire, and craters, blood, noise, shit, and smoke across northern France—became a killing machine.

To Albert, alone now in his big, empty apartment, enjoying his newfound quiet and solitude and the lack of responsibility, the war seemed at first like some catastrophe that was happening to somebody else, far, far away. "Europe, in her insanity, has started something unbelievable. In such times one realizes to what a sad species of animal one belongs," he wrote to

Ehrenfest. "I quietly pursue my peaceful studies and contemplations and feel only pity and disgust."[15]

His one fear, he said, was for Freundlich, who was already in the Crimea preparing for the solar eclipse. "Freundlich (my good astronomer) will experience the beginning of the war instead of the eclipse."

So what might have been the most exciting scientific experiment of the young century was instead among the first casualties of the young war. Freundlich was arrested as a spy, and he and his assistants were sent to a POW camp in Odessa. They were subsequently traded back to Germany for a group of Russian officers. He was back in Berlin by Christmas, but his equipment stayed behind. A second expedition, led by E. E. Barnard from California's Lick Observatory, also succeeded in reaching the Crimea but fared little better once there; on the day of the eclipse it rained. Their specially built eclipse camera was then impounded by the Russians, and it took three years for Barnard to recover it, ruining Barnard's chances to detect the Einstein effect at the next eclipse, in 1917.[16]

AFTER A BRIEF STAY in the Augustinerhof Hotel across from Zurich's train station, Mileva and the boys moved down Bahnhofstrasse into a pension. There, jobless and angry, she had no alternative but to wait for money and word from Albert. Hans Albert later recalled it as the worst time in his life.[17] Nobody knew what the future would bring. Were they going to stay in Switzerland? Would they return to Berlin? Was Papa going to join them? When would they see him again?

By September, having exhausted her funds, Mileva asked Albert for more money. Otherwise, she averred, she would be forced to "seek out the assistance of other people."

Albert read her note as a threat and replied in an exasperated tone, ticking off all the sums—100 marks here, 150 francs there—that he had already sent her and the means by which he had dispatched them. He had also paid for the move, he reminded her, as well as an operation for his mother. He had no money himself, but promised her 400 francs out of his next paycheck in October. Meanwhile, he told her, "I know very well without this from your earlier behavior what I can expect from you.

"You took my children away from me and saw to it that their feelings for their father were poisoned. Other people who are close to me, you will take away from me. You will attempt to poison everything in any way possible that has remained my life's pleasure. This is the well-deserved punishment for my weaknesses, which have chained my life to yours.

"Let me repeat: whatever you do, it will not be a surprise to me."[18]

As quickly as the money trickled into Zurich, it seemed to disappear like snowflakes in Mileva's palm. As a stopgap measure, she borrowed some sheet music from Lisbeth Hurwitz, figuring that she could earn a little by giving music lessons. Mileva and the boys moved into a second-floor flat in her old neighborhood on Voltastrasse, just up the hill from the university. It had its charms, for at its front, a small balcony with wrought-iron balustrade looked out over the hillside and the lake. Soon, in addition to teaching music, Mileva began a new career as a math tutor while she waited and winter clouds darkened the distant shining lake.

Within a few weeks of his family's departure Albert moved out of the Dahlem flat to a smaller bachelor apartment in the center of the city. A visitor recalled it as a spartan landscape, barely furnished and littered with stacks of papers, with Albert forever shuffling through the piles in search of some scientific article he needed. It was in these magnificently lonely rooms that he took up the battle again with his much-troubled and hopelessly ambitious theory of gravitation.

"As his mind knows no limits, so his body follows no set rules," reported his friend Janos Plesch, another doctor. "He sleeps until he is wakened; he stays awake until he is told to go to bed; he will go hungry until he is given something to eat; and then he eats until he is stopped."[19]

Although in academic and artistic circles Einstein's name was already familiar, with Elsa as his social guide, his circle of acquaintances expanded quickly. One of Albert's closest new friends was a doctor by the name of Hans Mühsam, who worked at the Jewish Hospital. His brother Eric was a well-known writer who had been sent to prison for his pacificist views shortly after the war began. Mühsam, who followed developments in science, had by chance been introduced to Elsa one day in 1915. He remarked that she had a famous name and asked if she had ever heard of Albert Einstein. When Albert heard this story he was curious and sought out Mühsam. Soon they were spending Sunday afternoons together hiking and talking about physics and biology.

Minna Mühsam later recalled that for a long time Albert came to see them daily. "First comes your husband," he told her once, "then for a long while comes nothing, and only then come all other people."[20]

Hans and Minna and Albert and Elsa became a close-knit group. They spent Friday nights reading the Bible after Minna scolded Albert for knowing almost everything about everything except the Bible. These sessions—and the Mühsams in general—were a window for Albert into a world he had ignored or may never actually have known. His own upbringing had been so solidly secular, he told Minna, that theirs was the first consciously Jewish household he had ever experienced. It suddenly dawned on him that he had been surrounded by strangers throughout his entire life.

Even among Jews, though, Albert would always be alone. He once admitted to Minna, "I can't bear it when a person comes too close to me. Then I withdraw back into myself."[21]

ONE BY ONE Albert's colleagues were drafted to help support the national war effort. Otto Stern was soon sending Albert postcards from the Eastern Front. Max Born, who had wandered through the Göttingen art museum with David Hilbert rapt with relativity and had recently been recruited to Berlin, found himself studying the physics of artillery trajectories. Ludwig Hopf, Albert's first assistant in Zurich, joined the German Air Ministry and helped design aircraft. Karl Schwarzschild, the promising mathematical astronomer in Potsdam, was sent to the Eastern Front as well. Walther Nernst, temporarily suspending the promotion of his heat theorem, accepted a commission as an officer and began advising on chemical payloads for shells.

The ever-eager Fritz Haber volunteered for military service immediately, but failed the physical. The rejection left him depressed, but his chemical inventions proved to be so crucial to the war effort that he was soon in uniform as a consultant on ammonia production. Haber supervised the making of chlorine for the first German gas attack in 1915 and then went on, as head of the chemical warfare service, to develop mustard gas.[22]

Not even Max Planck was immune to the patriotic fervor that surged through the ranks of German society. Besides his usual enthusiasm for anything that required discipline and sacrifice for the greater good, Planck was gratified to see his oldest son, Karl, who had been drifting and depressed, rally to the cause. "And yet what a glorious time we are living in," Planck wrote to his sister. "It is a great feeling to be able to call oneself a German."[23]

And so it was that Planck's name, along with those of Haber and Nernst, appeared on the list of ninety-three scientists, artists, and scholars who signed a "manifesto to the Cultured World" that was published in all the major German newspapers on October 4, 1914. In the document, composed by a pair of writers and a vice mayor of Berlin, Germany's intellectuals made a ringing announcement of their solidarity with the German army, and attempted to repudiate lies about the alleged brutality of the Belgian occupation spread by Germany's racist foes.[24] This was, after all, the nation of Beethoven, Kant, and Goethe. The manifesto and others like it—two weeks later there appeared a "Declaration of German University Teachers" —was instrumental in straining relations between German scientists and their foreign colleagues. Planck's old friend Wilhelm Wien drew up his own manifesto demanding that German physicists have nothing to do with British scientific journals because British physicists were always stealing German discoveries. Although he expressed his sympathies with Wien's sentiments, Planck declined to add his signature to the statement, and in fact was already having second thoughts about having signed the first one. Some of his misgivings were inspired by Lorentz, who had begun sending Planck journalistic and other accounts of German military ruthlessness. Planck initially protested that there had been suffering on both sides; by then his nephew had been killed and his younger son captured, and both of Nernst's sons were dead. Within a year, however, Planck was apologizing privately to his scientific friends for signing the manifesto. He had only meant, he said, to stand up for his country, not necessarily to defend every act of its army.[25]

Albert also refused to sign the manifesto, as did his Göttingen cohort David Hilbert. As a Swiss citizen, Albert ran little risk of being called a traitor, or even raising much of a ruckus. That was not quite the case, however, for the manifesto that he did sign, an antiwar screed drafted by Elsa's sometime physician and friend, Georg Nicolai.

Thus entered onto the world stage and into Einstein family affairs a man who was almost a parody of the Nietzschean *Übermensch,* brilliant, courageous, handsome, ruthlessly egocentric, sexually voracious, and utterly opposed to the interests of the German nation as presented by its alleged leaders. Born Georg Lewinstein in Berlin, the son of a pair of intellectuals and political liberals, he had been expelled from boarding schools and colleges all over Germany for fighting and dueling, leaving illegitimate children and broken hearts in his wake. He eventually achieved degrees in medicine and zoology, lived in Paris, was a drama critic in Leipzig, traveled in the Far East, worked with Pavlov in St. Petersburg, and became one of the world's foremost experts on the electrocardiogram.[26]

Until 1914, war and pacifism were not significant issues in Nicolai's life. Presumably by using family connections, he had somehow managed to avoid any kind of military service or training at all. At the outbreak of the war, Nicolai was a professor at the University of Berlin, and he arranged to become chief of cardiac services at a large military hospital in the suburb of Tempelhof.

Nicolai had been one of the first to recognize that on the new killing fields of the Western Front, sheer quantities of bullets, armor, and rations would count for more in the long run than the traditional warrior virtues of valor and genius. He sent a letter to the General Staff that consisted of a simple graph with two curves, one representing the industrial output of Germany, and the other, that of its allied enemies. Germany's curve would initially rise faster, he explained, because of its policy of war readiness, but then would soon fall as the war cut off its access to resources. Meanwhile, the Allies' capability would rise until the curves eventually crossed. At that point, he argued, Germany would lose the ability to negotiate peace.[27]

Nicolai was outraged by the manifesto of the ninety-three and its smug appropriation of Goethe, Kant, and Beethoven. To him it was simply an act of intellectual treason. Soldiers and the masses could celebrate war and the sundering of nations, but artists and scholars were part of a greater, more sacred cultural tradition. Nicolai titled his countermanifesto "Appeal to Europeans." In it he described the war as a calamity that would destroy the ties between nations just when technology was making those ties more essential and unavoidable than ever. There would be no victors, only victims. And in a particularly prophetic note, he warned, "It is therefore not only desirable but a dire necessity that educated men of all nations make their influence felt, so that whatever the outcome of the war, the terms of the peace do not become the source of future wars."[28]

Nicolai brought the document to Albert and to Wilhelm Förster, an elderly astronomer who was chagrined at having been one of the "ninety-three," and both agreed to sign as coauthors. Only one other person was willing to sign the document, however: Nicolai's friend Otto Buek, a writer and private scholar.

Not surprisingly, Nicolai was unable to get his "Appeal" published in Germany, and only several years later did it appear as part of a book published in Switzerland. The appeal's most immediate effect was that Nicolai, who was also delivering antiwar lectures to his biology classes, was transferred out of Berlin to the garrison town of Graudenz, the first of many such relocations. Although he was a civilian, he was under contract to the army and thus subject to their orders.

Nicolai took his lecture notes with him and began to expand them into a book, *The Biology of War,* while going on foxhunts with his erstwhile jailer and political opponent, the camp commandant. Meanwhile, the various Einsteins—Albert, Elsa, and Ilse—sent him encouraging postcards. One such missive featured the heads of the so-called Central Powers, the kaiser, the sultan of Turkey, and the emperor Franz Joseph; above the triple portrait, Albert had scrawled sarcastically, "Union is Strength."[29]

Nicolai's opinions continued to get him in trouble, however, and after a string of misadventures, court cases, and countercases, involving a cast of characters that included the emperor's wife, whom he had once treated, Nicolai's contract was canceled. In 1916 he was drafted into the army as a private and assigned humiliating work as a medical orderly.

Elsa, of course, had been a longtime friend and client of Nicolai's.

Whether their relationship was ever anything more—"*You can see Nicolai as long as he behaves himself*"—is lost along with a Russian notebook of his conquests.* Elsa's daughters had also known him, and their reactions to him were as starkly different as their personalities. Margot, the youngest, the artistic and shy one, hated Nicolai, whom she recalled as beautiful but not attractive, and couldn't understand what so many women saw in him. She remembered a frightening experience when she was twelve and Nicolai took her on his lap and ran his hands all over her body. On another occasion Nicolai had tried to hypnotize her, and when he had no success, became annoyed and locked her in the basement, beneath the noisy X-ray machine. She had to scream for help to be let out.[30]

Margot's older sister Ilse was much more vulnerable to his charms.

Ilse was seventeen when the war broke out, "a spirited idealistic young woman," in the words of Wolf Zülzer, Nicolai's biographer, and possessed of a certain swanlike beauty.[31] A childhood accident had deprived her of sight in one eye. Ilse's politics were decidedly left-wing. One of her best friends, Fanja Lezierska, was a member of the radical *Spartakusbund* led by Karl Liebknecht and Rosa Luxemburg. Ilse was also remarkably level-headed, for she managed to negotiate the shoals of close relationships with two of the century's most dangerous men, as we shall see, without apparent damage.

Ilse became smitten with Nicolai during his political crusades and travails, and the two became lovers for at least a while, according to Margot.[32] Ilse's admiration for Nicolai was little short of hero worship, and she became an accomplice and comrade in some of his adventures—typing up his protest letters, playing courier around the Berlin underground, introducing him to socialist comrades, and lobbying Albert in behalf of his causes. Perhaps equally important, she served as a direct conduit between Nicolai, who was forever hatching antiwar projects, and Albert, who might be convinced to participate in some of them.

IN EARLY 1915, Albert and Elsa began attending the Berlin meetings of a genteel antiwar club known as the Bund Neues Vaterland—Organization for a New Fatherland. The name was a piece of political camouflage. The

---

*Nicolai had a mesmerizing effect on women, even those smart enough to know better. In a notebook that he kept during a later sojourn in Russia, Nicolai compiled two lists—one chronological and the other alphabetical—of some 106 women with whom he had slept. According to Wolf Zülzer, Nicolai's biographer, the list was not complete, but it nonetheless included two mother-daughter combinations.

founders of the Bund were a diverse, pacific-minded collection of academics, writers, politicians, businessmen, lawyers, and retired diplomats who hoped to work within the system not only to bring the war to an early end but to promote universal suffrage, democratic reforms, and workers' rights. They circulated petitions and lobbied members of the government with long memoranda. The Bund also published pamphlets on its own printing press. The group was active, subject to increasing censorship, from its first meeting on November 11, 1914, until the military closed it down early in 1916.

Albert and Elsa showed up as "guests," according to the BNV minutes, at a meeting in March of 1915, perhaps at the urging of Georg, count von Arco, a physicist, inventor, and businessman.[33] They later joined as formal members. It was not long before Albert made a new friend. Through the international peace grapevine, the Bund had been corresponding with the famous French novelist Romain Rolland, whose pacifist convictions had led him into exile in Switzerland, where he was working for the Red Cross (and was about to win the Nobel Prize for literature). While duly denouncing German atrocities, Rolland held the ruling classes of all nations responsible for the war; imperialism, he said, was the real culprit.

Albert, sensing a kindred spirit, wrote to Rolland comparing the current nationalist frenzy to religious lunacy. "Even the learned men of the different countries are behaving as if their cerebrum had been amputated eight months ago," he complained. "I put my humble forces at your avail in case you think that I can be instrumental to you—whether by my location, or by my connection with German and foreign representatives of the exact sciences."[34]

ONLY A FEW months later Albert got his chance to help the cause. In June the Bund, perhaps at Albert's suggestion, decided to arrange for an appeal by intellectuals and academics in support of international understanding. Albert and his friend von Arco were part of a committee charged with writing a declaration. The plan was to have colleagues in various countries sign it and then issue it in their own names. Officially, the Bund would not be involved, because it was already in a vulnerable position. A letter by a Bund member criticizing the violation of Belgian neutrality, which had appeared in a confidential BNV circular, had somehow wound up being reprinted in newspapers, causing its author to be branded a traitor.

Albert, naturally, wrote first to Lorentz, who, having better sense, turned him down. That was the last anyone heard of the international intellectuals' appeal. It had been a naive idea, Albert concluded, although he continued to muse about what he could do to unite the colleagues of different "fatherlands." The small flock of industrious thinkers, as he told Ehrenfest, was the only "fatherland" for which he could generate any enthusiasm.[35]

Despite his pacifist connections, not even Albert could escape at least the fringes of the war effort. Early in 1915, as a physicist and veteran patent examiner, he was enlisted to resolve a patent dispute between Hermann Anschütz-Kämpfe, an industrialist who had designed a compass based on gyroscopic principles—the forerunner of today's internal guidance systems—for a proposed submarine expedition to the North Pole, and the American inventor Elmer Sperry, who was competing to supply the German navy with gyrocompasses very similar to Anschütz-Kämpfe's. The fee for Albert's services was 1,000 marks—money he could now certainly use.[36]

Perhaps because he was used to imagining devices from reading descriptions of them in patent applications, in his first fumbling court appearances Albert relied on the patent documents and blueprints, and ruled in favor of Sperry, but the court was not satisfied and commanded him to inspect the rival compasses personally. In July, therefore, Albert took Sperry's compass to the Anschütz-Kämpfe factory in Kiel and ran a series of tests, which convinced him that Anschütz was right.

The court subsequently revoked Sperry's contract and awarded it to Anschütz-Kämpfe. In the process Albert and Hermann Anschütz-Kämpfe became great friends, and thereafter Albert often vacationed in an apartment at the Kiel factory, which was conveniently located for sailing.

The compass episode had an interesting effect on Albert's scientific life. A gyrocompass tries to maintain its orientation in space because it is spinning. Albert wondered if such an effect on the atomic scale might explain the phenomenon of ferromagnetism, the puzzle that had first enchanted him at the age of six. According to the new quantum notions, each atom was a little electromagnet with electrons spinning around in it, and magnetism resulted when the spins of all the atoms in a substance were aligned. If that were the case, it might be possible to measure the angular momentum produced by the currents in a magnet, a notion that had first been suggested by the French physicist André Marie Ampère in 1820.

To that end, early in 1915, Albert and Wander Johannes de Haas, a Dutch physicist and Lorentz's son-in-law, conducted a series of experiments at the Physikalisch-Technische Reichsanstalt, where Albert had a visiting appointment, using a giant iron cylinder that hung like a stalactite from the ceiling. Albert, who joked that he had discovered a passion for experiments in his old age, reported that the cylinder engaged in a series of "adventurous motions," in response to pulses from an electric solenoid. But the molecular currents remained undetected, and in the end the experiments provided only a brief respite from Einstein's gravitational endeavors and the turbulence that was engulfing his life.[37]

THE EINSTEINS' private war was not going much better than the public one. As the end of 1914 and the holiday season neared, it was clear that Albert wasn't coming to Zurich anytime soon. Dividing their possessions had taken time. The last to go was some lace belonging to Zorka, after which Albert declared that he didn't want to be bothered over such trifles. He promised that he would support Mileva with 5,600 marks a year, in quarterly installments, "at least as long as my income does not sink considerably below its current state." He sent the children some parlor games that arrived in time for Lisbeth Hurwitz's New Year's Day visit.[38]

Meanwhile, Hans Albert had staged his own protest by refusing to answer his father's chatty letters. The boy was only ten, but his childhood had already ended. The first year in Zurich was beginning to cast what would be a long shadow. Fifty years later he would snap at his daughter Evelyn, "Nobody has a right to expect a happy childhood."[39] Albert believed he knew whom to blame for the break in correspondence. "It seems that the greetings I send the children are not being passed on, since they would have surely sent me greetings after such a long time. It seems useless to keep sending greetings."[40]

Mileva pressured her son, and the correspondence resumed in a tense, lopsided fashion, with Albert sending geometry puzzles, cheery promises of summer hikes, and admonitions to keep up the piano lessons, and Hans Albert ("Albertli" or "Adu") only grudgingly surrendering bits of news interspersed with bursts of rudeness and spite that left his loving father in a state of shock.

Such incidents reinforced Albert's fear that Mileva would turn the boys against him. He missed them, he told her, but he would not tolerate a situation in which they were being used, crushed by some feminine agenda. If she really didn't want to destroy his relationship with the children, Albert advised, she should not discuss his letters to the boys with them, but should let Albertli (and eventually Tete) write back on their own without supervision or censorship. If he sensed any indications to the contrary, Albert warned, he would break off writing at all. Mileva responded by having new photographs made of the boys and sending him a set.

Mileva's complaints were of a more familiar nature. She made the mistake of asking Albert for a portion of an extra thousand francs he would soon receive, only to meet with a furious response. As his wife, he reminded her, she would always have a claim to his "little bit of money," but he was not about to hand her the proceeds of his scientific awards. "I have provided for your support sufficiently and I find your persistent attempts to wrest for yourself everything in my hands extremely unworthy. If I had known you twelve years ago as I know you now, I would have judged my responsibilities toward you quite differently."[41]

It was an unusually snowy winter in Zurich, with deep drifts persisting through March. With her tutoring jobs, Mileva's finances had stabilized, though the family's situation was hardly comfortable. Sporadic entries in Lisbeth's diary portray a game but threadbare existence: Mileva on the streetcar laden with birthday packages for her distant husband, wondering if she should take up teaching Italian after borrowing a book from Lisbeth. She announced that there would be no summer vacation this year. Mileva's brother Milos had also joined the army and was now missing on the Russian front, leading her mother to become ill from grieving.[42]

Tete, enjoying a streak of good health, was beginning to come into his own. His astonishing intellectual precocity continued unabated, and when he began to show a particular affinity for music, Mileva arranged lessons for him, straining her resources even more. "The children are so dear," Lisbeth recorded. "Especially the little one, who is so lively and funny."[43]

Meanwhile, Albert began pressing Mileva for a suitable time when he could visit Zurich and the boys and take Albertli on a hiking trip. She equivocated. Against her own backdrop of privation, uncertainty, and unconfidence, Albert, the lion of Berlin, suddenly loomed as a potentially disruptive force to what little she had left of her own. "I expect to hear from you soon as to whether I will be allowed to take a short trip with little Albert," Albert reminded her. Albert imagined he was being punished, and he worried that his old friends might be judging him harshly for the breakup of his marriage.

He was eager to explain to Zangger, in particular, what outrageous things had happened to him. "I'm burning to take my dear boys into my arms again," Albert told Zangger. "As hard as the separation from them is for me, it was a question of my existence."[44]

At least twice, a date was set for his visit and then postponed. On the second occasion it was because Hans Albert suddenly refused to go hiking with his father, whereupon Mileva took her cue and scheduled a trip with the boys out of town.

His ongoing domestic problems did little to increase Albert's appreciation for, or expectations of, his fellow man. Whether by coincidence or not, a dose of pragmatism or pessimism had recently begun to undermine the almost cosmic optimism that had informed his antiwar activities. "Why shouldn't one let the servants live happily in the madhouse?" he asked

Zangger, explaining that he was beginning to feel content in his separation from the "crazy tumult" that occupied the public.[45]

As a foreigner and a Jew, Albert could get away with not supporting the war as long as he wasn't too publicly outspoken about his position. Also, none of the manifestos to which he had lent his support was having any effect. In fact, the sinking of the ocean liner *Lusitania* in July by Germany's aggressive and controversial submarine warfare program had brought the United States into the war, bringing further pressure on Germany and hastening the day when the lines on Nicolai's graph would cross.

So Albert trudged from the Academy, where his peers were busy exchanging waistcoats for officers' uniforms, to the Bund's quiet little coffeehouse gatherings, where he could gripe safely about the destruction of Europe while avoiding being dragged into the vortex of martyrdom that attracted people like Nicolai and Adler. He was content to admire that martyrdom from a distance. As a holy fool devoid of organizational ability or inclination, he knew he was no threat to the establishment. Within these confines he would sign anything—including a petition opposing the annexation of Belgium (but not the Baltic or certain African states) that incredibly and patriotically predicted total German victory—and join any organization, especially ones that never apparently got around to meeting, like the "Large International Council," and later in the war, the Vereinigung Gleichgesinnter, or Association of People of the Same Opinion. "Take me on your list such that I am comforted: *dixi et salvavi animam meam* [I spoke and saved my soul]," he said upon receiving one such solicitation.[46]

He expressed his feelings most frankly to Besso and Zangger. "In this time one sees that the only thing really worth pursuing in the world is the friendship of superb and free people who can't be led to the most perverse mental attitudes by any printed bullshit they read," he wrote Zangger in the spring of 1915. "People are not as consequent in anything, including hate and unfriendliness, as they seem. A little impartiality regarding others might help you get over it with ease.

"I always fare the best with my innocuousness, which is up to 20 percent conscious. This is easily attained when you're indifferent to the feelings of your dear fellow humans—but you are never as indifferent to them as they deserve."

He went on, "Personally, I have never been so calm and happy as I am now. I live completely withdrawn and yet I'm not lonely, thanks to the kind care of a cousin who was the one who drew me to Berlin."[47]

AS THE SUMMER passed Mileva ran out of money for the rent and had to avoid her landlord. Albert finally arrived for his long-awaited visit in Zurich in September, but this reunion with his children was a bust. Mileva was apparently uncooperative about letting him take the boys away. Hans Albert, in any case, didn't want his friends to see him with his father. The ambiguity and tension of the family's situation was crushing. Was the family moving back to Berlin? Albert didn't know. Were he and Mileva getting back together? When it came to the future, Albert was as silent as the glaciers.

While he was in Switzerland Einstein and Zangger took a train to Vevey on Lake Geneva, where Romain Rolland was holding court in exile. In his account of the meeting the great pacifist described Albert as a young man, not tall, "with an ample figure, great mane of hair a little frizzled and dry, very black but sprinkled with gray, which rises from a high forehead."

They spent the afternoon by the lakeside discussing the war and German politics. Rolland was surprised by Albert's independence and outspokenness. "Einstein is unbelievably free in his judgments on Germany where he lives. No German has such a liberty. Anyone other than he would suffer by being so isolated in his thoughts during this terrible war. Not he, however. He laughs." He was also shocked by Albert's fatalism. Germany, Albert reported, was too drunk on recent successes on the Eastern Front for the war to end soon. Only a crushing defeat would bring them to the negotiation table now. As he was leaving, Albert confided his hope for exactly such a defeat, which would bring down the government and dismember the Prussian dynasty.[48]

Back in Berlin, Albert got a chance to speak equally straightforwardly to German society. The Goethe Society, which was planning to publish a book called *Goethe's Country 1914–1916,* invited him to contribute a no-holds-barred essay containing his opinion on the war. That was too good an offer to pass up, and as it turned out, it was too good to be true—the book was never published. Albert's first draft of his essay began by evincing the belief that war was the enemy of human evolution and that in the future an international organization of European states would be required to prevent it. That notion was acceptable to his editors, but the following passages, which raged against nationalism as a vessel of "animal hatred and mass murder, taken out and used, in case of war, by the obedient citizen," proved to be too much.

"I tried to suit your purpose without becoming dishonest and without being obliged to say something about patriotism," he wrote to the editors in November, acceding to the cuts they demanded. "In case your fine feelings resulting from contact with humanity as it exists around here find the changed text offensive, let me know, please."

The final version concluded with a ringing declaration of Albert's outsiderness. "Why make so many words when I can say everything in one sentence; moreover, in a sentence fitting to me as a Jew: 'Honor your master Jesus Christ not only with words and songs, but, above all, by your deeds.'"[49]

It was a declaration for which he would eventually pay dearly, but for the moment, he had every right to be full of himself. It was only that month, November 1915, after eight years of mathematical torture, false hopes, wrong assumptions, and misleading answers, in full view of the academy of war criminals and the world, that Albert had finally cracked open the problem of gravitation, space, and time. On his lonely stage on Wittelsbacher-strasse littered with scribblings and calculations Albert stood erect, his black eyes blazing, his frizzled black mane atangle amid the broken eggshells of Newton's universe. The militarist paymaster Koppel had gotten his money's worth. The golden hen had come through. A new world, a new universe was dawning.

20

THE NOVEMBER REVOLUTION

"THERE ARE TWO WAYS THAT A THEORETICIAN GOES ASTRAY," ALBERT wrote to Hendrik Lorentz early in 1915, as he contemplated the twists and turns that had led him to a crucial turning point in his long quest for a new theory of gravity. "1) The devil leads him around by the nose with a false hypothesis (for this he deserves pity) 2) His arguments are erroneous and ridiculous (for this he deserves a beating)."[1]

Over the last decade, Albert had certainly courted both pity and a beating many times over. He had begun back in 1907 simply looking for a way to modify Newtonian gravitational theory so that it would fit into relativity. The happy realization that a falling body feels itself as weightless had led him to enlarge his ambitions. Perhaps relativity itself could be generalized so that the laws of physics would look the same not just for people privileged enough to live in so-called inertial reference frames but for observers in any state of motion—covariant laws.

Using the thought experiment of a physicist being pulled upward in an elevator to create an analogy or "equivalence" between gravity and acceleration, Albert was able to deduce some of the relativistic properties that gravity should have: Light would bend and clocks would slow in a gravitational field. Even the speed of light would be variable; in fact, Albert had attempted briefly and futilely to construct a theory in which the gravitational potential was represented by the speed of light—that was the devil's first appearance in this drama. This elevator analogy eventually led to Albert's second big breakthrough, in the summer of 1912, when he concluded that gravity could be defined as a warp in the normally Euclidean geometry of space-time.

But how to weave this collection of insights and hunches into an actual, full-blown theory? Mathematical considerations led him and Grossmann to the Riemann tensor as the possible basis for their gravitational equations. An equation based on the Riemann tensor, or its slimmed-down cousin, the Ricci tensor, would yield laws that were covariant, good for any observer—a mathematical miracle, but did it have anything to do with gravity? Apparently not, Albert concluded when the Ricci tensor gave what he believed—the devil again—were the wrong answers for some simple gravitational problems. In response he formulated another tensor theory, the *Entwurf,* that did give the right answers, as far as he was concerned.

Although it was consistent with relativity, as Einstein had defined it in 1905, the Einstein-Grossmann theory was not covariant, which was disappointing (not to mention that it did not even completely satisfy the equivalence principle, which was downright embarrassing, nor did it soon give the correct answer for the perihelion advance of Mercury). Albert soon found good reasons—such as the conservation of momentum—why he had to choose the tensor that he did, but despite his assurances to Lorentz, he was still troubled that his theory lacked covariance, and he kept poking at it, like a suspicious diner.

Finally, in September of 1913, Albert found an answer that satisfied him: a deep and fundamental reason why a covariant theory of gravitation was impossible. The problem lay not in tensors or Newtonian physics, but in the idea of covariance itself, which, Albert now thought he could prove, violated the notion of cause and effect, without which neither physics nor the world held any promise of sanity. Albert first mentioned the possibility of this new argument at the Vienna science congress at the end of the month.[2] Two months of mathematical churning later, he reported to Ehrenfest, "The gravitational affair has been clarified to my *complete satisfaction. . . .* It can be proven, namely, that *generally covariant* equations that determine the field *completely* from the matter tensor do not exist at all."[3]

Albert arrived at this radical and far-reaching conclusion by means of a subtle paradox he called the "hole argument." In it, he managed to turn the meaning of covariance against itself to conclude that in a covariant universe, there could in principle be an infinite number of different gravitational fields for the same configuration of masses, or, conversely, that many different arrangements of mass would result in the same gravitational field.

This was an absurd and unsatisfying result, especially if you believed, as Mach had taught and Albert was endeavoring to prove, in the relativity of inertia, that is, that the properties of the universe were uniquely and completely determined by the relationship between its contents.

Einstein's reasoning raised fundamental questions about the meaning of space-time, and philosophers and physicists are still arguing about it. What follows is not an exact rendition of the hole argument, but merely is intended to supply some of the flavor of a difficult and subtle issue.

In principle there were an infinite number of ways to label and organize the points of space and time. Imagine, for example, that we are laying out a city. We could choose to organize it on a grid, like Manhattan, with avenues running north-south and streets running east-west, or we could organize it in a hub-and-spoke arrangement, with avenues running outward from some central point and the streets forming concentric rings around it. We could have streets that intersect on a slant or that undulate like waves, or combinations of any or all of these ideas. We can even sit on top of a tower in a revolving restaurant at the center of town and keep track of outside events by where and when they appear through the window by our table, which is somewhat analogous to the situation of astronomers on earth trying to keep track of the action in the sky.

The lure of covariance was that the laws of physics would have the same form regardless of the coordinate systems in which they were expressed. At any point in space-time, regardless of how it was labeled, Einstein's equation would give the value of the gravitational field—the metric—based on the density of mass and energy there. It didn't matter how we organized the city of space-time; reality should be the same, Albert thought. And therein lay the paradox.

Consider the universe as a city laid out as an infinite grid with streets and avenues—or their Cartesian equivalents, x and y—extending regularly and rectangularly forever out into the trackless void, or at least the outer boroughs. Among the permissible coordinate systems, Albert demonstrated, was another grid that was identical to the first, except that somewhere, say between 59th Street and 110th Street and 5th Avenue and 8th Avenue, there were no buildings—just raw dirt—and the imaginary streets departed from a perfect grid—the blocks getting larger and then bigger, or the streets curving concave and convex around a central region. On the outside of this "hole," as Albert called it, the imaginary streets would line up with the grid again, but inside it they would be different.

In modern-day mathematical parlance, the kind of remapping that Albert was talking about—in which the straight streets are bent into curves and blocks grow or shrink—is called a diffeomorphism. Lee Smolin, a quantum gravity theorist at the Pennsylvania State University who has studied Einstein's argument, explains that this is different than simply changing addresses of the points of space-time. "The best way to think about this is to imagine that there is only a finite set of points. Think of alphabet blocks with letters painted on them. A change of coordinates is like painting different letters on the same set of blocks. A diffeomorphism is like exchanging the positions of the blocks." The meaning of covariance was that the laws of physics should not change when the alphabet blocks are rearranged (as long as the blocks are rearranged in some mathematically defined pattern, and not just arbitrarily jumbled). And it was there that Albert thought that he spotted a glaring contradiction.

Suppose now that this is the universe, with stars instead of buildings scattered through it, which have combined to produce some complicated gravitational field. There were an infinite number of different maps for this universe that would match outside the hole and disagree inside it. Since the hole was empty, Albert argued, changing from one map to the other should not change the gravitational field in the hole—that was the meaning of covariance.

Albert began by imagining that we solve the field equations for the gravitational field at some address in the hole, say 76th Street and 6th Avenue, according to one such map, and then asked what this solution would look like from the point of view of a second map, or a different coordinate system. By switching back and forth between the maps and making adroit use of the properties of covariance, Albert showed that the value of the gravitational field assigned to 76th and 6th in one map could get reassigned to the point corresponding to 76th and 6th in the other map, which was a different point in space-time altogether.

This was a disaster, Albert argued, since it now meant that there were two completely different and equally valid gravitational fields that matched outside the hole but diverged inside it. If there could be two fields there could be an infinite number—all arising from the same distribution of stars.

It meant that the field outside the hole didn't determine what happened inside it. And what, after all, was the point of laws of physics except to be able to determine effects uniquely from their causes, or force fields from their sources? In principle this imaginary hole could be anywhere, and of any size—as small as an atom or as large as the solar system, as long as it was empty. This meant, therefore, that determinism in principle failed everywhere or at any given time. Gravity in a covariant world was a question with a million answers, which was the same as having no answer at all.

Albert was certain that the hole argument spelled an end to the quest of truly generalizing relativity. Covariance was just a little *too* magic, if it eliminated cause and effect, and so it could safely be jettisoned. In the fog of argument, he concluded that he could not write the laws of gravitation in a form that would be true for everybody and everywhere in the universe.

Satisfied, or at least reconciled to his theory's limits, at last, during the winter of his impending move to Berlin and throughout 1914 Albert turned his energies to exploring the features his theory *did* have. He was hoping to establish, for example, that the notion of relativity at least extended to rotating observers. By the fall he had completed a series of calculations that seemed to show that the latter was the case—in the

Einstein-Grossmann world, rotation, at least, was relative. This was a happy circumstance; the theory was covariant enough to support the equivalence principle and thus the identity of gravitational and inertial masses, but not so covariant that it would undermine cause and effect.

Elated, Albert booked himself into the Academy in October for a talk that would be the definitive exposition of the Einstein-Grossmann theory complete with elegant and formal explanations of the hole argument.[4] The response from his colleagues was, as usual by now, muted, with an undercurrent of uncomfortable grumbling. But Albert was undaunted.

"Now I am completely satisfied and no longer doubt the correctness of the entire system, regardless of whether the observation of the solar eclipse will succeed or not. The logic of the thing is too evident," Albert declared to Besso in the flush of apparent success. "The general theory of invariants was only an impediment. The direct route proved to be the only feasible one. It is just difficult to understand why I had to grope around so long before I found what was so near at hand."[5]

A bit later, in January, Lorentz made a typically perceptive observation, namely, that the aether was alive and well in Einstein's new theory. The fact that it was not totally covariant meant that Einstein's theory favored some coordinate systems over others. Lorentz did not oppose that, for it suggested at least the possibility of some sort of a universal reference frame after all. "You are right in what you say," Lorentz stated, "only because you do not wish to hear of an ether at all."[6]

It was in replying to this wicked but understandable confusion that Albert made his famous comment about the devil and erroneous arguments, as an illustration of the treacherous and laborious path, haunted by mathematical and epistemological demons, that he had traversed to arrive in the clear light of his present understanding.

But the devil would not let him alone. Gravitation, a theory he thought was now settled, turned out to be no such thing. Like hunters closing in when the fox gets tired, the critics of the *Entwurf* were gaining ground on him.

One of those was Max Abraham, his old adversary, who had long since moved to the University of Milan and had finally achieved a measure of stability, if not serenity. The Italians still believed in the aether, and though Albert kept sending him letters and papers, Abraham continued to snarl in his characteristic manner that his theory made no sense. Italy was also the home of Tullio Levi-Civita, one of the inventors of the tensor calculus on which Albert's theory was based. In fact, Levi-Civita, a professor at the University of Padua, had been instrumental in recruiting Abraham.[7]

In February 1915, Abraham, having received yet another mystifying Einstein missive, passed it on to his friend Levi-Civita, expressing his consternation.

Levi-Civita indeed saw things in Albert's mathematics to be concerned about. In his 1914 paper and in presentations to the Academy, Albert had employed a powerful and elegant mathematical technique known as the variation principle to defend his theory. Levi-Civita wrote Albert and criticized his calculation, potentially destroying the foundation of the entire *Entwurf.*

Albert was grateful for the attention, up to a point. "But when I saw that you attacked the most important demonstration of my theory, which I obtained with streams of sweat, I became not a little alarmed," he said.[8] Albert replied with a counterargument, but by then Levi-Civita had already come up with another criticism, and so it went for the next two months: point and counterpoint.

April 2, 1915: "For one and a half days I had to reflect unceasingly until I understood how your example could be reconciled with my proof. . . ."[9]

April 8, 1915: "I must really admit that my deeper reasonings provoked by your interesting letters have strengthened my conviction that the proof . . . is correct in principle."[10]

For once Albert was too engaged by an opponent to resort to his usual vitriolic asides. "Only Levi-Civita in Padua had his wits about him, because he's familiar with the math used," Albert reported to Zangger. "The correspondence with him is unusually interesting; it's my favorite preoccupation right now."[11]

"At the moment there is a very modest interest in this topic," he told Levi-Civita himself. "It is strange how little our colleagues in the field feel the inner necessity of an *authentic* theory of relativity. Unfortunately, the attitude of our fellow human beings who do not work in the field is incomparably more strange. Therefore, it is a double pleasure for me to know better a person like you."[12]

The debate ended after two months in mutual mental exhaustion, with Albert still insisting that his theory was right but granting that he could not prove it completely. "I willingly acknowledge that you have touched the sorest spot of my proof," Albert admitted. "Here my demonstration is lacking in acumen. I also think that we have exhausted our subject up to the limits allowed by our present level of understanding." For the moment, then, the theory held, but, like a bridge that had withstood a tremendous flood, the integrity of its mathematical foundation was now in doubt.[13]

A month later, in June, Albert went to Göttingen for a week to deliver a series of six two-hour talks. Göttingen was David Hilbert's town, and there was probably not a scientist in Germany who was more in tune with Albert's way of thinking. Hilbert had been an early aficionado of relativity, a friend of Minkowski's, and one of the few fellow signers of Nicolai's manifesto. To Albert's delight, Hilbert understood the generalized relativity theory immediately. In fact, as events would prove, he understood it only too well.

Albert returned from his trip enthused about Hilbert's support and friendship. "I had the great joy of seeing that in Göttingen everything is understood to the last detail. With Hilbert I am enraptured. Hilbert is convinced of the general relativity," he told Zangger. "Sommerfeld is also beginning to agree."[14] Among the Göttingeners, however, there was widespread condescension for Albert's mathematical abilities and doubt that he, as a product of the "Zurich tradition," could bring his theory to completion. Typical was Felix Klein's remark that "Einstein is not innately a mathematician, but works rather under the influence of obscure physical-philosophical impulses."[15]

Albert, in fact, was already beginning to have his own doubts. When Sommerfeld asked him to write a couple of chapters on general relativity for a new book, Albert begged off, grumbling that none of the current expositions of the theory was correct.

After spending most of the summer planning and putting off his trip to Switzerland and vacationing on the Baltic with Elsa and her daughters, it was not until after he had returned from his meeting with Rolland on Lake Geneva that Albert turned back to his relativity calculations. A growing suspicion that his field equations might be wrong was gnawing at him. He was having a "rotten time," as he told Lorentz.[16]

There was much to be disgruntled about. In addition to Levi-Civita's criticism, which still worried him, a Dutch physicist, Johannes Droste, had used Einstein's theory to calculate Mercury's perihelion shift and gotten the same wrong answer—18 seconds of arc per century instead of 45.[17] But these alone were not enough to make Albert abandon the *Entwurf.*

It was only when Albert redid the calculation that had made him so happy a year earlier and discovered that he had made a mistake that he suddenly and finally realized that the *Entwurf* was doomed. The Einstein-Grossmann equations did not work with rotating observers after all; rotation was still not relative. In fact, the theory, he realized, probably did not work for any kind of accelerated motion. Finally, no doubt with a sinking heart, Albert repeated the entire process by which he had derived the field equations and found another error. Where he had previously been so proud that the field equations seemed to spring directly from first principles of mathematics, he saw now that they were riddled with tacit assumptions about gravity and the appropriate coordinate systems to use. Without those assumptions, the choice of equations was completely arbitrary; they could say anything. Alas, Levi-Civita had been right to be skeptical. All Albert's elegant manipulation of covariance requirements and coordinate systems, all his delicate and intuitive shadings of gravitational theory, three years of skull-wracking intensity, had gotten him nowhere.

Moreover, he was not alone in recognizing that his theory was now in trouble. David Hilbert had been enthusiastic enough about general relativity to go off and ponder it on his vacation, whereupon, according to Sommerfeld, he too had discovered unspecified flaws in the theory.

The only course left to Albert was to swallow his pride and go back to the Ricci tensor, the one he had rejected three years before. Being truly covariant, it had all the properties Albert needed, and then some, but for the moment all the arguments by which he had convinced himself that covariance was physically impossible or gave the wrong answer would have to be set aside. In one respect the switch in strategy was an easy one. The Ricci tensor was the beast in the alley that he had spent three years avoiding, and by now he knew it reasonably well. Mathematically, he simply had to do the one thing he hadn't yet tried.

THE STAGE WAS set for one last four-act drama, a final, heedless, month-long Götterdämmerung of tensors, proofs, transformations, and calculations that would finally ring down the end of the Newtonian universe, at least on paper. The audience was to be the Prussian Academy of Sciences, its members graying, distracted, half of them in uniform or dispatched to the nether realms of the embattled empire, and one mathematical genius three hours away down in leafy Göttingen, waiting by the mailbox.

David Hilbert, as it happened, had embarked on his own quest for the final magical field equations, his own "axiomatic solution to your grand problem," as he described it to Albert.[18] In this project he had two advantages over Albert. He was, first of all, a superior mathematician. Second, he was not a physicist, and, because he knew nothing about gravitation, he wasn't restricted by preconceptions about how the solutions should blend in with classical Newtonian gravity, or even about cause and effect. He could therefore go straight to the mathematically logical, most covariant theory. Hilbert liked to joke that physics was too complicated to be left to the physicists.

THE CURTAIN WENT UP on November 4, a Friday, when Albert stepped before the assembled gentlemen of the Academy to announce that he had "completely lost confidence" in the theory he had presented to them only a year earlier. It was based on an illusion, he admitted, an erroneous argument. And so, he explained, he had been led after all to seek new field equations that were generally covariant, "a requirement which I had abandoned only with a heavy heart in the course of my collaboration with my friend Grossmann three years earlier."

Albert outlined a new theory based finally on the Ricci tensor. It was a giant step toward a true generalization of relativity, but it still relied to some extent on special reference systems. "No one who has really grasped it can escape the magic of this theory," he declared.[19]

A week later Albert was back before the Academy with a modified version of the equations that, he believed, brought them closer to the ultimate goal of covariance. "Meanwhile the problem has been brought one step forward," he told Hilbert the next day. "If my current modification . . . is justified, then gravitation must play a fundamental role in the structure of matter. Curiosity makes it hard to work!"[20]

Armed with these new equations, Albert set out to recalculate the consequences for the two great experimental predictions of general relativity: light-bending, and the perihelion shift of Mercury. The first was a relatively simple calculation: The amount of light-bending turned out to be double the amount predicted by the old theory. A ray of light just grazing the limb of the sun would be deflected by about 1.7 seconds of arc.

Computing Mercury's shift was in principle a much more complicated process, but Albert, unbeknownst to the world, had the advantage of having already done it once with Besso under the aegis of the Einstein-Grossmann theory. The result on that occasion had hardly been a victory for the *Entwurf.* Mercury's orbit was shifting due to a variety of planetary influences, but after every other factor was taken into account, a change of 41 seconds of arc per century in the position of the planet's closest approach to the sun remained unaccounted for. The Einstein-Grossmann theory, according to Albert and Michele's calculations, had only predicted 18 seconds.

Primed by his earlier experience, Albert began methodically redoing the calculations. We can imagine him sitting alone at his writing table in his spartan, paper-strewn bachelor apartment amid a haze of tobacco smoke, a cup of coffee nearby growing cold, cookie crumbs underfoot. He might very well have been scrawling away on the unused backs of old letters—the computations have not survived. He might work at any hour of the day or night, for living alone, Albert had drifted into a kind of free-running state. He later told Besso he had worked so strenuously that it was strange that one could endure it.[21]

One of the neatest and yet most daunting aspects of this particular calculation was that there was nothing to adjust: The theory had no parameters that could be fiddled with to make the answer come out right once the mass of the sun and Mercury were inserted in the equations—the new, nearly covariant equations. This time, the answer came out to exactly 43 seconds of arc.

When he saw that number Albert had heart palpitations. He later said that he felt as if something had literally snapped inside him.[22] "For a few days I was beside myself with joyous excitement," he told Ehrenfest.[23] God had spoken in a way that few mortals are ever privileged to hear, and in a way that Albert would never forget. An adventure in pure thought had penetrated to the very heart of reality. In that instant, with what Zangger later called a "savage certainty,"[24] general relativity was cemented in his mind as the truth about gravity. Albert was thirty-six and a half years old.

The moment was still in his mind, ringing as clearly as a clock-tower bell eighteen years later when he told an audience in Scotland, "The years of searching in the dark for a truth that one feels but cannot express, the intense desire and the alternations of confidence and misgiving until one breaks through to clarity and understanding are known only to him who has himself experienced them."[25]

Albert duly read his results to the murmuring Academy on November 18. There was little response, but down in Göttingen Hilbert was amazed that Albert had been able to solve Mercury's orbit so rapidly. "If I could calculate as quickly as you," he wrote the next day, "then the electron would have to capitulate in the face of my equations and at the same time the hydrogen atom would have to offer its excuses for the fact that it does not radiate."[26]

The work was not quite done, however. In the course of carrying out the Mercury calculations, Albert had realized that he no longer needed to impose any restrictions or special conditions on his gravitational equations at all. Many of these accommodations had been designed back in the early days in Zurich, as a way to make the solutions appear to match the predictions of classical Newtonian gravity in special simple cases.

So in the midst of his Mercury calculations, Albert bent himself to the task of recasting his equations one last time. Meanwhile Hilbert was getting close to concluding his own calculation, and letters were flying between him and Einstein. A few days earlier, on November 13, Hilbert had written, excited about his impending solution, and invited Albert to hear him lecture in Göttingen on the sixteenth. Albert declined, pleading a bad stomach, but encouraged him onward: "The indications on your postcards lead to the greatest expectations."[27]

Too great, in a way. On the eighteenth, the same day that Albert announced his Mercury results to the Academy, he received a copy of Hilbert's new paper, a dense thicket of theorems and mathematics containing a set of gravitational field equations based on the selfsame Ricci tensor. Albert looked it over and was shocked by its familiarity, even though he and Hilbert were pursuing different trains of thought utilizing different mathematical tools. It seemed to him that Hilbert was trying to steal his theory. "The system [of equations] given by you agrees—as far as I can see—exactly with what I found in recent weeks and submitted to the Academy," he wrote tersely back to Hilbert.[28]

On November 20 Hilbert sent a paper based on those equations, modestly titled "The Foundations of Physics," to the Gesellschaft der Wissenschaften in Göttingen.

Albert kept forging ahead and recasting his own equations. It took him most of the next week to arrive at the final answer. On November 25 he reappeared one last time before the Academy to announce that he had found the final, fully covariant, fudge-free version of the gravitational field equations.

In the compact language of tensor notation, where superscripts and subscripts denote whole universes of parallel expressions, the resulting equation looked deceptively simple—not even sufficient, really, to fill a modern-day T-shirt. On the left-hand side of the "equals" sign was a capital R for the Ricci tensor, representing the curling and warping of geometrical space-time. On the other side were several terms combining the stress-energy tensor (capital T), representing matter and energy, and the metric, g, representing gravity. What the expression said was equally simple, in a sort of yin-yang way: Matter, energy, space, and time were tied together in an endless elegant loop.

The presence of the metric g over on the "source" side of the equation symbolized the fact that gravity itself was energy, and so the bending of space was itself the cause of further bending of space, and so on. It meant that, the more that space bent, the more it *wanted* to bend. Gravity was like an itch in space-time, that got worse the more it got scratched. Among other things, this itchiness implied that gravity and space-time could rarely be in equilibrium, which was to have profound consequences for the fates of stars and for the universe itself. It also made Einstein's gravitational equations, simple as they looked, devilishly tricky to solve.

General relativity was written in the language of God: one law, without restrictions or favors for special coordinate systems, good for anyone in the universe, no matter the condition of their motion or acceleration—no matter what. "Finally," Albert announced, "the general theory of relativity is closed as a logical structure."[29]

"I am satisfied but kaput," Albert later told Besso, adding that he was surprised the work had come out so well.[30] In the next few days the cards and letters fanned out to the Friends of Einstein announcing the triumph of the revolution that Riemann and Gauss had initiated a generation before in a euphoria of fourth dimension enthusiasm. Besso passed his announcement along to Zangger: "I enclose the historical card of Einstein reporting the setting of the capstone of an epoch that began with Newton's apple."[31]

IT WAS NOT all champagne and celebration, however. The day after his big announcement Albert wrote indignantly to Zangger about his sense that Hilbert had plagiarized his talks in Göttingen. The theory was a thing of beauty, he said, "But only *one* colleague has really understood it, and this one, in clever ways, tries to 'nostrificate'* it (an expression by Abraham). In my personal experiences I have rarely learned better to know the shadiness of people than in connection with this theory. However it does not trouble me."[32]

The Hilbert affair only played into an increasing testiness, even a bit of a persecution complex, regarding general relativity, that Albert had begun to develop. Writing to Sommerfeld soon afterward, he complained that only the intrigues of a few paltry people were preventing the astronomer Freundlich from carrying out an important test of the theory by measuring the deflection of light by Jupiter—a difficult measurement at best. He told Besso, "The colleagues behaved atrociously in the matter. You will have your fun when I tell you."[33]

To Hilbert, with whom he had exchanged cards and notes almost every other day, Albert suddenly said nothing. Finally, Hilbert wrote to apologize, saying that he hadn't meant to intrude on Albert's turf, and had in fact forgotten the Göttingen talk.[34] Albert wrote back to say there had been a certain "pique" between the two of them, "the causes of which I do not want to discuss." But now, he explained, he had conquered his bitterness. "I think of you again with untroubled friendliness and ask you to try to do the same regarding me. It is really a shame if two real fellows who have freed

---

*The word "nostrificate," says Heinrich Medicus, a retired physicist and former neighbor of Zangger's daughter Gina's, and the man who found the letter, means "to make one of ours." Max Abraham used it as a code word for taking somebody else's theory and changing it just enough to make it look like your own. In fact, until recently some scholars have argued that Einstein might have plagiarized Hilbert. When the paper that Hilbert had submitted on November 20 was published the following March, it contained the exact same field equations that Albert had finally arrived at on November 25, five days later. In 1997, however, a trio of historians, Leo Corry, Jürgen Renn, and John Stachel, reported that they had found galley proofs of Hilbert's paper, showing that Hilbert had inserted the correct field equations in December, after Einstein's dramatic announcement.

themselves to some extent from this shabby world should not enjoy each other."[35]

Albert may in fact have been mollified by reading Hilbert's actual paper, which was published in March and which had been modified on the press to include Einstein's final field equations. It began: "The mighty issues raised by Einstein, the ingenious methods he devised to solve them, and his original and penetrating concepts . . . have provided new approaches to investigate the foundations of physics."[36]

By the following spring Einstein and Hilbert were visiting each other again. But if the greatest mathematician and the greatest physicist in the world resumed their friendship, they were never able to bridge the intellectual gulf between their worldviews. While Hilbert claimed to understand Einstein's work better than Einstein himself, Albert still didn't understand Hilbert's work at all. He found it obscure, even after Hilbert had coached him through it, and he complained to Ehrenfest that Hilbert had "pretensions of being a superman."[37] Hilbert was too audacious; he wanted to draw a picture of the entire world, but, as a physicist, Albert knew that there were mysteries yet to be confronted—in the quantum realm, for example.

WITH THE ADVENT of general relativity the old Newtonian certitudes died a second death. In his new theory, Albert liked to say, time and space were finally deprived of the last vestiges of objective reality; they were merely artifacts of the gravitational field, that is to say, the metric that imposed geometry on formless possibility. It was the inevitable conclusion of the mental gymnastics that Albert had to go through to extricate himself from his infamous hole argument.

That proposition had applied a glossy philosophical sheen to Albert's original failure to come up with a covariant theory, but once he decided to continue his pursuit of covariance after all, he was not about to let philosophical scruples stand in the way.

In fact, a way out of the hole argument had been proposed to Albert earlier by Paul Hertz, another Göttingen mathematician, during the fateful summer of 1915. Hertz suggested that the different gravitational field solutions in the hole were not really distinct, but were in fact mathematically related.[38] They could be mapped onto one another by a mathematical transformation, and so in some sense represented the same *physical* situation. If one system could be changed into the other, in other words, they both represented the same reality, and the hole problem was therefore no problem at all. But Albert, still thinking he could save the *Entwurf* theory, was hardly ready for a solution to the hole argument, and rejected Hertz's proposition rudely.

There was, in the argument attributed to Hertz, an echo of the philosophy of Gottfried Wilhelm Leibniz, Newton's greatest rival, coinventor of calculus, and Mach's inspirator. Leibniz had dismissed the notion of absolute space by asking why the universe was where it was and not a foot to the left. It was demonstrably the same universe, whatever arbitrary and patently artificial spatial coordinates were assigned to it. Leibniz's point was that the important thing about the universe was not where it was but what went on inside it. A difference that could not be detected had no meaning, and was no difference at all.

This line of thought was carried forward in the summer of 1915 by a young physicist named Erich Kretschmann, who was spending time in his hometown of Königsberg after receiving a Ph.D. under Max Planck. Coordinates are abstract entities that we don't observe, he argued. In physics, what mattered was not the address of some point in a coordinate system but what happened at that point. Using the analogy of New York City, if a musician named Garland lives on the West Side of Manhattan and a writer named Kate lives on the East Side, and they cross paths in Central Park every morning on their way to work, it doesn't matter what address is assigned to their intersection or where the streets run. All that matters is that they meet. Nothing else matters; everything else is artifice.

In the midst of a long, rambling essay on coordinate systems and measurement Kretschmann made this point in a more formal way, saying that what scientists observed was the "spatiotemporal coincidence of parts of the measuring instrument with parts of the measured objects."[39]

Kretschmann's paper, which appeared in the *Annalen der Physik* on December 21, 1915, apparently struck a chord with Einstein. By now, of course, the hole argument was an embarrassment, and he was eager for an answer. Five days later Albert wrote back to Ehrenfest, who had been pestering him about the hole problem, with an answer almost identical to Kretschmann's. Space-time points, he said, gain their identity not from coordinates but from what happens at them. The phrase he used was "space-time coincidences."[40]

"The physically real in the world of events (in contrast to that which is dependent upon the choice of a reference system) consists in *spatiotemporal coincidences* . . . and in nothing else!" he told Ehrenfest. Reality, he repeated to Besso, was nothing less than the sum of such point coincidences, where, say, the tracks of two electrons or a light ray and a photographic grain crossed.[41]

In his magnum opus on the new general relativity theory early in March

1916, Albert paralleled Kretschmann almost word for word: "All our space-time verifications invariably amount to a determination of space-time co-incidences. . . . Moreover, the results of our measurings are nothing but verifications of meetings of the material points of our measuring instruments with other material points, coincidences between the hands of a clock and points on the clock dial, and observed point-events happening at the same place at the same time."[42]

What about those points where light rays or the world lines of electrons were not crossing, the empty lots on the street map of New York? What about those regions where there was no gravitation, no light, no matter? The idea that space and time are there waiting to be filled and gridded, that there is in fact some empty stage waiting to be occupied, however, was the Newtonian delusion. The meaning of general relativity, Albert believed, was that, in the absence of the metric—that is, in the absence of gravity, matter, and energy—there was no geometry and thus no space, no time. In the eyes of many modern scholars, the hole argument provided the final demolition of the classical view of space-time as an independent thing, a container waiting to be filled. Rather, it arose, as indeed Leibniz and Mach had preached, from the interaction of material objects. There is a certain irony in this, says Janssen, the Einstein Papers relativist. Einstein set out to eliminate the concept of absolute motion, through general relativity, but wound up proving instead that there was no absolute space. In discussing this issue some philosophers often use a rather homespun analogy in which they compare the universe to a pizza, in which the crust represents space-time and the toppings represent the contents of the universe—matter and energy. In Einstein and Mach's world, the crust cannot exist independently of its toppings; moreover, each arrangement of pepperonis results in a different crust, a unique space-time. The city does not exist independently of its buildings and people.

Empty space was no space, or the "manifold," as it was technically called. You could think of the "points" of this nonspace as being like children mobbed on a sidewalk on the day before school classes begin. Without classrooms or grade assignments there is no way of knowing who should be next to, ahead of, or behind whom. The metric, or the gravitational field, was like the school system itself telling the students where to go and sit. Without the metric, however, the points of the so-called manifold, those space-time children, were nothing and nowhere. As a home-plate umpire once famously explained about judging baseball pitches: "Some are balls and some are strikes, but they ain't nothing till I call 'em."[43]

"In some sense," the great Princeton physicist John Wheeler has revealed, "general relativity tells us that space and time are not fundamental aspects of the universe." What could be more fundamental than space and time? As of this writing, eighty years after Albert tore down the temple, physicists still don't have much more than a few wild guesses.

Curiously, the normally generous Albert never mentioned Kretschmann's paper as the source of his inspiration about space-time coincidences.* There is no record of Kretschmann's complaining that his work had been co-opted but he was certainly aware of his place in history. In a later article about Einstein's hole argument, Kretschmann listed his own 1915 paper as the main reference.[44]

IN THE SPATIOTEMPORAL scheme of things, it was now late November in Berlin. As the Great War entered its second year, snow began falling in the trenches of Europe. The gray apartment blocks of the city were lit with thoughts of sons Out There. An exultant but kaput Albert was looking forward to a trip to Zurich and spending some time with his boys, and had already begun softening up Mileva to let him take Hans Albert for a month. Hans Albert, however, was again being sullen, leading Albert to plead, "Please write soon! You are almost as lazy about letter writing as your old father."[45]

Albertli responded with a stony demand for an outrageously expensive— 70 Swiss francs—Christmas present. Furious and once again suspecting Mileva's hand, Albert canceled the trip. "I will not visit you again until you request," he wrote his son.[46]

Alarmed on Hans Albert's behalf and angry at Mileva's lack of support, Zangger stepped in and promised to help unite Albert with his son, if he would only come. Albert relented. "There is still a glimmer of possibility that Albert would be pleased if I came to visit," he told Mileva. "Tell him that and see to it that he is at least somewhat cheerful when he meets me."[47]

Albert left for Zurich a few days before Christmas, but the Swiss border was closed, and after waiting for a few hours and watching people getting refused entry, he turned back himself. He promised Hans Albert that he would come at Easter, even if he had to camp out on the border.

---

*The close timing of Kretschmann's paper and Albert's mention of "space-time coincidences" to Ehrenfest led Howard and Norton to conclude that Albert had appropriated Kretschmann's idea. As late as December 14, just before the paper came out, they point out, Albert was repeating to Moritz Schlick his mantra that space and time had lost their last vestiges of reality, without mentioning the space-time coincidence argument. "The facts make almost irresistible the conclusion that Einstein read Kretschmann's paper or learned of its content when it appeared, found the ideas on coincidences extremely congenial, and turned to refine and exploit them to explain to his correspondent Ehrenfest where his hole argument had failed."

21

QUANTUM TIMES

AS THE WAR GROUND ON, BACK IN ZURICH THE COFFEEHOUSES AND bars were filling with artists, revolutionaries, pacifists, and refugees of every stripe, exulting in what Albert wistfully called Switzerland's "freedom of the mouth." Vladimir Lenin was encamped in the Zurich Public Library. Across the Limmat in a Niederdorf bar a group of painters, poets, and other performers began gathering regularly for bizarre evenings that they dubbed the "Cabaret Voltaire." Hugo Ball, one of its organizers, randomly thrust a knife through a dictionary to find a word, and thus came up with a name for this movement: Dada.

Up the hill on Glockenstrasse, Mileva and the boys entered their second winter of estrangement from Albert. Mileva was still just getting by, supplementing Albert's payments by teaching music and math. Luckily she had many old friends in the Eastern European quarter from her student days. Tete was approaching school age and had emerged as a delicate and brilliant boy. Indeed, his precociousness had so startled Mileva, according to Trbuhovic-Gjuric, that she wondered if she should try to slow it down. Despite her financial difficulties, Mileva had hired a special music teacher for him. His older brother, meanwhile, was commencing a love affair with geometry.

There is no indication that Mileva was following the great discoveries and breakthroughs of November 1915, unless she heard about them from Zangger and Besso. In the Zurich household they had become only "interesting and beautiful things about science" that Albert was promising to tell the boys one day, outweighed and overshadowed by the ongoing war of nerves about Albert's intentions and his plans for Albertli. Mileva remained terrified that her boys would fall into the clutches of Albert's venomous relatives, but in the fall of 1915 she acquired an unlikely pair of champions, namely Michele Besso and Heinrich Zangger.

Despite Albert's occasional attempts to bring them together, his two best friends, Besso and Zangger had never met before that fall. Albert and Heinrich had first grown close when Albert started teaching in Zurich. Just then, however, Besso had quit the patent office and moved to Trieste, which was then part of Austria on the Italian border, to work in a textile factory. When Italy finally joined the war against Germany and Austria, putting Trieste in play, Michele bundled up his family and took them back to Zurich. There were no jobs, and they wound up living in the Winteler summer cabin about thirty miles east of Zurich while Michele taught physics part-time at the Winterthur *Technikum*. Later he would get a *Privat-dozent* position at the Polytechnic and rejoin the patent office. His dark beard was in the process of turning snow white, which, combined with his olive skin, high cheekbones, and coal-like eyes, imparted an exotic fierce-ness to his visage. High on Besso's agenda upon his return was meeting Zangger at last, and the two became friends immediately. The same eccen-tric amateur-physicist minds that had endeared them to Einstein bound them to each other.

One of the first things they agreed on was that their dear friend Albert was behaving badly with regard to his family, especially his children.

"I have a frightening unease about it," wrote Zangger, confessing that Albert's increasing involvement with Elsa was tormenting him. "He stands between the destiny of two young children and possibly has scarcely any notion of it, and is so dependent." Albert's reckless and insensitive demands for Hans Albert, particularly to have the boy visit him in Berlin, offended Zangger's natural sense of propriety.[1]

Besso responded that they should take a stand: "It is important that Ein-stein knows that his truest friends . . . would regard a divorce and subse-quent remarriage as a great evil, indeed as a calamity." Moreover, Besso was steeling himself to draw the line at visits by the boys to the love nest in Berlin. "If I had to break with my old, dear friend Einstein because of this, I would have to find consolation that I did my utmost duty toward him."[2]

Albert assured them that he had no intention of marrying Elsa, despite the pressures being exerted by her parents, which he attributed to bourgeois vanity. "If I let myself be captured, my life will become complicated, and my boys would probably be very much grieved." But no matter how much he loved them, he admitted, he could not endure living with Mileva any longer.[3]

In February of 1916 he broke the news to Mileva: "I hereby apply to you to change our, by this time, well-tested separation to a divorce, whereby we can in the essentials follow the plan of our friend Haber. I believe that it is in both our interests that in this way the obligations and rights of each other are clearly established by us so that each can arrange the rest of his life independently of the other as far as the situation allows it." The details could be worked out once she agreed.[4]

Albert followed up with a visit to a lawyer, who advised him that the process would go more simply and smoothly—and presumably more cheaply—if Mileva were to sue *him* and the whole affair were handled in Berlin.[5]

Mileva had no intention of doing anything of the sort. She responded by playing the only card she had, Albert's access to the boys, which brought a yelp of outrage from Albert. It was an "unjustified vexation," he said, that she wouldn't let him take Albertli to Berlin. In the interests of compromise, and perhaps a smooth divorce, he decided not to press the issue, but warned Mileva not to attempt a similar ploy when he came to Zurich for Easter. Otherwise, he vowed, it would be a long time before he visited again. "I don't want to have to beg for the right to be with my boys," he told her.

"The fact that I personally was unable to live with you has nothing to do with my relationship to the boys and has no influence on it. . . . The fact that I am leaving them in your care is a great sacrifice on my part rather than an absence of fatherly feelings."[6]

By the time he arrived in Zurich in the first week of April it had been a year and a half since Albert had seen his children. The trip began well. To Albert's relief, the boys showed up at Zangger's house on time, politely, and even cheerfully.[7] But when Albert and Mileva finally got together disaster ensued. Albert apparently brought up the divorce, and Mileva became hysterical. In the end nothing was decided, and Albert retreated in despair back to Germany, doubting whether he would ever be able to face her again.

After he left, Mileva went into a tailspin that culminated in a complete emotional and physical breakdown. In July she had a series of heart attacks (she was only forty-one years old) and was confined to the Theodosianum, a hospital run by Catholic nurses. The boys were sent to the Bessos and then to Lausanne, where Helene Savic and her husband were spending the war.[8]

Nobody was quite sure exactly what was wrong with Mileva. Her symptoms ran a confusing gamut from depression to mysterious sores on her fingers. According to Pauline Einstein, who heard it from Maja—who supposedly heard it from her in-laws, the Bessos—Mileva was faking a large part of her maladies. "Mileva was never as sick as you seem to think," Pauline told Elsa.[9]

Albert was kept informed of her condition by regular bulletins from Besso and Zangger. At first he believed that her problems were psychological. "Mileva really seems sick in the head," he told Michele, thanking him for being a "true helper." Albert begged Besso to keep him apprised of Mileva's situation, even if her actions seemed to make no sense.

"In the time of quantum theory," he added, "this is not so outrageous."[10]

By now, Anna Besso was fed up with Albert. She had been through a similar situation back in 1896, when Albert had dumped her sister Marie. She added a postscript to Michele's next report, expressing her anger in no uncertain terms. Albert, who didn't notice that it was Anna and not Michele writing, was shocked not only by the harsh tone but the fact that he was suddenly being addressed by the formal *"Sie"* instead of the more intimate *"Du."* Had their friendship fallen so far?

Albert tried to defend his actions in a long letter that amounted in part to a treatise on male-female relations, as he saw them. "Dear Michele," he sighed, "we men are pitiful, dependent creatures, I'll admit that to anyone with pleasure, but compared to these women each of us is a king: Because a man stands at least halfway on his own feet without always waiting for something outside himself to cling to. But they are always waiting until someone comes along (a guy) to dispose of them according to his own discretion. If this doesn't happen then they simply fall apart."

Mileva, he maintained, had a carefree life, with two strapping boys, a home in a splendid neighborhood, and plenty of free time, and she now stood "in the halo of her departed innocence," lacking only someone to lord it over her. Was it so terrible that he had finally fled as if from an odious smell?

Finally he addressed what he thought was Michele's angry postscript. "Dear Michele," he pleaded, "we've been good friends for twenty years. And now I see in you a growing fury against me because of a broad who means nothing to you. Defend yourself against it. She wouldn't be worth it even if she were a hundred thousand times in the right."[11]

Albert had scarcely mailed this screed when he finally realized that it was Anna and not Michele who was the author of the offending postscript. He quickly wrote again to apologize, "To think that this should happen to a scientist."[12]

Zangger performed his own examination of Mileva meanwhile and told Albert that she might be suffering from tuberculosis of the brain. If that was so, Albert thought, a quick end would be preferable to a long, drawn-out suffering. Nevertheless, it was now clear to him that she was genuinely ill and wasn't going to get well soon, which left him in a quandary. Should he go back to Zurich in order to be with his children when they returned from Helene Savic's? He had no pressing business in Berlin, beyond the universe and the enigmatic quantum. The fact was, heading to Zurich was the last thing in the world he wanted to do. If he went to see his boys, he explained to Zangger, Mileva would undoubtedly demand to see him as well. Under the circumstances he couldn't refuse her, and she might exploit her advantage. "In this situation I could be forced to make promises regarding the children," he said mysteriously, "whereby in the event of my wife's death the boys would be wrested from me. I have a deep fear of this, but if you deem it advisable, then I will come." In the meantime he told Zangger to give her whatever she wanted and to procure a nurse.

"My boys must not get the sad feeling that they have no support from their papa," he declared. If Mileva died, he would bring them to Berlin and tutor them himself at home. Without his friends' help, Albert said, he would be losing his mind. "I feel sorry for my wife and I also believe that her difficult experiences with me and through me are at least partly to blame for her serious illness."[13]

In the end, Albert remained in Berlin, spending the summer puttering around with relativity and quantum theory.

Mileva's condition gradually improved, and she even tried to take Tete off to the German seacoast for a kind of joint convalescence, but the living conditions were bad and food was scarce. They left after three weeks more exhausted than when they had arrived. Her illness had had an effect on Einstein, however. By the end of the summer Albert had given up on the divorce, at least temporarily. His relatives could wail and complain all they wanted; he told Besso that he had long since learned to withstand tears.

Meanwhile, Hans Albert stopped writing again. "I believe his feelings have sunk below the freezing point," Albert moaned.[14] When Helene Savic sent him a friendly and reassuring report on how the boys were doing, he complained to her that they seemed to harbor a grudge against him, and believed the time had finally come for him to give up.

"I think that, although it is painful, it is better for their father not to see them anymore." He would be satisfied if they managed to develop into honest and well-liked people, which they showed every indication of becoming, even if he was not going to be around to supervise. Knowing that in speaking to Helene he was probably speaking indirectly to his wife as well, he went out of his way to emphasize what he could not say and what Mileva could or would not hear in the tumult of face-to-face conversation. "The separation from Miza was for me a matter of survival," he explained. "Our life together had become impossible, indeed, depressing. Why, I am unable to express. . . . Despite this she is and will always remain an amputated part of myself. I will never get close to her again. I will end my days without her."[15]

Mileva would need help, he said. She would suffer, at least for a while. As for himself, he asked for no pity. Despite the war and his family problems, his life, he declared, was going in perfect harmony. "My mind is completely devoted to thought," he wrote. "I am like a person, amazed at broad horizons, who only gets disturbed when an opaque object prevents their view."

ONE OF THE PATTERNS that emerges from a long-term view of Albert's life—and one of the problems of writing coherently about that life—is that his successes seemed to feed on one another. Discoveries in one area of physics seemed to accompany discoveries in other areas; advances came in rushes across the scientific spectrum. The result is a series of historical scientific traffic jams. He invents relativity at the same time that he invents light-quanta, developing mutually contradictory notions about the nature of nature as if he were keeping his bets covered. One such traffic jam occurred in the summer and fall of 1916. At the same time that he was reinventing the universe in accordance with general relativity, he was also incubating a new and devastating idea about the enigmatic quantum.

In the time of the quantum, this is not so outrageous.

Ten years after Albert had formulated the idea that light was composed of discrete little energy bundles, or *Lichtquanten,* the concept was still appalling to most physicists. Typical was the comment of Robert Millikan, a University of Chicago scientist who spent ten years shining lights on electrodes in order to disprove Einstein's prediction that the velocities of the electrons emitted would be proportional to the frequency of the light. In 1915 he announced: "Einstein's photoelectric equation . . . appears in every case to predict exactly the observed results . . . yet the semicorpuscular theory by which Einstein arrived at his equations seems at present wholly untenable."[16]

Albert's interest in the quantum question had never really waned, even during the long period of work on general relativity. In January of 1916 he had come across an intriguing calculation by the theorists Hugo Tetrode[17] and Otto Sackur, which apparently inspired him again, although it wasn't until summer that he wrote to Besso that a "splendid idea" about the emission and absorption of light had just struck him: *"Alles ganz quantisch* [Everything is quantum]."[18]

His revelation concerned how to calculate the famous blackbody radiation spectrum, the problem that had started it all. Planck had obtained the right answer by assuming that radiated light was only emitted or absorbed in discrete lumps, but his calculation had been flawed and Planck's formula for the blackbody spectrum had never been properly derived. Albert now saw a way to do the calculation free of metaphysical assumptions about entropy and the like. He imagined a roomful of atoms bathed in radiation. According to Bohr's new theory, one of these atoms could absorb energy from the ambient radiation only by capturing just enough to raise an electron inside it from one permitted orbit (or energy level) to another. Likewise, an electron dropping from one orbit to a lower one would emit precisely that amount of energy as light. Albert's calculations showed that if each of these hops up and down the atomic energy ladder corresponded to the emission or absorption of a single light-quantum, the spectrum of radiation in the room would be Planck's spectrum, the blackbody spectrum. The same outlandish theory now explained both great mysteries of optics and matter, blackbody radiation and spectral lines, in one neat flash of the pencil. "An amazingly simple derivation of Planck's formula," Albert commented to Besso. "I should like to say *the* derivation."[19]

Albert did not stop there, however, and what he achieved next opened a door that would never close for him again. The quantum experience was about to stand before him for the first time in all its random, unclothed disagreeableness. The idea that a light-quantum might carry momentum as well as energy had always been at the back of his mind, though he never explicitly addressed it in any of his papers or talks. Abraham Pais, in his biography, *"Subtle Is the Lord . . .": The Science and the Life of Albert Einstein,* points out that the simple mathematical manipulation involved could have been accomplished as early as 1905. But perhaps even Albert didn't take his *Lichtquanten* seriously enough as physical objects, as opposed to calculational fictions, to do the final math. In classical electromagnetic theory, light waves exerted pressure, as had been so spectacularly demonstrated only a few years before by the tail of Halley's comet—which consisted of minute dust motes blown by sunlight—as it swept halfway across the sky. Momentum was different from energy, however, in that it had a direction as well as magnitude, which raised the issue of what happened when an atom emitted or absorbed light. Was there a recoil? Did a quantum squirt out in some particular direction? Or did light, as in classical theory, pulse out from the atom in a spherical shell of energy in all directions at once, and thus with no net momentum, no recoil?

The answer, Albert knew, was already in front of him, encoded somehow in Planck's intricate and now solidly founded formula that described blackbody radiation. It was by studying the statistics of Planck's formula back in 1909 that he had teased the mathematical signatures of both particles and waves out of expressions for fluctuations in the radiation and concluded that light had some qualities of both. Now Albert proceeded to use similar tools to sift the Planckian fluctuation statistics for the signature of atoms being bumped about by quanta, much the way he had once sifted the movements of dust motes for molecular nudges. The equations were consistent, Albert found, only if the radiation from atoms was indeed directed—in so-called needle rays—as opposed to being released as a spheri-cal blob. The quanta had momentum and direction. If you threw one, you recoiled; if you caught one, it knocked you back. In short, they were real.

"With this, the existence of light-quanta is practically assured," he told Besso.[20]

But—and this was a qualification that would echo through history—there was no way of knowing which direction the quantum tossed off by an atom would go, nor, for that matter, exactly when this event would oc-cur. It was, Albert admitted in his paper, "a weakness of the theory . . . that it leaves time and direction of elementary processes to chance. Neverthe-less, I have full confidence in the route which has been taken."[21]

All roads led to this breakdown of the verities. Even general relativity, for all its strangeness, was a conventional theory of classical physics: It dealt in causes and effects; one thing could be predicted from another as long as all the relevant clocks and measuring rods were set properly. Within those con-straints of prior logical thinking, the whole history of the universe and every particle in it from the infinite past to the infinite future could in prin-ciple be deduced. But if this new result was right, the quantum stood as an exception, a mockery, of this deterministic dream. For in the quantum's world, important elements, like the timing and direction of a quantum shot, were left to chance. Chance had already invaded physics in the notion of radioactive half-lives. Half the atoms of any radioactive sample would decay in a half-life, but no one could say which ones, when. In earlier cor-respondence with Besso, Albert in fact had fancifully compared an atom about to emit light to a radioactive atom.[22] In 1916 it might have been said that physics itself was radioactive, awaiting some mystical invisible com-mand to self-destruct. How could physics penetrate the order of the uni-verse if on the most fundamental level of reality there *was* no order?

"I feel that the real joke that the eternal inventor of enigmas has presented us with has absolutely not been understood as yet," Albert remarked.[23]

In the summer of 1916 Albert was like a man standing with one foot each on a pair of galloping stallions. The quantum was Einstein's baby or it wasn't anybody's baby at all. And so was general relativity. It would be many years before the most radical and far-reaching implications of either theory were discovered, but it was already clear to him that they were headed on different paths, that they in fact described completely different universes and worldviews. Together they embodied some of the warring paradigms of Western thought. In relativity nature was continuous and causal; its grav-itational and electromagnetic fields could easily be described by those won-derful differential equations that had entranced physicists from Newton to

Maxwell. In atomic and quantum theory, nature was discrete and could only be described statistically, by counting. These two views of the world were fundamentally mathematically incompatible. Which was right? How to reconcile them? It was the kind of conflict that had sent him into nervous spells as a young man in Bern.[24]

THE LUNACY of the world outside physics continued to assert itself. In October of 1916 a pair of Albert's old acquaintances collided, with fatal results. Fritz Adler, his old socialist neighbor and self-sacrificing friend, walked up to the prime minister of Austria in a hotel in Vienna and shot him dead for refusing to convene parliament and thus put the war to a democratic test. (The man he killed was Count von Stürgkh, the man who as education minister had recruited Albert to teach in Prague six years before.) Adler was apprehended unharmed, and his trial was scheduled for the spring.

Needless to say, Adler's action did nothing to halt the giant slaughter machine. The Germans and French each lost a third of a million men in a six-month standoff at Verdun on the Western Front. More than a million others were killed in the fall before the Allies abandoned a counteroffensive at the Somme. Austria overran Serbia and Rumania. A million Russians died on the Eastern Front. Having forsworn all-out submarine warfare in the wake of American outrage at the sinking of the *Lusitania,* Germany was being strangled by a naval blockade while Britain enjoyed relatively free access from the sea. A year later the submarines would return to the shipping lanes, their abilities honed by the Anschütz compass for which Albert had served as patent referee.

Strangely, as it turns out, that gyrocompass was not Albert's only contribution to the German military-industrial complex. In 1916, in an incident belying his antipathy to the war, he had tried his hand at designing an aircraft, which went far enough in its development to be flight-tested. Aero- and hydrodynamics were a subliminal family interest, dating back to Albert's sailing boats in a barrel as a boy in Munich. Vero Besso later recalled going kite-flying in Bern, with Albert explaining in great detail how the kite worked. Albert might also have been inspired by Conrad Habicht's attempts to design a flying machine back in his Bern days. The physics historian Hans-Jürgen Treder even argued in 1979, on the basis of unspecified material in the Prussian Academy of Sciences archives, that Albert was recruited to Berlin at least in part for his expertise in aviation technology.[25] If that was indeed the case, this little-known episode shows that his recruiters misjudged badly.

The story begins in June 1916, when Albert delivered a lecture to the Deutsche Physikalische Gesellschaft about waves and flight. The work stemmed from his old love of statistical mechanics—in particular, a feature known as Bernoulli's principle, which states that moving air exerts less pressure than still air. (A way to see this effect at work is to blow between a pair of apples suspended by strings a few inches apart; the apples will fall together.) "Where does the lift come from that allows planes and birds to fly?" Albert began, in a published version of his talk in *Die Naturwissenschaften*, saying that he had failed to find even "the most primitive answer" in the literature.[26] He went on to argue that Bernoulli's principle was the answer to the mystery of flight—the outward curve of a wing or airfoil makes the air flow faster over the top of it than along the bottom, creating a pressure difference (in fact this theory had already been elucidated in Russian and German aeronautical journals).

Shortly thereafter, the story goes, a long letter thick with equations arrived on the desk of Paul Georg Erhardt, head of the experimental department of Luftverkehrsgesellschaft (LVG), an aircraft company. In it Albert proposed using a novel humped wing, shaped like cat's back, to increase lift. Erhardt was a pilot, one of the first heroes of German aviation, who had once paid 500 gold franks to fly with Orville Wright. He had never heard of Bernoulli, and passed the letter to his engineers, who invited Einstein in for a talk. Albert explained that his design would produce the maximum lift from a minimal amount of thrust.

Within a few weeks a pair of "cat's back" wings were being fitted to the fuselage of an LVG biplane. As he watched the device taking shape, Erhardt became more and more dubious. "We'll see how this hare is going to run," he grumbled to Albert as he took his place in the cockpit.[27]

The pilot's skepticism proved to be well-founded. After a long takeoff the plane's tail dipped, and Erhardt found himself hanging in the air like a "pregnant duck." He was overjoyed to make it back safely down to the ground again just short of the airway fence. A second pilot had the same hair-raising experience. In its original configuration, the plane was not even safe enough to attempt a turn, and further tinkering improved its maneuverability only slightly.

Albert later admitted to being embarrassed at his folly. "That is what can happen to a man who thinks a lot but reads little," he commented in a bemused letter to Erhardt. His only regret was that he had not done his homework.[28]

Was this a violation of Albert's pacifistic principles? Scholars have pointed out that airplanes played a relatively insignificant role in the Great War. To Albert, it seems, flight was a technological problem, not a military one, and like many scientists before and after him, he could not resist the lure of a sweet technological fix. The possible wartime applications were a factor he simply walled off from himself, as he managed to wall off so much, for fear of being engulfed. His principles, after all, didn't preclude his staying good friends with Fritz Haber, who was busy inventing poison gas, or Hermann Anschütz-Kämpfe, who was profiting from submarine warfare, or having his Academy salary paid by Leopold Koppel, the ringmaster of the military-industrial complex. As far as Albert was concerned, as long as money was being passed around, it was better off being used to pay for his research rather than someone else's.

IN NOVEMBER Mileva wrote to Milana Bota to say that things weren't going well. Her heart had stabilized, but she was still plagued by health woes great and small, from sore fingers and hay fever to problems with Tete's ear. One of his mysterious maladies was a recurring pain in his ears. "So one continues to have drudgeries," she said. "I'll play the hero as long as I can, but there'll come a time when I have to submit . . . for every illness, even an insignificant one, is a great disruption.[29]

It was not long in coming. Shortly thereafter Mileva collapsed again. According to Besso, the new onslaught coincided with Hans Albert's refusal to show her a letter he had just received from Albert. She was once again confined to her bed and very depressed. Zangger and another doctor were prepared to call in a neurologist. Hans Albert, meanwhile, went to the Bessos' house, where Michele gave him an algebra lesson, showing him how much fun it was to get the right answer. Afterward they all sat around talking about science. Unfortunately these get-togethers could not be repeated very often, because of their unpredictable effects on Mileva's disposition.[30]

Albert was grateful that Michele, as usual, was stepping in, but he was under no illusion that his assistance was anything more than a stopgap. As the winter wore on, for the first time he began considered taking Hans Albert away from Mileva and bringing him to Berlin, where he could look after the boy himself. Although he soon realized he could not take the boy away against Mileva's will, the issue of the children's plight in the event of Mileva's incapacity continued to gnaw at Albert.[31] For now at least they were all right. Hans Albert had friends and seemed to be doing well enough in Zurich, and in any case, Albert's assuming custody would force him to take time from his work to do housekeeping, just when the universe was getting more interesting than he had ever dreamed.

PART FIVE

Master of the Universe

A man can do what he wants,
but he can't want what he wants.

Arthur Schopenhauer, quoted
by Albert Einstein in 1929

22

MACH'S REVENGE, OR,
THE WAR OF THE WORLD MATTER

AS OF THE BEGINNING OF 1916, THE NEWS OF THE DEATH OF NEWTON'S universe had not spread beyond a relative handful of German and Dutch scientists, namely the Prussian Academy of Sciences and Albert's friends. But the question of what kind of universe would succeed Newton's was already preoccupying a few brave minds. In Berlin, Albert was busy in the post-holiday vacuum grudgingly writing up a formal account of the new theory for the *Annalen der Physik* after having failed to convince Lorentz to do it for him, and grimly mulling over a book offer. Lorentz and Ehrenfest, meanwhile, were sitting in their respective studies ten miles apart racing to understand exactly what it was that Einstein had wrought. Lorentz's announcement that he had finally seen the light elicited an envious sigh from Ehrenfest. "Your remark, 'I have congratulated Einstein on his brilliant results,' has a similar meaning for me as when one Freemason recognizes another by a secret sign."[1]

Not among the theory's disciples was Ernst Mach, who passed away in February 1916 at the age of seventy-eight, adamant and unapologetic to the end in his scorn for atoms. He did, however, leave behind a surprise for Albert, although not until 1921, when his book on optics was posthumously published, would the world learn that relativity, as well, had earned Mach's contempt. "I gather from the publications which have reached me, and especially from my correspondence, that I am gradually becoming regarded as the forerunner of relativity," he had grumbled in a preface dated 1913. "I must, however, as assuredly disclaim to be a forerunner of the relativists as I personally reject the atomistic doctrine of the present-day school, or church."[2] He promised to explain his detailed reasoning in a later book, which was never written.

In the obituary of Mach he wrote for the *Physikalische Zeitschrift*, Albert didn't mention the atom problem: "In him the immediate pleasure gained in seeing and comprehending—Spinoza's *amor dei intellectualis*—was so strong that he looked at the world with the curious eye of a child until well into old age, so that he could find joy and contentment in understanding how everything is connected."[3]

Even in death, however, Mach cast a long shadow.

No idea had been more of a guiding influence for Albert in generalizing relativity than the concept of the relativity of inertia, which he would later call "Mach's principle"—the notion that the inertia of an object was some- how determined by its interaction with everything else in the universe. In the language of general relativity, Albert concluded, this principle meant that the curvature of space-time should be determined solely by the matter and energy in the universe, and not by any outside influences or initial con- ditions—what physicists called boundary conditions. One consequence of this, Albert thought, was that it should not be possible to solve the field equations for the case of a solitary object in the universe—an atom or a star—because there would be nothing out there for it to interact with. But Albert was in for a few surprises once other theorists started working with his equations, and his subsequent attempts to build Mach's principle into natural law would lead him to reinvent the entire universe.

IT BEGAN on the Eastern Front, in Russia, where Karl Schwarzschild, the brilliant director of the Potsdam Astrophysical Observatory, found himself computing artillery trajectories after having volunteered for the military at the start of the war. As a younger man Schwarzschild had dabbled in Reimannian geometry and had followed the early stages of the develop- ment of general relativity. When the reports from the November Academy meetings arrived from Berlin, Schwarzschild seized on Albert's new equa- tions as a means to solve a simple astronomical problem—namely, what was the gravitational field surrounding a point mass in empty space?

By January he had gotten the answer and sent it to Albert. The work, Schwarzschild commented, "permits Mr. Einstein's result to shine with in- creased purity."[4] Albert read Schwarzschild's report into the record at the Academy. Several months later Schwarzschild struck again, this time solving the equations for the field *inside* a particularly simple model of a star.

Aside from the fact that they existed at all, there was something odd about Schwarzschild's solutions. In both cases there was a certain radius, a "magic circle," within which the mathematical calculations became non- sense. The metric, which prescribed how to calculate the distance between two space-time points, became infinite; the pressure at the center of the star rose to infinity. It was as if this "Schwarzschild radius" marked some kind of inner boundary to space-time. This was obviously not a happy result. "If this result were real, it would be a "true disaster," Albert warned, "and it is very difficult to say *a priori* what could happen physically, because the formula does not apply anymore."[5]

The operative word was "if." Opinion among the nascent relativity experts was divided about whether this feature was a mathematical artifact, caused by the choice of coordinate systems in which such calculations were done, or whether it represented real potential misbehavior on the part of nature. The radius at which this disaster manifested itself depended on the amount of mass that was enclosed within it, and all real objects, even atoms, were known to be too large to be affected. For the sun, it was pointed out, the magic radius was 1.47 kilometers, which meant the entire solar mass would have to be squeezed into a ball only 3 kilometers across. For the earth, the required gravitational radius was a mere 5 millimeters. Johannes Droste, a student of Lorentz's who obtained the same results in a simpler mathematical form, suggested that theorists simply ignore the territory inside the Schwarzschild radius.[6]

Among those who subscribed to the notion that nature would not permit such extreme conditions to exist was a young astronomer and relativity enthusiast in England by the name of Arthur Stanley Eddington, who summed up his objections to the prospect by enumerating their absurd consequences in his book *The Internal Constitution of the Stars.* Among his thoughts about these massively condensed bodies was that "the force of their gravitation would be so great that light would be unable to escape from it, the rays falling back to the star like a stone to the Earth." Moreover, "the mass would produce so much curvature of the space-time metric that space would close up around the star, leaving us outside (i.e., nowhere.)"[7]

Schwarzschild himself would have no more to say on the subject, for he contracted an autoimmune disease and died that May at the age of forty-three. Like Minkowski, who regretted that he had had to die just as relativity with being born, and Riemann, whose mathematics was fifty years ahead of the physics of his time, Schwarzschild passed away on the frontier of the promised land, having been favored with only a glimpse of the strange and beautiful landscape within, another genius mowed down by indifferent, banal circumstance.

SCHWARZSCHILD'S RESULTS initially took Albert by surprise—not because of the so-called singularity, but because there was any solution at all that described a single, solitary body in an otherwise empty universe. "I had not expected that one could formulate the exact solution of the problem in such a simple way," Albert confessed to Schwarzschild. According to what Albert now grandly called "Mach's principle," a body by itself had no properties. Everything we know is relative—no atom is an island, a single hand does not clap. This was the germ of the vision that he had been following for ten years. Neither Albert nor Mach had spelled out any theory of how this mysterious interaction worked, but Albert had been hoping and presuming that general relativity had somehow incorporated or would at least account for this mechanism. "One can express it this way as a joke," he said as he explained to Schwarzschild how things were *supposed* to work. "If all things were to disappear from the world, then, according to Newton, Galilean inertial space remains. According to my conception, however, *nothing* is left."[8]

And yet here, according to the equations, was a star bending and defining space all by itself, a whole universe in its own little nutshell of not-quite nothingness.

Of course, the universe did not come in a nutshell. Or did it? The universe at large was a mystery. Was it infinite? If not, what was on the other side? Heaven? God? Dragons? A nutshell indeed. Underlying this rather frivolous figure of speech is a deep problem about doing physics in the universe, one that was about to command Albert's full attention. The fact is that the universe either did or did not come in *something,* and this something, Albert realized, was the missing ingredient that could perhaps restore a Machian order to the cosmos.

In physics, Albert knew, it was not sufficient simply to solve the laws of motion as a way to describe the evolution of some system—be it the solar system or the universe. You also needed to know its initial setup, its so-called boundary conditions. The boundary conditions for the solar system might be the positions and velocities of all the planets, moons, and aster-oidal and cometary chunks back at some particular moment on the very brink of their gravitational dance together. In general relativity, determining the boundary conditions of the universe involved making assumptions about what happened to the metric, the geometry of space-time, at large distances and remote times from the here and now under immediate consideration. Nobody could do a calculation like Schwarzschild's without having to make some choice along the way about the outside universe, just as one would use sea level as the reference point in describing the elevation of a mountain.

In 1916 the universe seemed deceptively simple. It consisted locally of eight planets and their ancillary moons, asteroids, and comets swirling around the sun (although alleged discrepancies in the orbit of the outermost planet, Neptune, hinted at the possibility of a ninth). The sun, in turn, was but one of millions of stars in a disklike cloud known as the Milky Way. Recent ingenious observations by the American astronomer

Harlow Shapley at Mount Wilson Observatory in California put the diameter of this galaxy, as it was called, at about 250,000 light-years, with the sun about 50,000 light-years from its center.[9] Whether anything existed beyond the Milky Way, nobody knew. Among the stars were thousands of small cloudlets of light called nebulae, some amorphous and tangled, others shaped like little pinwheels. No less a personage than Immanual Kant had suggested that these pinwheels were other Milky Ways sprinkled at incredible distances across the void. The majority of astronomers, however, subscribed to the notion favored by the Frenchman Pierre Laplace that these whirlpool clouds were local and represented new stars in the making. Indeed, in 1885, a new star[10] had suddenly flared in the center of the giant Andromeda nebula, right where the whirlpool theory said a new star was supposed to appear, and then dimmed to a presumably quiescent and mediocre life. The controversy raged on, particularly in America, but Albert was not an astronomer and didn't know the details.

The idea of an infinite universe was troubling to Albert for a number of reasons. If it was filled with stars, a simple argument showed that the gravitational force exerted on any one star by all the others would be infinite; the universe would collapse. On the other hand, if the Milky Way were just an island in empty space, there was nothing to stop a star from escaping the Milky Way now and then and drifting off alone to infinity, where, according to Mach's principle, it should lose all its inertial mass. Albert concluded that, in a sensible universe, this could not happen. In a sensible universe space must be fenced in, somehow.

The first hint of where this chain of thought was heading came in March 1916, shortly after Schwarzschild's discovery, when Albert finished his general relativity paper and started sending it out. In it, Albert illustrated the notion of the relativity of rotation—a presumed feature of the theory—by imagining scientists who lived on a sphere spinning in empty space. Viewed from afar, this sphere would be seen to be bulging at its equator, due to centrifugal forces. But the inhabitants of this world, believing themselves to be at rest, would attribute the bulge in their world (or, more technically, the extra mathematical terms in their local space-time metric) to another factor, namely the gravitational pull of "distant masses."

What were these masses? Albert didn't say, but since the normal cosmic stuff of stars, nebulae, dust, and planets had already been excluded by definition from his *Gedankenexperiment,* these masses had to represent something else, something cosmological, and, hopefully, Machian. Something to do with boundary conditions of the universe. Perhaps they were the nutshell that would make Mach's idea work at last.

In the spring of 1916 Albert set himself to the task of incorporating the notion of distant masses into a theory of the universe that would be consistent with Mach's principle. His scheme, expressed in letters about the proposed mathematical form of the metric, was that these masses would form the boundary of the universe, crumpling up space out beyond the Milky Way and thus corralling the stars, like a tablecloth rolled to keep crumbs on the table. There would be no space without matter, and no matter without some point of reference, no chance for any atom or star to be alone and so escape the influence of its neighbors and the mystical Machian network of all for all.

This was not the first time, nor would it be the last, that some physicist would posit a completely new and unsuspected phenomenon—like that of masses forming the boundary of the universe—in order to preserve an elegant theory. It was a rather un-Machian thing for Albert to be doing in the name of Mach's own principle, but he was on a winning streak, still giddy after his experience with Mercury's perihelion, over the power of theoretical thought. But he had not gotten far with this notion before it ran into friendly criticism from a *real* astronomer.

With a pointy white beard and a hypochondriacal disposition, Willem de Sitter, forty-four, director of the Leiden Observatory, was seven years and a generation older and more conservative than Albert. He had gained his knowledge of the universe the hard way, measuring stars for a Dutch survey of the Milky Way and observing Jupiter's moons at the royal observatories in Cape Town and Groningen. De Sitter had been one of the first scientists to receive a copy of Albert's general relativity paper. Not only was he part of the Dutch scene along with Lorentz and Ehrenfest, but he also had connections in England, where Albert was eager to get his theory recognized. De Sitter did exactly what he was supposed to and sent the paper on to Eddington, who was secretary of the Royal Astronomical Society. Eddington was enchanted and commissioned de Sitter to write an article for the Society's *Monthly Notices*. De Sitter's papers there and other places subsequently served as the English-speaking world's introduction to curved space and related topics.

De Sitter and Einstein had been corresponding since early 1916, and in the give-and-take that followed between them the universe was made and remade. They first met—and de Sitter first heard about Einstein's cosmological proposals—that fall, when Albert visited Leiden for a couple of weeks.

De Sitter loved general relativity but regarded Einstein's latest proposal with a jaundiced eye. These "supernatural masses"—or "world matter," as he sometimes referred to them—were just Newton's old idea of absolute space back again in new clothes, he argued. By definition they could never be part of the visible universe, and therefore they would have to keep shifting outward as telescopes got more powerful and saw farther into space. It was okay to say for now that the source of inertia was outside the Milky Way, but what would their grandchildren say if some new invention were to extend the boundaries of the known universe as dramatically as the telescope had three hundred years before?

"I am convinced—but this is only a *belief* that cannot be proven, of course," de Sitter wrote to Albert in a typical comment shortly after the Leiden visit, "that these masses will go the way of the "aether wind.""[11]

If Albert was stung, or bemused, at being pronounced the new avatar of absolute space, he didn't show it. Writing back to de Sitter in an attempt to clarify his ideas, he explained that although this universe he was proposing would be finite in space (although unimaginably larger than astronomers had previously conceived), it would still be infinite in time. Moreover, the distant masses, Albert said, were not anything supernatural or exotic, but rather were ordinary stars, like the ones we observe in the Milky Way, strewn through the far reaches of the universe. "This is compatible with the facts," Albert admitted, "only when we imagine that the portion of the universe visible to us must be considered extremely small (with regard to mass) against the universe as a whole."

He went on to apologize for his Machian obsession with the origin of inertia, saying it was a matter of taste. Still, de Sitter shouldn't scold him, Albert said, for trying to imagine a world in which inertia was relative. "I in no way demand that you share this curiosity."[12]

De Sitter did not, and criticized Einstein's distant masses in the first of his *Monthly Notices* articles, in December 1916, but his tone was gracious. Even if Einstein had not convincingly explained the origin of inertia and gravitation, de Sitter acknowledged that he had made giant strides in physics. "Perceiving the irrelevance of the representation by coordinates in which our science was clothed, he has penetrated to the deeper realities which lay hidden behind it . . . and has thus made an important step towards the unity of nature."[13]

All this was a prelude.

There was a way to "construct" a universe that had no boundaries and yet was finite, so that no star could wander off to inertialess infinity. In fact, we live on the analog of such a world. A man could sail the globe for millions of miles and thousands of years without ever encountering an edge and yet without ever being more than a few thousand miles away from the teeming centers of civilization. Riemann had pointed the way in 1854 when he imagined the universe as a hypersphere with three-dimensional space curved back on itself. It had occurred to Ehrenfest back in 1912 that this might be the way to describe the universe, but he had not pursued the idea. In the winter of 1916–17, however, he remembered it and mentioned it to de Sitter. Whether he ever passed the idea on to Albert as well is not known, but by the end of January 1917 Albert had set aside his previous efforts and was hard at work reinventing a universe that had no need for boundary conditions. He was anxious to see what Schwarzschild would think of it and meanwhile told Ehrenfest, "I have . . . once again perpetuated something about gravitation theory which somewhat exposes me to the danger of being confined to a madhouse."[14]

In brief, Albert realized that general relativity gave him a prescription for taking a leaf from Riemann's book and constructing a spherical universe. If the gravity of all the mass and energy in the universe was sufficient to bend space back on itself, the result would be a cosmos that was finite but without boundaries. A light ray shot off in any direction would eventually return from the opposite direction. There would be nothing "outside" the universe to dictate the conditions or properties of whatever was inside. The universe would have only itself.

With a little help from a mathematician friend, Jakob Grommer, Albert proceeded to dash off a short paper entitled "Cosmological Considerations on the General Theory of Relativity."[15] He read it to the Academy on February 8, nearly a year to the day after Mach's death, and it is fair to say that since then, the universe has never been the same.

As he often did when he was dealing with complicated issues, Albert began by reviewing the problem of an infinite universe in terms of familiar Newtonian physics. Given Newton's laws of motion, and enough time, a cloud of stars like the Milky Way would eventually evaporate like the molecules in a perfume bottle, light would radiate away to infinity and be lost. A star lost to infinity would have no inertia. But the stars were still here; their density had not dwindled to zero.

Albert found that he could keep the stars from evaporating simply by adding an extra term that he called lambda, after the Greek letter, to the Newtonian equations. Thus adjusted, the equations yielded a universe with a constant average density of stars. "A universe so constituted," he explained, "would have, with respect to its gravitational field, no center."

Of course, we don't live in a Newtonian universe, but Albert now had a road map of sorts to follow when he turned to the task of creating a theory of the universe as curved space. That road map would lead to what he called a "slight modification" of his cherished field equations only three months after he had finally discovered them. The change was to be the inclusion of that funny term, lambda.

Having failed to invent a universe with boundaries or one in which the inertia of a star vanished at infinity, the only recourse, as Albert said in his paper, was a universe with no boundaries at all, a spherical space. In exchange for this freedom from boundaries, however, another condition had to be fulfilled. Mathematically, the mark of a sphere was constant curvature. In general relativity, the "curvature" at any given point in space was determined by the density of matter and energy there. So in order to be spherically closed, the universe had to have on average the same curvature and thus the same density of matter everywhere. On one level, this was an absurd proposition. Locally, for example, matter was clumped into giant conglomerations like our own solar system, with vast voids in between. On the scale of the entire universe, however, only densities averaged over enormous distances counted toward the curvature. It was reasonable to assume, Albert argued, that this overall distribution of matter was uniform throughout the universe. If it was not, he pointed out, the density differences would create anomalous gravitational fields, which in turn would cause great rivers of starflow across the sky. As far as astronomers could determine, however, the velocities of the stars in the Milky Way were random and small; on average, they weren't going anywhere. The inescapable conclusion was that the universe was stable; the way it looked now was pretty much the way it had looked eons ago and would look eons from now. The framework of the stars was the new aether.

Such stability, however, turned out not to be a feature of general relativity. In order to remain static, the universe required something that the original field equations did not provide, namely, a force that would counterbalance the attraction of gravity. Otherwise the universe could deflate, and gravity would keep rolling it into a tighter and tighter ball.

Luckily, the fix was easy, and it was the same fix as before. Albert added lambda, which he now dubbed the "universal constant," to the field equations. Lambda did not change the properties of the equations that Albert already knew and loved; they still gave practically the same answers for the bending of light and the shifting of Mercury's orbit, and they were still as covariant as they ever were. Its only effect was on the universe as a whole, where it provided a kind of long-range cosmic repulsion that balanced the mutual gravitational attraction of the contents of the cosmos.

"Admittedly," as Albert wrote in his paper, the addition of lambda "was not justified by our actual knowledge of gravitation."[16] Technically, lambda could be interpreted as representing the energy content of empty space, a curious concept, since Machian thinking presumed that space *had* no properties in itself. By now it was just as well that Mach was in his grave. Albert was perhaps only dimly aware that he was destroying Machian principles in order to save them. In the context of this new work, the original field equations had amounted to a tacit assumption that empty space had *zero* energy. But maybe that was a bad assumption; why *shouldn't* space have energy? Once introduced, lambda, or the cosmological constant, as it came to be known, would turn out to have a stubborn life of its own.[17]

The end result of this trickery and tinkering was what is sometimes called a "cylindrical" universe. If we imagine a stripped-down universe with one dimension of space and time, time would run down the long axis of the cylinder while space ran around it. One particular moment in time would be a slice through the cylinder: a circle. Another frequent analog is to imagine space at any given moment of cosmic time as the skin of a soap bubble or the surface of a balloon. People invariably ask what is inside or outside the skin of the balloon, and the always frustrating answer is that from the standpoint of a two-dimensional flatlander living on the balloon, "inside" and "outside" are no more sensible or reliable than Charles Hinton's famous fourth dimension, the favorite space of artists and mystics. An arrow shot in any direction would puncture nothing, but rather would return from the opposite direction after following the curve of the cosmos around. A light beam would have the same trajectory. In fact, one of the more curious aspects of Einstein's universe is that light from a distant star could have more than one route around the cosmos to our eyes. For a star on the opposite side of the cosmos from us, Albert noted uncomfortably, the whole universe would act like a giant magnifying lens, and its light would fill the sky, however faintly.

It was an odd vision, but in its way, cloaked in its natural raiment of mathematics, Einstein's universe was as solid as an oak barrel. The world made sense, at last. The world was static, as everyone believed, and its geometry was strictly determined by matter. The latter, as Albert later explained to de Sitter, was the core of his interpretation of Mach's idea of the relativity of inertia. "To me, as long as this requirement had not been fulfilled, the goal of general relativity was not yet completely achieved. This only came about with the lambda term."[18]

Of course the joke was that Albert did not need a static universe in order to have a Machian one, as Michel Janssen points out. "Einstein needed the constant not because of his philosophical predilections but because of his prejudice that the universe is static."

"I have naturally constructed a spacious castle in the sky," Albert announced to de Sitter.[19] As far as Albert was concerned, his cosmological considerations were more of an exercise in logic and philosophy than in astronomy, but now that relativity worked on paper, he was anxious to see if it was working in the real world. Although in his paper he had modeled the universe as a sphere, Albert had quickly realized that other shapes were possible and that it could even be lumpy like a potato. Einstein's new theory defined a relationship between the density of stars in the universe, the radius of the universe, and the value of the cosmological constant, and he had already prodded Freundlich to start working on the statistics of stars in order to measure those quantities.

The so-called radius of the universe, Albert had calculated, based on the density of stars in the Milky Way, was about 10 million light-years. The most distant known stars, by contrast, were a mere 10,000 light-years away. If the chasm between those two numbers could be bridged, it would mean, Albert thought, a new epoch in astronomy.

"One can illustrate our question by way of a beautiful comparison," he told de Sitter. "I compare space to a static cloth floating in the air, a certain part of which we can perceive. That same part is curved like a small part of a spherical surface. We philosophize about how to extrapolate the cloth where equilibrium is achieved through its tangential stress/tension, whether it is enclosed at the edges, endlessly expanded, or finitely large and enclosed upon itself. Heine gave the answer in one of his poems:

"And a fool waits for an answer."[20]

DE SITTER WAS laid up in a sanitarium when Albert's new universe hit him. As usual he found a lot to be suspicious of.

First and foremost, he questioned the assumption of a static universe. As an astronomer, de Sitter knew that there were hints that the situation was more complicated than the lack of commotion among the Milky Way stars might indicate. In 1914 an American astronomer by the name of Vesto Slipher, working at the Lowell Observatory in Arizona, had made spectroscopic observations of thirteen of the controversial spiral nebulae and found that the characteristic lines in their spectra were shifted in wavelength to the red, as if by a Doppler effect. The amount of the shift suggested that these nebulae were all moving at speeds that dwarfed the speeds of stars. And, even stranger, they were all moving *away* from us: If nebular motions were as random as those of the stars, some of them should be approaching. But thirteen was perhaps too small a sample. Slipher had reported these results to the American Astronomical Society in 1914, but had not pursued the subject further.[21] He had been hired by Percival Lowell to look for extraterrestrial life.

Another factor that dismayed de Sitter was the preposterous scale of Einstein's cosmos. Its estimated mass was equivalent to 10,000 Milky Ways, which implied that astronomers so far had discovered barely a trace of the contents of the universe. Where were all these extra galaxies? Einstein, de Sitter, concluded, had simply brought the world matter back in from the edges of the universe and spread it around uniformly inside. There was no longer necessarily any distinction between the world matter, responsible for inertia, and ordinary matter, responsible for gravitation.

In fact, the world matter, de Sitter noted, was no longer needed at all. The cosmological constant itself contributed to the curvature of space. When it was added to Einstein's equations, the space that resulted was curved even if no matter was present at all. All the mathematical conditions that Albert had imposed to ensure the relativity of inertia were met without any matter being present. And with that, Mach's vaunted principle vanished into the air.

As if to show how it should have been done, in the last of his three articles in the *Monthly Notices,* de Sitter proceeded to offer his own solutions to the Einstein equations for the universe. The first of them, which he labeled type A, was similar to Einstein's universe, with an even distribution of matter and a cosmological constant. The second solution, which he called B, had no matter at all—only the cosmological constant. Whereas Albert in his own calculations had essentially ignored time, in his cosmological calculations, assuming the universe as a whole was static, de Sitter had included time and wound up with a four-dimensional universe that was shaped like an ellipsoid.

De Sitter used a quirky coordinate system, and what resulted was an empty and presumably unchanging cosmos with strange features. Clocks ran slower, for example, the farther away they were from you. At the maximum distance, on the other end of the ellipsoid, time would stand still, and all motion and energy would come to zero. "All these results sound very strange and paradoxical," de Sitter admitted, but they weren't as bad as they seemed, because it would actually take an infinite amount of time for a light beam or a particle to reach that standstill point.[22]

Strangest of all, when de Sitter tried placing test objects in this universe he found that due to this dilation of time they would appear to be flying away; the farther away they were, the faster they would appear to be going. "The frequency of light vibrations diminishes with increasing distance from the origin of coordinates," he explained. "The lines of very distant stars or nebulae must therefore be systematically displaced towards the red, giving rise to a spurious velocity." In fact, he noted, Slipher had recently found that several nebulae did indeed appear to be receding. Perhaps this spurious velocity, which soon became known as the "de Sitter effect," had already been detected. In which case, with more data, it might be possible one day to choose between universe B and universe A.[23] In the space of a few weeks the large-scale shape of the world had gone from a philosophical abstraction to grist for the analysis of marks on photographic plates.

As with Schwarzschild's solution the year before, Albert was dismayed that de Sitter had managed to invent an empty universe in apparent contradiction to Mach's principle. "In my opinion it would be unsatisfying if there were an imaginable world without matter," he explained. "The [metric] should on the contrary *be determined by way of the matter.* That is the core of what I understand under the claim of the relativity of inertia."[24]

As far as the redshifted nebulae were concerned, Albert stuck to his conclusion, based on the low velocities of stars, that the universe was "quasi-stationary."

"This chain of reasoning is completely erroneous," de Sitter scrawled on the margin of Albert's letter.[25]

"We have only a snapshot of the world," he warned Albert in his reply. He personally doubted that even the Milky Way was a stable system. "And then must the great world be stable? The supposition which you silently make, that the mean star density is the same everywhere in the world . . . is justified by nothing and all our observations speak against this."[26]

Eager to prove that one could not make a universe out of nothing, Albert argued through the summer and fall of 1917 that de Sitter's universe was not empty at all, but had a band of matter concentrated around the periphery of the world, where time stopped and the mathematics broke down into gibberish—a so-called singularity. It was this band, then, that supplied the inertia to the universe and fulfilled Mach's condition, in effect turning de Sitter's world back into Einstein's.

"That would be distant masses again," de Sitter wrote dryly in the margin. "It is a *materia ex machina* to save the dogma of Mach."[27]

De Sitter suspected that the singularity was probably just an artifact of the strange coordinate system he had used, but he declined to commit himself. It didn't matter what inhabited the periphery because, as he had said in his earlier paper, it would take an infinite amount of time to get there from here. And thus these two opposite ends of the universe could have no effect on each other.

As other mathematicians like Eddington, Herman Weyl, and Felix Klein from Göttingen entered the fray and came to agree with de Sitter, at least about the supposed singularity, Albert saw his dreams of a Machian universe beginning to evaporate. In 1918 he denounced de Sitter's universe to the Prussian Academy as anti-Machian. If the cosmological constant allowed this kind of universe to exist, then maybe there shouldn't *be* a cosmological constant, he said.[28] It was the first of many occasions on which he was to express his regret at having let that particular genie out of the bottle. "If there is no quasi-static world," he grumbled to Weyl in 1923, "then away with the cosmological term."[29]

BEFORE 1916, the concept of the universe as a coherent mathematical entity had simply not existed. Now, thanks to general relativity, it did, even if in the process the cosmos had become like some aethereal ball of putty being swatted back and forth over the walls of war, or like a pair of dice tumbled on luminous green felt emblazoned with mystical symbols. Around the table the bets were being placed on relativity, momentum, cause and effect, atomicity, and entropy as the dice were handed from player to player, each dreaming of a natural. As far as the rest of the world was concerned, Einstein, de Sitter, Eddington, Weyl, Freundlich, and the rest could have been arguing about fairies dancing on the head of a pin. England, the land of Maxwell and Faraday, was full of old aether mechanics who hated relativity and hated Germans even more. The British journals were full of diatribes about the shoddiness of German science; there was talk of a boycott after the war. Except for a monograph by the precocious Eddington that was published in 1918, there were no further articles about general relativity published in English. In the Western world Einstein was just a rumor, his theories known mainly in the mirror of de Sitter's strange universe and by the ceaseless lobbying of Freundlich for some astronomer to go to an eclipse and measure light-bending. But in California a few young astronomers with nebulae in their eyes and frost on their breath were hunkered down on desert peaks in California where large new telescopes were being erected by dreamers like George Ellery Hale, founder of the Mount Wilson and Palomar observatories, little aware that Einstein's new universe would be their destiny.

23

THE BELLY OF THE BEAST

IF THE SPRING OF 1905, ALBERT'S *ANNUS MIRABILIS,* IN WHICH HE HAD discovered relativity and *Lichtquanten,* represented one of the most remarkable bursts of brilliance in the history of physics, the stretch from 1915 to 1917 in Albert's life represented arguably the most prodigious effort of sustained brilliance on the part of one man in the history of physics. During those years of war and heartbreak, of working like a horse, smoking like a chimney, and living on coffee, cheap sausage, and cookies, he had finally completed the general theory of relativity, reinventing gravity, space, and time, used relativity to reinvent the universe itself, and raised quantum theory to yet another bewildering level of mystery and weirdness—and the effort took its toll. In February 1917, as Albert was putting the finishing touches on his new cosmological theories, he collapsed with stomach pains.

The breakdown had been a long time coming. As far back as 1899 he had remarked in a letter to Mileva how an upset stomach—his poetic ailment—on a train ride home to Milan had reminded him *in harsh tints how closely knit our psychic and physiological lives are.* During his visit to Leiden the previous fall, it had been remarked upon that Albert looked sickly and pale.

Albert told Freundlich he had cancer, but that it didn't matter if he died, now that he had finished the general theory of relativity.[1] Freundlich convinced him to see a doctor, who supplied the first of what would be a cavalcade of diagnoses: a liver ailment, a stomach ulcer, jaundice, gallstones, and general weakness.

Within two months, Albert lost fifty-six pounds, and his doctor placed him on mineral water and a strict diet. The situation was exacerbated by wartime conditions in Germany, where food was scarce and tempers were short. A series of strikes in the spring of 1917 undermined the economy even further. The Berlin Einsteins were at least able to supplement their meager rations with produce from their cousins in the rural south.

In spite of his condition, Albert never stopped working. It was flat on his back eating what he called chicken food that he continued his correspondence with de Sitter, wrestling with the possibilities of a universe without boundaries.

He canceled plans to go to Zurich. "This winter I was so sick," he later explained to Hans Albert, who wasn't talking to him anyway, "I could not have gone on hikes with you or even have been allowed to eat in a hotel. In addition I doubted that you cared much one way or the other if I came."[2]

Meanwhile, Mileva and Tete were getting worse again. It was decided that Mileva should go back into the hospital; Tete would accompany her. That was no cure for his ear problems, though, and Zangger wanted him to go to a sanatorium in Arosa, high in the Alps outside of Chur. The prospect made Albert's already battered spirits even darker. "It's clearly impossible that he will ever be a whole man," he wrote in despair to Besso.

For the first time in his life, he told Besso, he blamed himself for his son's plight.[3] Usually he took things lightly and did not feel responsible, but it occurred to him now that Tete's and Mileva's conditions were probably connected. Tete had inherited his disease. Albert's guilt, it seemed, was in having consorted with an unwell woman: The limp. The goiter. The suspected tuberculosis—poison sprouts on the brain. The glandular problems. Not to mention the unhealthy and persistent melancholia.

When Albert and Mileva conceived Tete, on their Engadine vacation in the wake of the Anneli episode, Albert explained, "I didn't know anything about scrofula or that this tuberculosis is translatable to the children." He had thought nothing of Mileva's swollen glands when he lay with her.

Now he would have to live with it.

He agreed to send Tete to Arosa for a year, but that just raised another problem. "Who will bring him there?" Albert asked. "Shouldn't I come to the place myself and stay with him there a little so that the boy doesn't feel so lost?"[4] Zangger consoled him with the assurance that it was the right and only course to take. Though he himself was now sick and needed surgery, Zangger took charge of the financial and medical details, but because the sanatoriums were full of foreign children, the best he could do for Tete was a spot at 10 francs a day.

There was still the question of what to do with Hans Albert. Einstein continued to mutter to Besso and Zangger about taking the boy to Berlin, where he could teach him to be a comrade. Albert had a lot to give, he believed, and not just intellectually, but he continued to fret, "Do you think my wife would understand?"[5]

In April, Mileva, accompanied by Tete, went back into the hospital, while Hans Albert went to live temporarily with the Zanggers. Still undecided about whether to risk Mileva's wrath over a Berlin move, Albert thought that he could send Albertli to Maja and Paul instead. He asked

Besso to sound them out about it. If they agreed, Albert would contribute 2,000 francs a year to pay for Albertli's room and board. "Don't tell them it is my idea," he cautioned, "so they can refuse without being embarrassed."[6]

The Wintelers agreed, but the plan never went any further. In May a shortfall in Albert's income caused him to have second thoughts about how to support himself and his family as economically as possible. Out of an income of approximately 13,000 marks a year, he was sending Mileva 7,000, plus additional sums regularly to his mother, who was also ill. Any large expenses threatened to leave him with nothing to put away for the boys' future. He pleaded with Zangger to take this into account as he juggled the medical bills of Mileva and Tete.

Albert even mused about moving to Zurich if he was needed. "I'm not as useless in these things as you might think thanks to the turbulent life I've had in the past."[7] Meanwhile, Albert's doctors were pleading and then ordering him to go away to a spa himself for the summer. Tarasp in the beloved Engadine was suggested, but Albert refused; he wanted to see his children during the summer and couldn't afford to take them all to Tarasp.

IN THE MIDST of his illness and his family problems, Albert's attitude toward the war and Germany further soured. In a nihilistic frame of mind, he had concluded that no amount of words or reason would stem the conflict. As he told Zangger: "I wonder if it wouldn't be good for the world if degenerate Europe didn't ruin itself completely. Whenever I hear the new nauseating word 'training,' my guts turn over. All our praised progress in technology, and in civilization in general, is like an axe in the hand of the pathological criminal."[8]

One who encountered his bitterness directly was Georg Nicolai. After a series of misadventures, the outspoken doctor had been drafted and sent to Danzig to work as a medical orderly. He had brought with him the manuscript of his series of forbidden lectures, which he had rewritten into a book called *The Biology of War.* Just before being arrested again, he managed to smuggle a copy of the book to a publisher in Zurich, where it would be published in the spring of 1917 to acclaim in non-German Europe.[9] In the meantime, however, Nicolai had already embarked on an even more devious publishing adventure: an anthology of selected writings by Kant, Fichte, and other classic German authors in which they espoused values like liberty, peace, and brotherhood. *The Politics of the Classics* would demonstrate, then, that modern Germany had betrayed its own heritage by the war.

In order to publish the book, which no legitimate publisher would go near, Nicolai proposed to form a corporation composed of himself, Einstein, and his friend the writer Otto Buek, who had signed the "Appeal to the Europeans." But when his principal backer, a wealthy cousin of Elsa's named Moos, got cold feet, Albert, who was supposed to be the glue in this syndicate, also had second thoughts. "I consider the matter completely hopeless and likely to cause you much annoyance and disappointment," he wrote Nicolai from his sickbed.[10] Shortly thereafter, without telling Nicolai, Albert quit the project. Moos then told Elsa and Ilse that he was withdrawing as well. Nicolai appealed to Albert to no avail.

"Nothing is so difficult as to turn down Nicolai. The man who is so sensitive in other respects that the growing of grass is a loud noise to his ears seems almost deaf when the noise involves a refusal," Albert answered. "I therefore raise my voice with the strength of a young bull just reaching maturity and say—shout—solemnly, passionately, 'No!' (The music stops, two bars rest, follow by an elegiac *piano*.)"[11]

Every man had the right to a hobbyhorse, he said. "You therefore have the right to plant stale cabbages, fertilize them with modern manure, and sell them personally in the market. I have the right to stay away and devote myself exclusively to my own hobbyhorses. That is how it shall and must remain."

Nicolai felt deceived. Albert had never explicitly said he was quitting, even when Nicolai asked him directly. He agreed that Albert certainly had the right to withdraw, but protested that it wasn't fair to convince Moos to do so as well, and thereby undermine the entire scheme.

Albert apologized for his "ungracious simile," explaining that he meant it as a joke and insisting that he never intended to damage Nicolai's enterprise. "I dissuade no one who would participate without me. It is only that I do not wish Herr Moos to go along *exclusively for my sake*. Such a sacrifice on the part of a relative and personal friend would embarrass me."[12]

Nicolai was baffled and disappointed by this attitude on the part of his comrade and protector in Berlin scientific circles, but there was little he could do but be gracious, for he was preoccupied with yet another court-martial. *The Politics of the Classics* died a probably merciful death, and Albert made a typically astute call as he danced along the edge of personal freedom and outright treason, writing seditious letters to Rolland and Zangger from within Switzerland, where censors would not read them, and hobnobbing with plutocrats and war criminals at the Academy.

In April, although he was still weak, Albert ventured out to a meeting of the Berlin section of the German Peace Society at the Cafe Austria. About two dozen people were present, many of them members of the now-banned BNV, when the police invaded and broke up the gathering. Some of the participants, including Albert, straggled to a nearby apartment, where they drank tea and finished the meeting.[13]

There was perhaps no better object lesson for the dangers of political passion than Albert's old friend Fritz Adler, the physicist-assassin, whose trial was coming up that summer. Adler had whiled away the long wait in a succession of jails and fortresses by writing a lengthy and densely philosophical treatise on relativity, *Local Time, System Time, Zone Time,* which he sent around to various physicists, including Albert, who was appalled at Adler's action and had stayed in touch with Katya, Adler's wife, and offered to do whatever he could. "He is one of the most exceptional men whom I have known. I cannot judge his deed because I do not know the motives for it."[14]

Albert offered to be a character witness but was declined. He then deputized Besso to act in his name to mobilize support for Adler in the Swiss physics community. In a private aside, however, he warned Besso that Adler was "a stubborn sterile *Rabbinerkopf* [rabbi-head], out of touch with reality, selfless and self-tormenting. A real martyr."[15] As an example he cited Adler's stepping aside for him at the University of Zurich so long ago.

Besso succeeded in getting the *Physikalische Gesellschaft* in Zurich to take up Adler's cause. They agreed to issue a statement testifying to his admirable qualities and his important role in the scientific brotherhood. The defense could use it as they wished.

When Albert sat down to compose the statement, though, he found himself in a ticklish spot. Adler's family had circulated his manuscript widely to psychiatrists and other physicists, hoping to use it to show that Adler was deranged and thus not responsible for his actions. In effect, the physicists were being invited to save their friend by insulting him and saying his work was loony, when it was really only bad. "The experts, especially the physicists, were placed in a very difficult situation," wrote Philipp Frank, who was teaching at Prague and received a copy.[16] Nobody was in a hotter spot than Einstein, the expert of experts.

Albert was frankly embarrassed by Adler's work, telling Besso that it was full of "worthless subtlety" and Ehrenfest that it "stood on thin bones." It didn't help that Adler was a devoted disciple of Mach. Despite his reverence for what Albert was now calling Mach's principle, about the relativity of inertia, he had no more use for Mach's stamp-collecting approach to epistemology. The confirmation of general relativity—a theory motivated by pure thought if there ever was one—by Mercury's orbital permutations had been a revelation to him that would send him increasingly in search of truth through mathematics as time went on. Theories came not from facts, but from what he would call "free inventions of the mind."

"Adler rides Mach's nag to exhaustion," Albert complained.[17]

Besso shot back that Albert himself had ridden Mach's horse quite successfully back in the old Bern days and that he shouldn't give up on the little nag yet.[18]

Albert finally concluded that Adler, with all his father's connections, was in no real danger. There was no need to insult him by asserting he was crazy when he was merely wrong—as most physicists are most of the time. Albert's statement, "Friedrich Adler als Physiker" ("Friedrich Adler as Physicist"), was mostly descriptive and nonjudgmental: "One can see in these few sentences that the work which Adler did with enthusiasm and unshakable zeal during the days of his detention pending trial wants to lay firm the epistemological foundations of classical mechanics and its relationship to the modern theory of relativity."[19]

Adler was ultimately convicted and wound up spending eighteen months in a relatively comfortable prison, Stein am Donau, from which he carried on a philosophical dialogue with Albert. He joked that given wartime deprivations, he was probably better off in prison than out in the world.

By July Albert had begun to gain weight again, thanks to Elsa's cooking, but the disease had taken an otherwise heavy toll. In the wake of his cosmological breakthrough and the new quantum calculations, his scientific work, he felt, had fallen into a lull, and he was ready once again to proclaim an end to his laying of golden eggs. "Truly new things one finds only in one's youth," he lamented to Zangger. "Later one becomes more experienced, more famous, and dumber."[20] He did as much traveling as his stomach would permit. Declining an invitation from Ehrenfest to Leiden, Albert paid a brief visit to his ailing mother and then went to Zurich, picked up Albertli, and took him to Arosa to see Tete, whose condition was improving at his "vacation camp." Then he repaired to Lucerne and Maja and Paul's "sanatorium" for an extended stay. There, the treatment consisted of making Albert lie down at the first hint of attack and apply heat, thus lessening the pressure on his abdomen.

Feeling "surrounded by love," he vowed to follow the doctor's directions for his belly and to live a medically correct life from now on. During this period, poor Hans Albert left his stricken, depressed mother and took the train to Lucerne to find his father stretched out in pain being fed the equivalent of baby food. His uncle Paul managed to take him on a hiking trip before Mileva demanded him back. "He is well developed but he acts crudely around me sometimes out of old habit," Albert complained.[21]

Meanwhile, illness had accomplished what neither lust nor tears could bring about: the partial conquest of Albert's stubborn solitude. On his return to Berlin, he acquiesced in being moved into a newly vacant apartment across the hall from Elsa's at 5 Haberlandstrasse. Now Elsa could nurse and feed (if not boss) him in style. Elsa later remarked that it was handy having a physicist around the house, because he managed to open any can, no matter how irregular or strange it was. With sporadic shipments of food and medicine from Zangger, Albert slowly improved.

Mileva and Tete, however, were not as fortunate. Albert's finances had improved to the point that Tete's treatment was less of a burden, but he was hoping fervently to get Tete back home by the end of the year. He told Zangger, "I am against a child like him spending his youth in a kind of disinfection apparatus. Our life is not only technologically, but also medically contaminated—really just a version of technological contamination."[22]

Adding to the tension was the uncertainty of what it was that ailed Mileva. Albert was not about to go see her, however, and asked Zangger, "Do you know exactly what my wife's affliction is?" Was it an abscess in the spinal cord, or multiple sclerosis? Was there any prospect of recovery? At one point the doctors prescribed X rays, but, through Zangger, Albert declined.

Mileva spent the summer bouncing from her balcony overlooking the city to the hospital. A missive from Zangger later that year in his usual style gave a terse snapshot of her condition. "The spinal cord has remained healthy for certain, the inflammation crept, which is unusual, to the bone and further. . . . The plates, which offer elasticity, have shrunk, any excursion is applying pressure again now. There's been no fever for a long time, mobility hasn't been impaired but *any pressure* is dangerous for the spinal column and the cracking of the sheath."[23]

In August Mileva wrote to her mother asking her to come and help out; Albert had apparently also been in contact with Mileva's family. Her mother couldn't come, so they sent her sister Zorka instead. She reported to Albert that she arrived in Zurich to find Mileva lying on her sunny balcony. When Mileva laughed, Zorka began to cry, until at her sister's prodding she was soon laughing too.[24]

Like Mileva, Zorka had a limp, only worse, according to Lisbeth Hurwitz, who was as always a faithful visitor. Zorka wound up staying three years. At first she was a considerable help, but she eventually began behaving oddly, singing for no reason, and acting hostile and paranoid to outsiders. Zorka was finally institutionalized.[25]

THE LONG-AWAITED Kaiser Wilhelm Institute for Physics finally opened its doors to business in October 1917, even if the actual doors that opened, according to the formal announcement in Berlin's *Vossische Zeitung,*[26] were at 5 Haberlandstrasse. The promise of the directorship of his own physics institute had been one of the lures by which Albert was reeled to Berlin, but with the chaos of the war there was still no building to be had. Haber had offered space in his own institute, but Albert naturally preferred to work at home.

It was an extremely modest operation at first.[27] Ilse served as secretary. The announcement described the purpose of the institute as to "induce and support systematic work on important and urgent problems in physics," and invited proposals to be considered by a "Direktorium" of Einstein, Max Planck (whose name Albert forgot to include in the announcement), Walther Nernst, Emil Warburg, Fritz Haber, and Heinrich Rubens. The group convened once in November 1917 and didn't meet again for two years. In the meantime Albert demonstrated exactly what his notion of "urgent problems" was when he hired Freundlich as an assistant on a three-year contract to conduct research on testing general relativity. It was an uncharacteristic bureaucratic triumph for Albert, who had been trying for years to pry Freundlich free from his duties at the Royal Observatory.[28] Of the first wave of proposals and any number of crackpot schemes, the Direktorium granted money to only one: Peter Debye, who had been Albert's successor at the Polytechnic, was given funds to buy instruments for generating high-energy X rays. Unfortunately, he returned the money several years later when the manufacturer couldn't come through.

In the outside world, political, rather than scientific, revolution was the order of the day. Seeking to exploit the ongoing unrest in Russia, Germany had offered to transport the revolutionary exile Lenin back to his homeland. And so in April Lenin and his comrades had left their refuge in Zurich and traveled in a sealed boxcar across Germany to the Russian border. The czar fell, and in November Lenin and the Communists finally seized control of the short-lived republic that had ensued. Peace talks had finally begun. The fate of the war hinged on whether the Americans from across the Atlantic or the Germans released from fighting on the Eastern Front would reach the Western Front first and turn the tide of battle. As harbingers of things to come, Rosa Luxemburg and Karl Liebknecht announced the formation of the German Communist Party, and Albert was secretly placed on a blacklist of thirty pacifists who were not to be allowed out of the country without the approval of the military high command.

ALBERT'S STOMACH PAINS recurred at the end of the year. He took to his bed again, staring at the bill for Tete's first three months at the Sanatorium Pedolin: 1,200 Swiss francs. Zangger had calculated a tentative budget for Mileva's and the boys' care during the upcoming year, but Albert was feeling short of money again and angry about it. Zangger, he felt, was bossing him around as if he were a student. Over the course of the past twelve months he had sent some 12,000 marks to Zurich. He notified Besso that for the following year he would send what Mileva could reasonably claim—

6,000 marks, and no more. "The question is about a screw without an end," he complained.[29]

At least he was on good terms again with Hans Albert, who sent him a letter and a card. "I can see that you are not only my beloved son, but also that a truly affectionate relationship is developing between us," he wrote, informing his son that there was no chance that his stomach would let him visit Switzerland this year, or perhaps ever again. "Doubtless, I have had my ailment longer than you have been alive."[30]

In the next room, the secretary of the Kaiser Wilhelm Institute for Physics, Ilse Einstein, composed a carefully worded letter to a certain Fraulein von Kampen about a lady who might possibly take in a certain young girl. "She would especially have to see the girl herself; if it's at all possible please arrange *immediately* for her to travel here. I hope the lady can come to an agreement with her."[31] The "girl" in question was in fact the revised manuscript of Nicolai's outlaw manifesto *The Biology of War,* which Nicolai was arranging to smuggle to Switzerland for a second edition. It seems amazing that Albert would not have known, whether he approved or not, of Ilse's involvement in the matter, which would likely have left him vulnerable to criminal charges if their role had been discovered.

For once, though, Nicolai had outsmarted himself. After another court-martial, for insolence, he had complained so incessantly about his commanders in Danzig that his friends in Berlin had pulled strings to get him transferred out to an infantry company stationed in the small town of Eilenburg. It was safe, being far from the front, but it was a "provincial backwater," and there was nothing for him to do there. Characteristically, Nicolai fought the transfer, but lost. The one consolation was that Eilenburg was close enough to Berlin for him to have visitors. His wife soon moved to the town to be near him, and Ilse came frequently as well.

PACKAGES LARGE AND SMALL—from a sealed railway car to a stack of papers—were slipped around smoky, weary Europe with furtive intensity as the war ground through its fourth year—bundles of hope, gathered and sent on bloody trains through shattered and contentious stations of memory, some abandoned now, some never more than a gleam in some dreamer's eye. General relativity was one such package, a sort of coded message to the future working its way like a virus into the bones of de Sitter and Eddington. So was the quantum theory. Another package was safely delivered. Zangger wrote to say that Tete was back home from Arosa, telling Albert, "You need complain about foreign homes no more." Tete would be back in school by the summer.[32]

24

THE LAST SCOUNDREL

IN THE SPRING OF 1918—APRIL 23, TO BE EXACT—MAX PLANCK TURNED sixty. Coming amid these unsettled times, with Europe now wracked by war and hunger, it was a bittersweet occasion for Planck personally. His oldest son had been killed in battle; just the year before one of his two daughters had died in childbirth. By now this quietly steadfast man had received every scientific honor that Germany had to bestow. Within a few months he would win the ultimate recognition: the Nobel Prize. He was still dragging his feet on the matter of Einstein's light-quanta. Earlier in the year he had told Lorentz that the question of whether light was propagated as corpuscles or as continuous waves was the most important issue in physics.[1] For now he was sticking with Maxwell's formulation of light spreading out in waves because it gave definite, calculable answers.

As is the scientific custom, a symposium was held at the German Physics Society headquarters in Planck's honor, at which Albert was a featured speaker. His address, entitled "Principles of Research," has often been quoted, and was the first expression of themes that he would return to for the rest of his life.[2] Using Planck as an ostensible example, Albert argued that science was an emotional, not a logical, calling, a lover's quest. But he might just as well have been describing himself:

"In the temple of science," he began, "are many mansions, and various indeed are they that dwell therein and the motives that have led them thither." For some, he went on, science was a sport; for others a way to get ahead. "Were an angel of the Lord to come and drive all the people belonging to these two categories out of the temple, the assemblage would be seriously depleted, but there would still be some men, of both present and past times, left inside. Our Planck is one of them, and that is why we love him."

Who were these few pure souls who remained, and what drove them? "Most of them are somewhat odd, uncommunicative, solitary fellows, really less like each other, in spite of these common characteristics, than the hosts of the rejected. . . . I believe with Schopenhauer that one of the strongest motives that leads men to art and science is escape from everyday life with its painful crudity and hopeless dreariness, from the fetters of one's own ever-shifting desires." A sensitive nature would seek to escape the fog of personal desires into the realm of objectivity, the same way a city dweller would yearn to escape his noisy, crowded surroundings into the peace of the high mountains, "where the eye ranges freely through the still, pure air, and fondly traces out the restful contours apparently built for eternity."

This negative urge to escape, Einstein said, was balanced by a positive side. Like the painter, the poet, or the philosopher, the scientist was striving to create a world of his own. "Each makes this cosmos and its construction the pivot of his emotional life, in order to find in this way the peace and security which he cannot find in the narrow whirlpool of personal experience."

If the physicist wanted to live in a rational world, his job, then, was to find those universal laws from which all else could, in principle, be deduced. "There is no logical path to these laws; only intuition, resting on sympathetic understanding of experience, can reach them," he declared, driving a stake through Mach's freshly covered grave. This approach might seem like a haphazard enterprise, he admitted, but history proved that it had many successes. Nobody could deny that the laws of physics were derived from experience, even if there was no logical bridge between phenomena and principles—a miracle due to what Liebniz had called a "preestablished harmony" between the mind and nature.

"The longing to behold this preestablished harmony is the source of the inexhaustible patience and perseverance with which Planck has devoted himself, as we see, to the most general problems of our science, refusing to let himself be diverted to more grateful and easily attained ends." Neither discipline nor willpower—often wrongly invoked in Planck's case—could account for this attitude, Albert declared. "The state of mind which enables a man to do work of this kind is akin to that of the religious worshiper or the lover; the daily effort comes from no deliberate intention or program, but straight from the heart."

NOW THAT HE WAS living at 5 Haberlandstrasse with Elsa and the girls, and the address had even appeared in the newspaper as the home of the Kaiser Wilhelm Institute for Physics, Albert's unconventional living arrangements (or so he claimed to Mileva) were attracting attention. Ilse was twenty years old, marriageable, and thus susceptible to social embarrassment. Tears were flowing again in Berlin, and he decided it was time to renew his attempts to get Mileva to agree to a divorce.

Almost two years to the week after he had made his first ill-fated divorce proposal, Albert wrote Mileva at the end of January 1918: "A desire to finally achieve a certain order in my private affairs prompts me to suggest a divorce to you for the second time."[3] To make the offer more attractive, he added a number of guarantees. Mileva would receive 9,000 marks a year, with 2,000 earmarked for the children, an increase from the current 6,000. If and when Albert won the Nobel Prize, which was then worth about 30,000 to 40,000 marks, he would deposit the money in a Swiss account with the interest going to her and the boys, in which case the proposed support payments would be appropriately adjusted. (Mentioning the Nobel was hardly presumptuous, since by 1910, Albert had already been nominated for the physics prize seven times.) Finally, he would see to it that Mileva would receive his widow's pension.

"Of course I would naturally only make such colossal concessions in the case of an uncontested divorce," he warned her. "If you do not agree to a divorce, you will not see a penny over 6,000 marks."

Mileva showed Zangger the letter, which struck him as brutal. "In short, knife to the throat, without any advance notice," he summarized its contents to the Bessos. "Friend Einstein causes me sorrow. Irritatedly he writes rude letters, even to [Hans] Albert, that all this [Mileva's illness] is nonsense."[4] In the harshness of Albert's correspondence he detected the hand of the egotistical Elsa, who did not want to send money to Zurich, and who, when Einstein was still faithful to his wife and not a famous figure, did not deign to invite him to her home, and in fact, scarcely acknowledged him.

Zangger asked Anna Besso to talk to Mileva and then help her draft an appropriate reply, which Zangger and Besso would then show to Emil Zürcher, Mileva's lawyer and neighbor.

In her relapsed state, Mileva was, in fact, barely able to contend with anything, and she at first rudely rejected Anna's suggestion that they try to establish a budget to bring her expenses under control.[5] As if some entropic curse had settled over the Einsteins, things had only grown worse. When Zorka had to be sent off to an institution, Zangger hired a temporary nurse to fill in, but she was expensive. If cheaper help couldn't be found, Mileva would have to return to the hospital. Albertli could stay with the Zanggers, but Tete would have to be placed in a children's pension. "Everything costs and is insanely expensive if you view from the distance," wrote Zangger to Albert, "but unfortunately in an emergency nothing more can be done than what we are presenting to you now."[6]

In the meantime Mileva poured her heart out to Anna.[7] Albert had fallen under the spell of his cousin back in 1912, Mileva claimed, and had been lying to her ever since. Her children, she predicted, would never see a penny of whatever money Albert was saving. "Elsa is very greedy," Mileva said. "Her two sisters are very rich and she's always envious of them." Albert's cousin, she went on bitterly, cared little about her own children, and had even admitted once that Ilse and Margot were estranged from her, but it didn't matter because their grandmother—Albert's aunt Fanny—took care of them. (In fact, events would subsequently suggest that Ilse and Elsa were not alienated at all.)

Anna sent a long memo to Zangger detailing all this, and also a letter to Albert analyzing Mileva's financial situation, pointing out, among other things, that the 6,000 francs he was offering her was far too little to live on.

"That was the good old trusty Anna again," Albert wrote back.[8] He agreed to increase his payments to 8,000 francs. Then he went on to outline what he thought the best solution to all their problems would be, and it was a dire one indeed. Hans Albert, he said, should go to live with Maja and Paul permanently. "Miza, who will remain incapable of heading a household for the rest of her life, is cared for permanently in a sanatorium, in Lucerne, if she likes so that she can see Albert daily." Tete would have to be shipped off to a healthy mountain environment to avoid an early death.

Spinning out a tale of his own woes—his illness and Elsa's vulnerable social position—Albert begged Anna to see things from his point of view. He had no choice but to marry Elsa, he said, in order to lessen the embarrassment for her marriage-age daughters, but he promised that he would not let the new marriage hurt Mileva or the children. "Is it astonishing when I sometimes feel bitter and get the impression that all are united in making life unnecessarily hard for me?"

"If Elsa had not intended to make herself so vulnerable, she ought not to have run after you so conspicuously," Anna retorted, declining the role of bringing Mileva to heel.[9] Being sick was not such as a great reason to get married, she pointed out, and then reminded him that only a year before, he had assured everyone that he would not marry again. "I'm only telling you this so that you can compare your *present* standpoint with the one *then*."

Albert was shocked that anyone—especially a woman—would be so brazen with him. "She has written such shameless letters that I have forbidden myself further letters from her," he fumed to Mileva.[10] "One must however pardon her because she is not entirely normal." This last comment was almost certainly an allusion to the Winteler family's own strain of instability and melancholia—Marie's depression, Julius Winteler's double homicide of his mother and brother-in-law—with which he was intimately familiar.

Having confessed the truth to Anna, Mileva could no longer hide it from herself. This was the end; she no longer had the strength to fight this juggernaut. What made its way through this winding channel of advice and negotiation, from Mileva to Anna to Michele to Zangger, was a draft of a pitifully fragile white flag of surrender.[11]

"You will understand that it is difficult for me to come to a decision due to my present illness," she began. "I understand your desire for a free future; if it is necessary for your work I do not know, but I do not wish to block your way and prevent you from happiness. But it looks to me that everything will become much easier after the war, communications, everything. I have asked Dr. Zürcher to find out about the procedure. It does not seem that simple. I am afraid of getting too excited and must do everything so that we are secure and that nothing can take this security from me, because of the children. Let your lawyer write to Dr. Zürcher, how he thinks that all should be arranged, how the contract should be.

"Everything that could upset me I must delegate to people who are objective and are knowledgeable regarding the actual situation. You see that I will not stand in the way of your happiness, if you are decided, but due to my illness it is more difficult than you assume—in particular today—to do something definitive."

Some version of this letter was subsequently relayed to a relieved Albert, who could not resist chiming in with his own symptoms, as if there were still some strange sympathy between them. "My health is adequate but my sense of balance is still unstable; I'm lying down a large part of the day and then being fed like a child who's just been weaned. Despite that I vomit sometimes, and any excitement causes with certainty a pain."[12]

There was still plenty to haggle about—in particular the provenance and amount of the various funds, and contingencies for their use, and who had to travel where and when to see the boys. The subsequent negotiations were businesslike. Perhaps, Albert considered, he had made things harder than they had had to be earlier, with his fierceness and hardness.

As Einstein and Zürcher traded drafts of divorce agreements throughout the spring of 1918, the Germans launched a major offensive, one last push to win the war before the Americans reached the front and reversed its course. By early summer German troops had gained such a strategically superior position that the civilian government was confident that it could negotiate a favorable truce, but Field Marshal Paul von Hindenburg and his chief of staff, Erich Ludendorff, blocked the effort. Strikes and hunger ravaged the country. Albert ruled out any trips to Switzerland that summer on account of his health. Elsa asked Zangger to send him milk; when it arrived, Albert forbade him from doing so again.

"I am curious to see what will last longer: the World War, or our divorce," Albert wrote Mileva. "They both began essentially at the same time. This situation of ours is still the nicer of the two."[13]

As promised, Tete improved throughout the spring, and Zangger described him as growing up into a nice young man, a shy fellow with girlish movements, "not Einsteinlike at all."[14] Hans Albert, meanwhile, managed to find a new way to disappoint his father. As an outgrowth of his long childhood fascination with rivers, he had decided that he was going to study civil engineering.

That news occasioned a trip down memory lane. Albert himself had originally planned to be an engineer, like his uncle Jakob. But he found that the prospect of applying his brainpower to such tedious ends was unbearable. "Thinking for the sake of thinking, like music!" That was what appealed to him, he told Zangger.[15] Clearly Hans Albert was not cut out for this lofty life.

Well, it was nice that the boy was interested in something. "What he is interested in isn't really important, even if it is, alas, engineering," he told Mileva. "One cannot expect of one's children that they inherit a mind."[16]

On June 12 Albert and Mileva finally signed a formal divorce agreement. It consisted of two handwritten pages on brown paper, and among its major provisions was Albert's promise to deposit 40,000 marks' worth of stocks or bonds in a Swiss bank for Mileva, who could spend the interest as she pleased, but not the capital without his approval.[17] If she died or remarried, the money would go to the children. In the case of a Nobel Prize, Albert would give the prize to her and get the 40,000 marks back. As a hedge against *not* getting the prize he agreed to have another 20,000 marks deposited in a German bank, from which Mileva could draw interest after his death.[18]

The agreement also stipulated that Mileva would let Albert have the boys during their school vacations when he came to Switzerland. That was a victory for her, but one Albert was willing to cede. If Mileva was satisfied, he didn't want to tamper with the agreement.

In fact, 40,000 marks' worth of securities had already been transferred to a Swiss bank, more or less over the objections of the German central bank. In addition, later that summer, Albert mentioned to Mileva that she should have received a paper "regarding the deposit of 20,000 marks."[19] It is not clear whether the other larger chunk had been deposited or not, but Mileva seems to have requested that the interest be paid to her in Swiss francs, rather than marks, as Albert had arranged. Albert, however, didn't want to change the arrangement and confuse the bank just then. In the meantime she could collect interest and exchange it, if she wanted. "I believe the bad times are coming again for the exchange!" he said, like the finance student he once was at the Polytechnic.

It remains a mystery where Albert got the cash to finance his divorce. One source of funds could have come from Einstein's continuing role as consultant for Anschütz-Kämpfe. By now he did not lack for rich friends.

"I now ask you to submit the divorce immediately so that the matters finally fall into place," Albert wrote as the 20,000 marks clanked down into some metaphorical cash register of the soul. He had fulfilled his end of the bargain. His freedom was bought and paid for. The future he had vouchsafed to Elsa in all those cozy adulterous letters back in 1913 was seemingly within sight.

But there was only one slight problem, one nagging uncertainty to be resolved. According to one controversial document that has surfaced, Albert no longer knew whom he wanted to marry. He had fallen in love with someone else, and a rather inconvenient and embarrassing someone else. Of all people, it was Ilse.

ILSE WAS, of course, already in love with Nicolai and had spent a good part of the year shuttling back and forth to Eilenburg, where Nicolai was plotting and pacing in his self-styled prisoner-of-war cell. When at home, she helped Albert organize and administer the new physics institute. Working with her seems to have stirred up old longings. It was spring, and he was beginning to feel better.

Subtle signs of Albert's entanglement can be discerned in his correspondence with Ilse earlier in the spring. She had sent Albert a postcard from Eilenburg with a drawing of a goose. His response had a touch of the flirtatious childishness reminiscent of his glee at the receipt of cookies from his mother or goosecrackle from Elsa. "The power of movement only resides in my head," he wrote. "Seldom does it go astray down into my fingers. But my pleasure in your letter and the beautiful *Geflügelkarte* is bringing it to pass. The poultry pleased me proportionately opposite to its size. The little nest and its surroundings seemed to be very nice, veritable nature practically unsullied by people."[20] There was news regarding the institute, but he would save it until she was back "so that you will feel entirely in your element."

Ironically, it was Nicolai who perceived where their relationship was heading. When Ilse apprised him of Albert's plans to marry her mother after his divorce, Nicolai ventured the opinion that a marriage between Albert and *her* (Ilse) would be more suitable. Elsa, after all, was forty-two and had spent the six years of their relationship getting old. Ilse, who was twenty, could bear Albert children. (Like most Berliners in his circle, Nicolai probably wasn't aware that Albert had children back in Zurich.)

Although Ilse didn't take Nicolai seriously, the question soon came up in a discussion in the Einstein household: Whom exactly should Albert marry, Elsa or Ilse? To Ilse's consternation, what had started as a joke quickly escalated into a serious proposition, which now had to be considered completely and deliberately. As she later explained in a letter to Nicolai, Albert had simply declared that he could marry either Ilse or her mother; the choice was up to them. Elsa, uncertain how seriously to take this, or perhaps possessed of a mother's self-sacrificial instinct not to stand in the way of her daughter's prospects, left the decision to Ilse.

Confused and dismayed, Ilse turned to her best and most worldly friend and counselor, Georg Nicolai, and poured her heart out in a long, extraordinary letter dated May 22, 1918.[21] Across the top she wrote in large letters, "PLEASE DESTROY THIS LETTER IMMEDIATELY AFTER YOU HAVE READ IT."[22]

"I know that Albert loves me very much, perhaps more than any man will ever love me, he even told me himself yesterday," she revealed, saying that he had confessed he had trouble controlling his sexual feelings when he was with her. But his feelings were not exactly reciprocated. "If there truly is friendship and camaraderie between two beings of different types they would certainly be my feelings for A. I have never sensed the wish or the slightest desire to be close to him physically." Without passion, she could not sustain a marriage and was afraid that she would grow to no longer love him and even resent him. "Last, but not least, I would feel something like a slave who had been sold."

While she could imagine growing fond of a stranger through living with him (as in an arranged marriage), "with A it just won't go into my head." For one thing, she had already gotten used to thinking of him as a father. "You will reply that that lies in the past, but I would be reminded of it again daily through Mama's presence." It would seem unnatural and "not entirely clean." Albert, she said, maintained that these were just social prejudices.

"A is in no way trying to persuade me," she added, "because he does not want to assume the responsibility of chaining such a young thing as me to himself." As long as she kept on living in the house, Albert had told her, it wouldn't matter whether Ilse was married to him or not, because she would always be her own free person and would mean as much to him as her mother did. It's not clear here exactly what Albert had in mind. Did he have in mind a sort of domesticated ménage à trois in which he would enjoy the best of both worlds—cooking and general social *Gemütlichkeit* from Elsa, and sex with Ilse? In discussing it, Ilse said that she would not be jealous of the "external splendor" that would accrue to her mother in such an arrangement. Albert also told her that, if she didn't have the desire to bear him a child, then it would be nicer for her not to be married to him. "And this desire I surely do not have," she asserted.

But the fact that she did want to get married someday was a source of pain to him. "A said to me that that would cause a tear in his life, and that it would be painful for him to have to miss me. This is a sore spot. I feel like Heine's ass. On the one hand I would like to be with Albert my life long, but to be married to him would require in my opinion a different kind of love."

There was also the effect on Elsa to consider. Her mother would withdraw, Ilse said, if she believed that Ilse could only be happy with Albert. But it would be difficult for her, and Ilse didn't feel right about usurping the position that her mother had finally attained after all the years of patient struggle. Her mother would be disgraced; the philistine relatives would be horrified.

"The question is this: What is more beneficial for the happiness of the three of us and especially of Albert?"

All her instincts, Ilse said, cried out against marrying Albert. It would only be the worst for Albert himself if he had to watch their relationship, which had never been anything but pleasurable, become for her a pair of shackles.

"It will appear unusual to ancestors that I, small dumb twenty-year-old thing, must decide on such a serious matter; I myself can hardly believe it, and also feel very unhappy about it. Help me!"

Nicolai's answer, if he made one, has not survived. He probably delivered it in person, because he himself was soon to be smack in the middle of this confused household.

Time had run out for the dashing pacifist, despite a string of symbolic victories that frustrated and embarrassed the war bureaucracy.

In May, finally fed up with his antics, the military authorities ordered Nicolai to report for drill with sidearms. Contending that it not only was an immoral order that violated his rights as a physician under the Geneva Convention and the Military Code but that it also broke a promise by the minister of war to use him only in a medical capacity, Nicolai refused the command and wrote out a formal protest to the minister of war, which Ilse transcribed and delivered.

When the war minister turned him down, Nicolai deserted and took a train to Berlin.

He went straight to 5 Haberlandstrasse, where Ilse and the rest of the Einsteins took him in.[23] From there he drafted an ultimatum to the war minister: If the wrongs done him were not redressed within four weeks, Nicolai would leave the country. At the same time, a letter in Ilse's hand-writing was drafted on what passed for Kaiser Wilhelm Institute for Physics stationery, urging that Nicolai not be treated as a deserter and that he not be required to bear arms. Whether Albert ever signed or mailed this letter is not known.

After waiting, presumably in the shadows of the Einstein households, for the reprieve that would allow him to consider himself a free man, Nicolai and three members of the Spartacus underground stole a pair of biplanes from an airfield northwest of Berlin and flew to Denmark, where Nicolai became the toast of Scandinavia.

Back on 5 Haberlandstrasse, the Einsteins celebrated. Ilse composed a long ballad in honor of the heroic flight and sang it to the accompaniment of a lute.[24] We can only imagine Albert's reaction, chuckling as he always did at BNV meetings when he learned that another pamphlet or letter had slipped the Prussian dragnet, watching Ilse as if she were the moon, setting serenely over a distant sea far and foreign and forever denied to him.

FROM THE START, it seemed clear that Albert's attraction to Elsa was as much a maternal as a sexual one. After years of war, illness, tensor analysis, constant movement, and Mileva's chilliness, Albert's "personal life," as he referred to it, had dwindled to the point that he once admitted that Freud's theories were lost on him because he "had no live material at hand to work with."[25] In his own memoir, Janos Plesch, a doctor and playboy friend of Albert's, suggested that Albert satisfied his urges during those semibachelor years in brothels, although he offered no specific support for that claim.

Suddenly, like a crocus in the April snow, was a vivacious twenty-year-old, and the charm of the relationship with Marie Winteler seemed reborn, after all these years. It is not hard to imagine Albert sitting amid the over-stuffed furniture and the portrait of Frederick the Great, listening to Elsa and Ilse debate their futures, hoping no doubt that the decision would go his way, leaning into the frowns and silences with the rhetorical equivalent of body language, so that he could have Ilse without asking for her. In physics, Albert was a hero, proud of his willingness to tackle head-on the most diffi-cult issues of the day.[26] But in his personal life his curiously passive style was to hint that he was available and hope for an invitation. He rarely had to look very far: a virtual sister in the Winteler family, a classmate, a cousin, the cousin's daughter. When the situations into which he fell became too com-plicated, he simply withdrew. The deepest form of intimacy is to confess to the other, not your fears, but your hopes. In that respect Albert had no voice. For a man with no defenses, the only refuge was to admit no desire at all.

If an examination of Einstein's love life sheds no light on the origins of relativity or the meaning of the quantum riddle, it can at least help us appreciate the desperation with which he sought the clarity of physical law, cause and effect, and mathematical beauty. The genius of abstraction who needs a woman to ground him in the world is a cliché. But in keeping with Niels Bohr's aphorism that a great truth is a statement whose opposite is also a great truth, so in Albert's case the cliché is turned inside out. It was with physics that he needed to ground himself in the surreal fog of desire and the dizzying claims of the belly and the other organs that assaulted and threatened his autonomy. It was physics that offered a small, ordered closet in which to hide from what he called, in a letter to Eduard, the dominant feminine psychology, the world as it sprang from his mother's knee.

WHEN ALBERT was frustrated, there was always the quantum to ponder. Shortly after Nicolai's escape, Einstein and the women went to Ahren-shoop, a resort village on the Baltic, where he spent the rest of the summer in a truce with his belly. "Uncountable hours spent futilely thinking about the quantum question," he reported to Besso, "but I do not doubt anymore the reality of light-quanta," adding that he would continue to stand alone in his conviction as long as no suitable mathematical explanation of the quantum phenomenon had been produced.[27]

His equanimity was also shaken by a surprise from Zangger. The old Polytechnic and the University of Zurich had joined forces to offer him a position back in Zurich, his spiritual hometown. Apart from the question of whether his personal life could survive a move nearer to Mileva, he wondered what they wanted with an old ruin like him.[28] "Whatever useful thing I have thought up, that is now alive in fresher and younger heads, while I must be constantly tormented by a belly that tends to rebel and takes my every excitement or exertion badly. The spirit becomes lame, the energy dissipates, but fame hangs around the fossilized shell." He was better off in the Academy, whose quintessence lay more in mere existence than in work. He suggested that Hermann Weyl, a mathematician who was making important contributions to general relativity and the untangling of de Sitter's complicated universe, would be a better choice.

However, feeling too guilty to turn his old friends down summarily, Albert offered to lecture twice a year in Zurich instead. He still felt bad. That night he dreamed that he was shaving and cut his throat with the razor.[29]

Meanwhile, Mileva relapsed and went back into the Sanatorium Theodosium at Albert's recommendation; this time the boys were left with a nurse. What had been a friendly streak of letters from Hans Albert ceased again, and his father had to resume his nagging, jollying-up mode. The depths of Albert's cynicism about human prospects were revealed by a remark he made to Max Born. "Recently I read that the population of Europe grew in the last century from 118 million to almost 400 million . . . a terrible thought, which almost could make you friends with the war."[30]

It was a comment he might not have made if he could have seen the immediate future. In the fall a deadly influenza rose up and swirled like scythe wind through the rotting trenches and the battered cities of the world. In army camps the dead were stacked like cordwood. It was the worst epidemic since the Black Death of the Middle Ages, when plague had taken a quarter or more of Europe. By the time the influenza outbreak of 1918 was over some 20 million people worldwide had died. Albert was particularly concerned about Tete's vulnerability, but in fact, all the Einsteins were spared.

Back in Berlin, Albert, Elsa, and the girls were swept into the chaos of the final act of the Great War. At the beginning of September Albert was invited to a clandestine meeting to discuss forming a union of intellectuals to save Germany from its militaristic ways.[31] Albert declined on the grounds that the members of the group were aristocrats and not democrats. "You will get a collection of non- or 'realist' politicians," he explained. ". . . they will agree on the principle that private morals are insignificant for the relations between states and people; that only the right of the stronger ones is decisive and that human beings would degenerate without war."

By the end of that month American divisions had piled onto the Western Front and were driving the monster back. Germany's main ally, Austria-Hungary, was crumbling as Hungary, Poland, and Czechoslovakia declared independence from generations of Hapsburg rule. Virtually overnight Austria became a sliver of a nation populated by out-of-work bureaucrats. Ludendorff told the kaiser to sue for peace, and urged him to form a new democratic government to carry on the negotiations. In that way the military could preserve its honor; it would be the civilians who had officially surrendered. Woodrow Wilson, point man for the Allies, also demanded a democratic government to negotiate with.

A cabinet was formed under Prince Max of Baden, a liberal, and reforms began, but progress was slow. The BNV began to meet again and advocate the release of political prisoners. The kaiser was now seen to be an impediment, and pressure grew for him to abdicate.

On November 4, sailors based in Anschütz-Kämpfe's town of Kiel refused to return to sea with their ships, dismissed their officers, and took over the city. The mutiny spread to the soldiers and workers in other cities. A general strike was called for November 9. People poured into the streets demanding peace and democracy. Wilhelm II fled to Holland. The Reichstag was dissolved, and a new socialist republic was proclaimed. Albert's relativity class was canceled due to the revolution. The next day, the newly legal BNV revived itself in an open-air rally in front of the Reichstag; Albert was appointed to its fourteen-member Working Committee. On the day after that, on the eleventh hour of the eleventh day of the eleventh month, the guns fell silent.

Albert wrote exultantly to his sister and other friends that the military religion was finally overthrown. He did not believe that it would rise again.[32] In the meantime, he reported, the old folks at the Academy were in despair at their fates in the new democratic Germany and had begun to come around and chat him up. He had by then the reputation, as he put it, of being a "supersocialist." With their huge bronze stomachs the old heroes of Prussia reminded Albert, he said, of mammoths from the Bismarck ice age as they came wagging their tails to him in the hope that he could halt their descent into the abyss.

But the revolution rapidly began to eat its own. A revolutionary students' council took over the university and arrested its conservative rector, Reinhold Seeburg. The university was closed, and Albert was asked to intercede with the students. He, Max Born, and Max Wertheimer, a psychology professor, took a tram downtown to the Reichstag, where the students' council had been given offices, and the trio was granted an audience. In Albert's pocket was a short speech heavy with references to the end of the old class order and the will of the people, warning the young idealists against replacing tyranny of the right with tyranny from the left. Instead he delivered a rousing speech in favor of academic freedom. "I always thought that academic freedom is the most precious asset of the German university. . . . I should regret if the old liberties will end."[33] The rector was freed, and the university was soon reopened as well.

In the streets, however, the turmoil continued to increase. As Albert reported to Hans Albert, "Many people are trying to profit from the confusion. The money and property of the state is handled carelessly. Paper money is printed in vast amounts, so that its value keeps going down. The workers want enormous wages, even though there is nothing to do in the factories. All the factories are in danger of bankruptcy. Today the soldiers arrived. It was an exuberant moment. Many boys like you rode with the soldiers or followed them through the streets. Everyone turned out to greet the men who had faced so many dangers . . . for so long. But Tete should not play soldiers. That is not my intention."[34]

With the returning soldiers came rising tension between the exuberant left wing and the right wing who believed that Germany and its military had been betrayed by its civilian leaders. Not a shot had been fired on German soil, and yet they had lost. Soon it would be the turn of students on the right to seize control of universities and kick out professors.

Shortly after the New Year Rosa Luxemburg and Karl Liebknecht, the founders of the German Communist Party, were captured and killed by a member of a rightist militia. It was rumored that the leader of the Social Democratic Party, Gustav Noske, had approved the killings in order to save Germany from Bolshevism. The police started hunting down Luxemburg and Liebknecht's comrades from the Spartacus group, one of whom, Fanja Lezierska, took refuge in the Einstein house.[35] Ilse spirited her out the backdoor just as the police were ringing the front doorbell.

Nicolai returned home a hero.

IN THE END, the process of Albert and Mileva's divorce did outlast the war. Zürcher had begun the legal proceedings on behalf of Mileva in October, petitioning the *Bezirksgericht*, or district court, for a speedy hearing, noting that the defendant was in poor health. In his own statement in response to the suit, Albert blamed the dispute on incompatibilities of character. "In the marriage there were many scenes as a result of differences of opinion that came about whereby there were also cursing and physical violence on the part of the plaintiff which I in a situation of irritation also returned."[36]

On November 20 came a hearing at which neither Albert nor Mileva was present. Zürcher told the court that Albert had admitted adultery with a cousin in Berlin, "one of the better classes," as he put it.[37] The court issued a summons for Albert to be interrogated by a judge in Berlin. By the time Albert received the summons, however, the date on it had already passed, and the documents had been sent back to Zurich. "My divorce is a joke to everyone who knows about it," Albert remarked to Besso.[38]

A second court date was scheduled just before Christmas, and at it Albert testified that he had indeed been committing adultery. "My wife, the plaintiff, has known since the summer of 1914 that I have been in intimate relations with my cousin. She has indicated to me her indignation about this."[39] Albert and Mileva were officially divorced two months later, on February 14, 1919, one month shy of Albert's fortieth birthday. Albert was fined 100 francs plus court costs. He was declared an adulterer and forbidden to remarry—at least in Switzerland—for two years.[40]

Conveniently he was in Zurich at the time, fulfilling his agreement to hold a biannual lecture series. Albert remained in Switzerland for more than a month, from January into March, staying in a boardinghouse and lecturing in a drafty hall that the Swiss complained cost too much to heat.

In the meantime he spent a lot of time visiting Mileva and the boys. As with most married couples, the final resolution of their future slackened the tension a bit between them, and they were able to remember on occasion why they had once liked each other. He played music with his children and with Lisbeth Hurwitz. At her best, Mileva had mellowed into a kind of gray, stoical presence; but even then, Evelyn Einstein, who later visited as a child, recalled that Mileva scared her to death. Mileva looked forward to Albert's visits and pestered him for copies of his scientific papers. "As time goes by," Albert said at one point, "you will notice that there is hardly a better ex-husband to have than myself. For I am faithful, possibly in a different way than a young girl dreams of, but still and all faithful and true."[41]

BACK IN BERLIN Albert finally moved across the hall into the full bourgeois splendor he had denigrated his whole life. Elsa's living room was decorated mostly with the obligatory portrait of Frederick the Great and busts of Schiller and Goethe and yellow wallpaper. Albert wandered these rooms, it was often said, like a bohemian guest, maintaining the apartness he had come to prize. He had his own bedroom, a spartan space with a simple single bed. Up a flight of stairs was an attic with whitewashed walls and a bare wooden floor, which he claimed for his office, furnishing it with a round table, a pair of chairs, shelves for his papers, and a portrait of Newton. There he retreated for hours on end to bask with monkish joy in the hazy light from a dirty, sloping window, and in the glow of the curling universe rising like smoke from the shuffled equations, the thinking like music, the music that still played for his ears alone.

Albert and Elsa were married on June 2, 1919.

That spring an acquaintance—one who knew him well enough to enjoy his propensity for bad rhymes—had told Albert that he should lecture in verse about relativity. Presumably this would not detract from the general level of comprehension of the theory. By way of turning him down, Albert wrote a verse. In part it read:

I am as I always am the organ player
Who can do nothing more than turn and turn
Until the sparrow sings it from the roof
And the last scoundrel comprehends it.[42]

25

THE MELTED WORLD

BY ONE OF THOSE STRANGE COINCIDENCES WHOSE CONTEMPLATION IS better left to theology than to science, the sun and the moon appear to be the identical size in the sky, about half a degree of arc in diameter. Which means that, in the endless and repetitive, wobbly dance of the worlds, at least every two years on average, the moon will pass in front of the sun and cancel out its image precisely. The inhabitants of some almost invariably remote, trackless, and undeveloped stretch of territory maybe a few hundred kilometers wide and a few thousand kilometers long find themselves experiencing a strange noontime darkness in which the horizon glows red for 360 degrees around, the cattle and the birds panic, the temperature drops, a giant shadow sweeps across the land, the vermilion sky winks into stars, and behold! The sun is replaced by an inky hole feathered by the pale, pearly fire of the corona.

It takes about an hour from the time the moon first appears to take its first bite out of the sun for the day to die. In the last minutes before totality the world is like a dream that is already slipping away as you stumble toward wakefulness. The shadows of every blade of grass become as sharp and black as sharks' teeth, while the grass itself melts like butter and the landscape runs between your toes. It is hard not to stare down the tube of blackness that is the moon's shadow and feel the catch of cosmic machinery in your gut, as the worlds line up like an exquisite billiard shot and you stand at the end of the cue stick. Even for veteran astronomers, the only logical response often seems to be to scream or cry.

Astronomers have been chasing eclipses ever since they realized that the cosmic events could be reliably predicted. The crack between the worlds lasted from only a few seconds to seven minutes (in the rare, most cosmographically advantaged case)—all the time that an astronomer might have in his or her entire life to scan the vicinity of the sun for so-called sub-Mercurial planets normally washed out in the solar glare, or to take delicate measurements of the mysterious corona. To make the most of those precious minutes required months of preparation. Money had to be raised. Telescopes had to transported, set upon newly formed foundations, and aligned with the celestial poles. Calibration observations had to be made. Astronomers had to travel, be housed, fed, protected from wild animals, and treated for their jungle fevers, insect rashes, and sunburn or frostbite, as the circumstances dictated. And after all that, it might rain.

The eclipse predicted for May 29, 1919, was one of the more promising of the breed. It would commence in the eastern Pacific, where the sun would rise already blackened, sweep across South America and the South Atlantic, and end at sunset in Africa, engulfing those lands in darkness for as long as five minutes. In March of that year Arthur Eddington sailed from London with a boatload of hastily assembled equipment for the island of Principe, off the West African coast. It was not the landscape that he and his colleagues hoped to see melt as they set off into the brine, but the Newtonian world itself with all its certainties, whose hegemony of three hundred years ranked among Shakespeare and Camelot as one of the glories of England but had now been challenged by a German Jew.

History would record that Eddington was the right man at the right place. At the age of thirty-six he was well on his way to becoming arguably the most important and influential British astronomer since Edmond Halley. A Quaker and lifelong bachelor who lived with his sister and mother in Cambridge, where he held the title of Plumian Professor, Eddington was a reserved, studious man with a penchant for the outdoors and for mysticism. Born in 1882 in the county of Somerset, he had been raised by his mother. As a youth he showed a flair for sports, history, literature, and above all mathematics. Rising easily through the school system, he studied mathematics at Owens College in Manchester and then Trinity in Cambridge, where in only his second year he won the distinction of "chief wrangler" in the famous and grueling triptos exams, the first time a second-year man had been so honored.

Following Cambridge, Eddington went to work at the Royal Observatory at Greenwich as an assistant to the Astronomer Royal, Frank Dyson. His hands-on training in practical astronomy included leading an expedition to a rainy Brazil eclipse in 1912. The marriage of astronomy and high mathematics was Eddington's trademark. Soon after moving to Cambridge he published a book analyzing the distribution and motions of stars and the structure of the universe, declaring his belief that the spiral nebulae were indeed distant galaxies. By 1916, already a member of the Royal Astronomical Society and the Royal Society, he had embarked on an ambitious program to solve the equations of electromagnetism, hydrodynamics, gravity, and thermodynamics in order to compute the structures of stars. This work culminated in his book *The Internal Constitution of the Stars,* which graduate students read to this day, and which would have assured Eddington's place in history even if he had never heard of Einstein. Eddington spent much of his later years in philosophical attempts to understand the meaning of quantum theory and the relationship of the different forces and phenomena of nature to each other. "We have found a strange footprint on the shores of the unknown," he once wrote. "We have devised profound theories, one after another, to account for its origin. At last we have succeeded in reconstructing the creature that made the footprint. And lo! it is our own."

Eddington was enthralled by general relativity as soon as he was exposed to it by de Sitter, and he rapidly became Einstein's self-appointed evangelist in Britain. He once admitted that if it had been up to him, he wouldn't have bothered making an eclipse expedition because he was already certain that the theory had to be right. As a popular anecdote told around Cambridge went, after a lecture on relativity Eddington was congratulated by one of his colleagues (who did not accept the theory) on being one of the three people in the world who understood it. When Eddington did not reply, the colleague accused him of false modesty. "On the contrary," Eddington said, "I was trying to imagine who the third person could be."[1]

Eddington's enthusiasm made him a member of a very small club. In Britain Einstein's theory was even less respected than in Germany. In addition to the Newton factor was the very fact that Einstein was a German. By the end of the war, Germans had effectively been exiled from the international scientific community and weren't even being invited to scientific conferences, a circumstance that outraged the idealistic Eddington. As of the day he sailed, not a single paper on general relativity, save de Sitter's original review articles and Eddington's own report, had been published in English. Luckily, one of Einstein's few English-speaking admirers was Eddington's friend Dyson, the Astronomer Royal. It was Dyson who first pointed out in 1917 that the upcoming South Atlantic eclipse would present an excellent opportunity to test Einstein's theory of light-bending and then organized enough institutional and financial support to make an expedition happen. He had more than one reason, as it turned out.

In 1917 the British government declared a draft of all able-bodied men. Eddington made it known that as a Quaker he could not and would not serve. Anxious to avoid the embarrassment of having one of their own sent to a prison camp for refusing military service, the Cambridge dons managed to obtain for him a scientific deferment. A letter from the Home Office was sent for him to sign, but Eddington, who had a little Nicolai in him, added a codicil saying that if he didn't get the deferment he would refuse to be drafted anyway, as a conscientious objector. The offer was retracted.

Dyson intervened and managed to get Eddington's deferment reinstated, but at the price of promising that Eddington would undertake an important scientific task. That task, it was stipulated, was to lead an expedition to the 1919 eclipse in order to test Einstein's theory, should the war be over by then.

That Einstein's theory was still untested nearly a decade after he and Freundlich had begun to lobby astronomers to do eclipse experiments was a testament to the vagaries of war and weather. The bending of light was general relativity's most dramatic effect, and Freundlich had in fact done his missionary work well. A succession of expeditions had been mounted to the far corners of the world by astronomers eager for the chance to prove or disprove this great new barely comprehensible theory of the universe.[2]

Under its energetic director, W. W. Campbell, Lick Observatory in central California had become the leading center of eclipse studies. The observatory had built a special "intra-Mercurial" camera to search for planets close to the sun during a succession of eclipses in the first decade of the century. In 1912 Charles Perrine, the director of the Argentine National Observatory and a Lick alumnus who had helped run the "Vulcan" program, borrowed the camera to look for the Einstein effect during an eclipse in Brazil. It rained.

Campbell brought the camera to the 1914 Crimean eclipse, but clouds blocked the phenomenon at the last moment. "I must confess that I never before seriously faced the situation of having everything spoiled by clouds," Campbell wrote later. "One wishes that he could come home by the backdoor and see nobody."[3] Worse, the Lick expedition was obliged by the exigencies of war to leave its equipment behind to be shipped after it left. There was an eclipse two years later in Venezuela, to which Perrine sent a team to make rudimentary observations, but there were no resources available to conduct the Einstein test.

That left a June 8, 1918, eclipse in Washington State as the big Einsteinian opportunity. The war was still on, but Washington was practically in Lick's backyard. Campbell planned an expedition to Goldendale in the Columbia River basin. The only problem was that the observatory's special eclipse camera still hadn't been returned from the Crimea; by March it had gotten only as far as Yokohama. At the last possible moment Campbell borrowed some lenses from a student observatory in Oakland and cobbled together a camera.

Expectations were high, although Heber Curtis, one of the Lick astronomers, couldn't help noticing that they were running *thirteen* instruments and that the house they were living in had *three* black cats as well as a dog named Shadow. The eclipse appeared through a sudden, miraculous hole in an otherwise cloudy sky. "CLOUDS FALL AWAY FOR SOLAR ECLIPSE," reported the *New York Times*.[4]

Unfortunately, the improvised optics fell short of the so-called astrographic quality required for precise positional measurements. Moreover, in order to increase slightly its narrow field of vision Curtis had operated the camera slightly out of focus. The star images, about fifty in all on a dozen plates, looked like fuzzy little dumbbells. Curtis, later to be known for his role defending the notion that the spiral nebulae were galaxies in the famous Shapley-Curtis debate in 1920, had been the driving force behind Lick's entrance into the Einstein sweepstakes. He wasn't particularly enamored of general relativity—complaining once that it had been changed so much that the original relativity should now be called "old relativity"[5]—but he realized that the light-bending prediction was a golden opportunity for astronomers to help resolve a major issue in physics. Curtis had conceived and designed the light-deflection experiment, but in the meantime had temporarily forsaken astronomy for war work. He took a leave to attend to the eclipse work, but then went back to the Bureau of Standards, leaving the plates that could doom or glorify Einstein sitting unmeasured and unanalyzed in their vault at Lick.

When it came time to organize for the 1919 eclipse, none of the usual suspects—neither Campbell nor Perrine nor Freundlich—was able to raise funds for an expedition. Dyson and Eddington had a clear shot. To reduce the chances of being thwarted by bad weather, they planned to send two expeditions. Eddington and E. Cottingham would go to Principe, while another pair of astronomers, Andrew Crommelin and Charles Davidson, would travel to Sobral, a town about fifty miles inland from the coast of Brazil.

Meanwhile, in the wake of a new report by the Mount Wilson astronomer Charles St. John, general relativity's prospects were dubious. St. John had looked for the gravitational redshift by measuring the wavelengths of spectral lines from the solar surface, but he could find no discernible shift from their wavelengths in the laboratory. Eddington was also concerned by the fact that the situation with respect to light-bending had grown more complicated over the years. He and others came to realize that ordinary Newtonian gravity also predicted the bending of light, if one took account of the relativistic principle of the equivalence of mass and energy—light had energy, thus mass, and was thus subject to gravity. In effect this blending of Newtonian gravity and relativity was what Einstein had done in 1911 when he first calculated the bending of a light beam by the sun. A star near the sun, he had concluded, would appear to have been displaced by an angle of 0.86 second of arc outward from its regular coordinates. If Albert hadn't modified his theory in 1915 and doubled that number to 1.74 seconds of arc, it would have been impossible to distinguish his prediction from Newton's.

As Eddington saw it, there were three possible outcomes to the eclipse experiments: the full Einsteinian deflection of light, half that amount, or no deflection at all. There was room in his universe for only two competing theories of gravity, even though a small galaxy of variations had been offered up by Abraham, Mie, and Nördstrom, among others.

"What will it mean if we get double the deflection?" Cottingham asked during a Royal Astronomical Society briefing on all this. "Then," said Dyson, "Eddington will go mad, and you will have to come home alone."[6]

In fact, any result at all would be a small miracle. A second of arc—one two-thousandths the diameter of a full moon—was about as sharp as stars appeared to the eye under the finest, most stable mountain observing conditions; atmospheric turbulence usually turned them into bigger smudges than that, and the images on photographic plates were invariably even larger. So the job of the astronomers was to measure a systematic shift that was less than the width of the pale dots they were working with. In principle this could be achieved by superimposing two plates of the same field of stars—one with the eclipsed sun in the center and one without—and determining how much the stars in the former were smudged outward from the latter. In practice, merely the temperature difference between day and night could change the telescope's focus and shift the images the hundredth of a millimeter that would be sufficient to make the results useless. In all, there were a dozen different parameters that had to be accounted for by measuring at least six different star pairs on the original and comparison plates in order to determine the deflection. On the positive side of the ledger, the eclipsed sun would be directly in the middle of the Haydes star cluster, which meant there would be a large number of potential stars to measure.

If, indeed, there were any pictures at all.

There nearly were not. In Sobral the rain-forest clouds cleared only the evening before the eclipse. Crommelin and Davidson had two instruments, an astrographic camera from Greenwich and a large Irish telescope with a four-inch-diameter lens; they wound up with some two dozen plates showing seven to twelve stars each. Over at Principe, Eddington and Cottingham had another Greenwich astrograph. Although it stopped raining an hour and a half before totality, the skies never cleared. Eddington grimly kept on taking pictures, looking up only once, hoping that some light would show through the clouds.

Eddington was rewarded for his faith with two plates (out of sixteen) that showed five blurry stars. He telegrammed Dyson that he was disappointed, but that he hoped to gauge the displacement anyway. After a day of measuring the images, he thought he could tell, at least on one plate, that the shift had been recorded. He later remembered it as the greatest moment of his life. He turned to Cottingham and said, "Cottingham, you won't have to go home alone."[7]

He described his experiences later at a Royal Astronomical Society dinner celebrating his return, in verse borrowed from *The Rubáiyát of Omar Khayyám*:

"The clock no question makes of Fasts or slows,
But steadily and with a constant Rate it goes.
And Lo! the clouds are parting and the Sun
A crescent glimmering on the screen—It shows!—It shows!!

Five minutes, not a moment left to waste,
Five minutes, for the picture to be traced—
The Stars are shining, and coronal light
Streams from the Orb of Darkness—Oh make haste!

For in and out, above, about, below
'Tis nothing but a magic Shadow show
Played in a Box, whose Candle is the Sun
Round which we phantom figures come and go.

Oh leave the Wise our measures to collate.
One thing at least is certain. LIGHT has WEIGHT
One thing is certain, and the rest debate—
Light-rays, when near the Sun, DO NOT GO STRAIGHT."[8]

In the meantime, however, there was news from Lick. Campbell, who had been getting restive on the sidelines, had finally managed to bring Curtis back from Washington, D.C., and put him to work measuring the Goldendale plates. He wanted a result before the British could report their own findings. Limited by time, money, and most of all by the blurry quality of his images, Curtis devised a novel analysis scheme. He divided the stars on his plates into two concentric zones and measured the average "expansion" of the star positions in each zone. According to general relativity, there should have been a difference of 0.18 second of arc between the two groups, but Curtis only measured 0.05. He duly announced to a June 1919 meeting at Mount Wilson that the Einstein effect was nonexistent.

"I am confident that there is no Einstein effect whatsoever; I have gone over the results again, and you can make [the conclusion] as strong as you like," he cabled to Campbell, who was on his way with other American astronomers to Europe for a meeting of the International Astronomical Union. Campbell reported Curtis's results to a special meeting of the Royal Astronomical Society on July 11. "It is my opinion that Dr. Curtis's results preclude the larger Einstein effect, but not the smaller amount expected according to the original Einstein hypothesis."[9]

OUTWARDLY, at least, Albert never flinched, and Heinrich Zangger would later write to him about his savage confidence in his theory: "Your perseverance, your perseverance in the thought that light must travel in a curve around the sun . . . is a tremendous psychological experience for me. You were so certain that this certainty had a violent effect."[10]

While the astronomers jockeyed to read his fate in the blurry stars, Albert himself spent the summer sailing with his boys and rambling around Switzerland, making up for lost time amid echoes of his lost life. Whom should he run into at Lake Constance, for example, but Lisbeth Hurwitz, who listened patiently to his explanation of the whole dreary divorce scenario, including his fears about Mileva's tuberculosis, and then accompanied the clan to Arosa, where Tete was about to be interred again. There was a reunion with his old Olympian friend Conrad Habicht in Schaffhausen. Nearby in Zurich, Mileva was teetering on the edge once more, and his mother was staying with Paul and Maja and suffering from what would later be diagnosed as stomach cancer.

Albert returned to Berlin in the middle of August to find the German economy in shambles. The exchange rate had soared to five marks to one Swiss franc, which left Albert responsible for earning about 40,000 marks a year simply in order to make his payments to Mileva. He and Elsa decided to take in a boarder to help pay the rent. In fact, he was contemplating moving the entire family to cheaper quarters in Potsdam, and ordered Mileva and the boys to move to Germany, where he could pay them in marks. Food was finally available in the markets again, but at vastly inflated prices. The elevator in their building no longer worked; every trip out was a mountain-climbing expedition. In the streets and the universities, the conflict between the left and right continued unabated. Strikes and shutdowns were the order of the day, as predictable as the weather. "Life is no small matter," he wrote to his mother. "We nonetheless are doing well these days."[11]

The Dutch thought that they could perhaps take advantage of this turmoil, and shortly after Albert's return, Ehrenfest wrote to offer him a post in Leiden. Albert turned it down; after all this time, he was beginning to identify with the Berlin physics community, especially Planck, now that they were being shunned and boycotted by the rest of the world. He would not abandon Planck and he knew that Planck would never leave. In fact, the German people as a whole now looked less objectionable to him. "Ever since the people here have been doing poorly I like them incomparably better. Misfortune looks better on people than success," he told Zangger.[12]

Despite his apparent nonchalance, Albert was burning for reports about the eclipse. The Dutch were still his pipeline to the English, and in his reply to Ehrenfest he asked if there had been any word from Eddington.

There was news. By September Eddington had finished measuring his plates. He had reported to an astronomy meeting in Bournemouth that a preliminary analysis showed light-bending of between nine-tenths of a second of arc and twice that amount—confirmation of general relativity at last, if the results held up.

"Dear Mother, joyous news today," Albert promptly told Pauline, who was encamped in a sanitarium in Lucerne, and could surely use some encouragement.[13]

He also shared the findings with his colleagues, coolly showing off the telegram to a graduate student, Ilse Rosenthal-Schneider, who asked him what he would have done if the results had been otherwise. "Then I would have been sorry for the dear Lord," he answered. "The theory is correct."[14]

But when an article soon appeared in the *Berliner Tageblatt* trumpeting the verification of Einstein's theory as "the true constitution of the universe,"[15] Albert sobered up and fired off a note to *Naturwissenschaften* pointing out that nothing had really been settled yet: The data indicated 0.9 to 1.8, but the theory demanded 1.7.

Exactly. Precisely how much off a straight course the light grazing the sun on May 29, 1919, had gone was in fact becoming a thorny problem as the rest of the astronomers made their way back from the South Atlantic to England and began to measure their plates and compare notes. Eddington resolved the issue of disparate results with either a stroke of intuitive brilliance, if you are an Einstein biographer or relativity fan, or an embarrassment, if you are a historian or someone preoccupied with the formal niceties of the scientific process.

In the end, Eddington and his crew had three sets of data from the three telescopes they had taken to the eclipse. By far the best plates were the ones exposed in the four-inch Irish telescope at Sobral. Measuring the seven stars on those seven plates gave a deflection of 1.98 seconds of arc, with an uncertainty of about 0.12 second. That was higher than Einstein had predicted, and taken alone would actually have cast severe doubt on general relativity. The next-best plates were from the Sobral astrograph. They showed many more stars, but heat from the sun had affected the mirror and blurred the images, and may even have affected the focus. The same analysis on these plates yielded a value for the deflection of 0.86, almost exactly the Newtonian prediction, but with a bigger uncertainty.[16]

Finally, there were the two plates from the Principe astrograph, shot through the clouds, with five blurry dumbbell-shaped stars on each. They were the worst quality of the lot. With only five stars, Eddington could not simply measure the images and solve an equation for the deflection. He had to engage in a roundabout series of calculations: assume some amount of gravitational deflection, compute the other components contributing to the stars' displacements, and use those to recalculate the gravitational deflection, repeating the process until all the numbers were consistent. In such a manner he converged on a value of 1.61 seconds of arc for the gravitational deflection—very close to the Einsteinian prediction, especially when the rather large uncertainties were accounted for.

What, then, was the correct answer: 1.98, which was too high; 0.86, which was too low; or 1.61, which was just right, but unreliable? Was it an average of all three? Traditions and empires of the mind hung in the balance. The spread in the results should have been a big yellow caution sign that the experiment was flawed. Presumably there was only one right answer. Averaging the data from all three instruments, the philosophers and historians John Earmann and Clark Glymour have pointed out in a historical essay, would have led to an estimate of the deflection that was somewhere between the Newtonian value and the Einstein value, which was precisely what Eddington had reported to Lorentz in September. If Eddington wanted to exercise some judgment and keep only the best data— namely, that derived from the Sobral four-inch—then the answer would rule out general relativity.

As the American cosmologist Allan Sandage likes to say, quoting the British astronomer Sir Hermann Bondi, no experiment should be believed until it has been confirmed by theory. Bondi could have gotten that from Eddington. Eddington in 1919 already knew the truth: The truth was general relativity. Eddington looked into the fuzzy forest of results and saw the trees leaning and he knew—or thought he knew—that there were only three choices for the amount of that lean. He threw out the Sobral astrograph, which had given the lowest number, and kept the other two. Their average was 1.75, right on the relativistic mark. General relativity was confirmed.

The unofficial rumors of support for Einstein's theory began to spread, and they soon reached Albert himself. As it happened, Ehrenfest had invited Albert to Leiden; it would be his first trip there in three years. Ehrenfest's house had become almost a symbolic home away from home for Albert, a casual place where the climate was receptive to general relativity, where there was always food on the table to snack on, music, children to play with, canals to stroll, and nobody to complain if he sunbathed naked on the back terrace. Albert had been there two days when the astronomer Ejnar Hertzsprung pulled him aside and showed him a letter from Eddington quoting his final result of 1.75 seconds for the light-bending. Einstein promptly wrote to Planck, "It is a gift of fate that I have been allowed to experience this."[17]

Two nights later the Dutch Royal Academy met in Amsterdam. With Albert seated on the stage in front of a thousand students, Lorentz lectured off the record on Eddington's final results from the eclipse expedition. The audience cheered Albert like a hero, but there were no reporters dogging his footsteps then, and their cheers simply evaporated into the night. Albert came home aglow, like an author who has already read the reviews of a book yet to be published. At least one of his academy colleagues congratulated him on winning a great and needed victory for German science.

"So everything is gradually getting better for you, even light bends for you after several million years," Zangger said, reporting that Mileva too was improving and leaving the house again. "The stars are doing perihelia—perhaps you'll command other cabrioles of them. Given the circumstances, it is understandable that you would rather write to Isak [*sic*] Newton than me."[18]

A FOG OF DREAD and anticipation hung over the joint meeting of the Royal Society and the Royal Astronomical Society, held at Burlington House on November 6, 1919. All the old lions of the empire were present, lining the room like living busts to watch the world turn: J. J. Thomson, Alfred North Whitehead, Sir Oliver Lodge, James Jeans. In his account for the *Times* of London, Whitehead compared the meeting to a Greek drama:

"We were the chorus commenting on the decree of destiny as disclosed in the development of a supreme incident. There was dramatic quality in the very staging—the traditional ceremonial, and in the background the picture of Newton to remind us that the greatest of scientific generalizations was now, after more than two centuries, to receive its first modification. Nor was the personal interest wanting: a great adventure in thought had at length come safe to shore."[19]

Thomson, the president of the Royal Society and chair of the meeting, began this coronation by describing general relativity as one of the highest achievements of human thought. "It is not the discovery of an outlying island but of a whole continent of new scientific ideas."

The Astronomer Royal Dyson, Crommelin, and Eddington were called on in turn. Describing the eclipse and the data from each of their points of view, they sounded a little like the proverbial blind man describing an elephant. Dyson stressed the four-inch Sobral result of 1.98 for the light-bending, from which he was willing to pronounce Einstein correct. Crommelin was inclined to give some weight to the astrographic data with less light-bending. Only Eddington mentioned the Principe data. He ignored the Sobral astrograph, as he did in his later writings, quoting the results of the expedition as 1.98 from Sobral and 1.61 from Principe. The Sobral astrograph effectively vanished from history, a fact that would particularly flummox Campbell, who did not attend the November meeting. How could Eddington have thrown out the Sobral astrographic data, while keeping the Principe plates, which were of even worse quality? As Campbell remarked later, "The logic of the situation does not seem entirely clear."[20]

To these and other criticisms, Eddington argued that though the Sobral astrographic plates might have been clearer and contained many more stars, they were plagued by *systematic errors,* namely the heating of the mirror by the sun. That argument was a little disingenuous, according to Earmann and Glymour, who point out that the main effect of heat on the mirror would have been to change the focus, blurring the images. Although the images were indeed blurred, they were not as badly blurred as the Principe ones.

Eddington and Dyson won the day, partly because no one else had sufficient command of the data to form an independent conclusion, and partly because of the pair's own glowing authority. Oliver Lodge, who was still doing aether physics and had publicly predicted there would be no deflection of light, walked out without a word. In the discussion that followed, the strongest objection was raised by one Ludwig Silberstein. He complained that they could hardly say that Einstein's theory had been confirmed when the Mount Wilson astronomer St. John had failed to detect the gravitational redshift, the third in the triumvirate of general relativity's fabled predictions. The redshift measurements were much more precise than the eclipse measurements, he said, and therefore deserved more weight. "The solar spectrum can, even in this country, be observed many times a year," he pointed out. "If the shift remains unproved as at present, the whole theory collapses, and the phenomenon just observed by the astronomers remains a fact waiting to be accounted for in a different way."[21]

Eddington sidestepped that criticism by asserting that it was not Einstein's theory that was confirmed, but "Einstein's *law* of gravitation," describing the trajectory of a light beam. As if a light beam could have a trajectory without all the formal tensor apparatus of a theory. Eddington, as it would turn out, had nature on his side.

Back in Berlin Albert felt a powerful urge to communicate with his sons. "I am longing to hear from Albert and Tete. Tell them!" he commanded Mileva.[22]

REVOLUTION IN SCIENCE

NEW THEORY OF THE UNIVERSE?

NEWTONIAN IDEAS OVERTHROWN

Thus read the headlines in the London *Times* the next morning. Beneath them was a two-column article repeating Thomson's exclamations about Einstein's momentous achievement, and declaring that "our conceptions of the fabric of the universe must be fundamentally altered." After summing up the eclipse story it stated, "But it is confidently believed by the greatest experts that enough has been done to overthrow the certainty of ages, and to require a new philosophy of the universe, a philosophy that will sweep away nearly all that has hitherto been accepted as the axiomatic basis of thought." In another article a week later, the paper claimed that observational science had been led back to "the purest subjective idealism, if without Berkeley's major premise, itself an abstraction of Aristotelian notions of infinity, to take it out of chaos."[23]

The *Times* gave no indication who this "Einstein" was, other than that he was a Zionist who had not signed the infamous manifesto of the ninety-three back at the start of the war. It mentioned no first name, nationality, nor even the fact that he lived in Berlin.

The *New York Times* followed the next day, the eighth, with a story on page six that called the result "epoch making," and chimed in the day after with a cascade of headlines that told the whole story and then some: "Lights All Askew in the Heavens / Men of Science More or Less Agog at Results of Eclipse Observations / Einstein Theory Triumphs / Stars Not Where They Seemed or Were Calculated to Be, but Nobody Need Worry / A Book for 12 Wise Men / No More in All the World Could Comprehend It, Said Einstein, When His Daring Publishers Accepted It."[24] After the first day, the lead of almost every story and editorial seemed to be the incomprehensibility of the theory. "Efforts to put into words intelligible to the scientific public the Einstein theory of light proved by the eclipse expedition so far have not been very successful," the story on November ninth began.

There had been a time when allegedly only one other person in the world understood general relativity—Hilbert, who had tried to beat Albert to it. By the time of eclipse fever the legend of the magic circle had swelled to three and included Eddington. Now the *New York Times* enlarged that story to proper biblical proportions, declaring on the basis of an authority forever lost to history that Einstein had said only *twelve* people in the world understood it. Shades of the disciples, the Illuminati, the dark anarchistic cells that were dominating the front pages of the newspapers, yanking the cords of governments from below, the papal conspiracies steering from above.

When he was asked in a follow-up interview about those twelve wise men, Albert just chuckled in his rich baritone and tried to tell the story of how he had invented the theory, a man falling in his chair who did not feel his weight. In the newspaper account it came out as a twist on the story of Newton's apple: Instead of an apple, a man had fallen off a roof onto a garbage pile and lived to tell the tale.

The Enlightenment, with its dream that reason and science could both understand and improve the world, could have died that very day, if its corpse were not already stinking up the quiet muddy fields of France and Russia. All the science, the crystal skyscrapered cities, the machines, the electrical networks, the rationalizations of time, the armies, the bureaucracies, the plans, the colonies and the commerce so artfully managed, the missionaries, the explorers, the revolutions, the manifestos, and the world systems, all the schemes and all the ideals, all the hard-booted engineers and all the covenants, all the best intentions and learning had all amounted to an artillery shell in the chest. It was the biology of war, the soft tissue stuck in the cogs of society.

There could no longer be any pretense of running the world, of understanding the art. Not even the soul could be understood, said Freud. Why then, should anyone expect to be able to understand the science? Indeed, mystification was part of the lure. For a hundred years natural philosophy, a pastime for broadly educated gentlemen, had been evolving into physics and nestling into corners of mathematical specialization and hard-edged incomprehensibility. The world was slipping away from the ordinary man. It was Einstein, a Jew who deigned to feel sorry for God, who was the one with a telegraph line to nature, staring out from newspaper photographs with those soulful, sexy eyes, seeing deeper than you could imagine. It was the fourth dimension all over again—the mystery realm in which everything was possible, everything was transcended, where the spiritual lay naked and close. When the relief—the escape—from dreary rationality and death finally appeared, it appeared exactly where the artist-mystic followers of Lobachevsky and Riemann and Hinton had said it would be—in the mysteries of curvy, higher-dimension geometries. The bourgeoisie already knew they weren't supposed to *understand* the fourth dimension, only celebrate it. But Einstein had approached directly what everyone else could only glance at sidelong as metaphor. The man behind those eyes had touched the secret of the universe, the secret that made light rays bend and stars dance.

IN BERLIN, at least, they knew who Einstein was. Although the local papers had been covering the eclipse and relativity saga since the beginning, Albert was considered a local character, the pacifistic professor, a folk hero of the left, and he awoke to no particular fanfare on the morning of November 7. It was the second anniversary of the Russian Revolution, and the streets had been barricaded and the troops readied for the usual battle between the workers and the former soldiers. As the day wore on, however, a parade of reporters began to find their way to Haberlandstrasse, and from then on they would never stop coming.

On December 14 a giant close-up of Albert's face—chin supported by his fingers, thoughtfully lined, looking downward with slightly shaded eyes—stared out from the cover of the *Berliner Illustrierte Zeitung.* "A New Giant in World History," it said.

Albert was like an actor who becomes an overnight success after twenty years of playing bit parts and waiting tables. "The thing reminds me of the fairy tale 'The King's Clothes,' but it's a harmless idiocy," he told Zangger.[25] He enjoyed the kind of banter he could have with reporters and photographers too much to allow himself to indulge in much of it, not to mention his wishing to avoid the unseemliness of appearing to pander to the public. He soon learned to let Elsa answer the door, the phone, and the mail, while he retreated into the perfect freedom of an apparent captivity. Which only made him seem more desirable.

"By an application of the theory of relativity to the taste of readers," Albert joked in a London *Times* article that was part thank-you note and part victory lap, "today in Germany I am called a German man of science and in England I am represented as a Swiss Jew. If I come to be regarded as a *bête noire* the description will be reversed, and I shall become a Swiss Jew for the Germans and a German man of science for the English."[26]

It was a glorious moment for science. Eddington and Einstein, the tensor twins, the Quaker and the Jew, had saved the notion of science as an international brother- and sisterhood. At a time when that fraternity was in tatters, with the distinguished professors of one nation calling their learned foreign colleagues liars and baby murderers, Englishmen had ventured out onto the high seas, seas where only a year before Einstein's countrymen had been hunting them in submarines, to test a theory that would topple Newton himself. It was the beginning of a beautiful friendship. Albert had been studying English, and his earliest adventures in that language were letters to Eddington, who replied, of course, in perfect German.

In their first official exchange Eddington allowed that he felt sorry for Freundlich. Albert stressed that general relativity was not out of the woods yet. Discovering the gravitational redshift, which Freundlich was involved in, was more crucial now than ever. "If it were proved that this effect does not exist in nature, then the whole theory would have to be abandoned."[27]

Then he ordered himself a new violin.

PAULINE EINSTEIN moved back to Berlin in December. For the first time since 1901, when Albert had set off for Lake Como and over the Alps for his first job, mother and son were reunited under one roof. Her stomach cancer was now its final stages. "A tragic affair with a half year to go," Albert told Besso.[28]

They installed her in the book-lined study, where Albert could hear her moaning in pain despite the morphine that, besides the company of kin, was her only treatment. "She hangs with every thread on life," Albert reported. "It seems that her torture will still last long, for she still looks good."[29]

Back in Zurich the other family of the new genius of humanity was wondering where they were going to live. "It seems that a kind of gypsy life has been granted us," Albert said.[30] Inflation had ruined the mark to the point that he could no longer afford to keep the family in Switzerland. Albert was proposing to move them all to Baden, in Germany, when a little windfall of a few thousand marks came his way, a sum that would be sufficient to support them a while in Hungary. He relented to Mileva's pleas and put off for six months any decision about relocating them to Germany. With time would come wisdom, presumably. Meanwhile, Hans Albert had the mumps.

In January 1920, Mileva was summoned back to Novi Sad, where her family was in crisis. Albert advised her to travel with a collapsible stool so she could sit and rest during the inevitable delays, travel being extremely difficult and taxing these days. To Zangger he confided that he thought she was going to pieces, as he coped with the question of how much of his incredibly shrinking bank account he should send to cover the expenses of his floating family. Tete decamped to yet another sanatorium; Hans Albert was sent once again to stay with Zangger.

Mileva found the situation in Novi Sad to be dire. Zorka had moved back to Serbia earlier and had only gotten worse in the interim. She was crude and unapproachable by anybody save her beloved cats. The tenants were frightened of her. Zorka was especially abusive and insolent to her father, Milos, who for his own part was depressed and rarely left the house. Gone were the days when he held court in his old uniform in the coffeehouses telling tales of the old times. Zorka looked up to her older sister, the Rose of Hungary, and for the three months Mileva stayed her presence was a balm. Milos eventually died of a stroke in 1922, but not before Zorka had brought the family even lower by inadvertently burning up its fortune, which Milos had hidden in a woodstove.[31]

Back in Zurich Mileva resumed her pattern of checking in and out of the hospital while the children followed dutifully in her wake, resisting Albert's demands that she and the boys move to Germany: Baden, Freiburg, Durlach—the locations of potential Teutonic nightmares fell regularly from his pen. Her friendship with Lisbeth Hurwitz was a lifeline. They went on walks and to concerts and lectures together.

One entry in Lisbeth's diary records their attendance in 1921 at a lecture by a renowned biologist and pacifist, one Georg Nicolai. The talk elicited stormy applause, but Mileva and Lisbeth were disappointed, having hoped for more—of what Lisbeth did not say.[32] Mileva seemed preoccupied, and Lisbeth wondered if she was thinking about her sister's harsh attitudes toward the human race. But Mileva might just as well have been thinking of her brother, an example of the sort of blood migration that war created, swept into Russia and remaining, living out his life in a foreign land. The biology of war.

Tete grew fat and quiet and seemed to have inherited some of his father's knack for childish awareness and curiosity about the obvious. When he was ten, Mileva sewed him a new suit. Tete told Lisbeth that certain parts of clothing always seemed new to him—the collars on sailor suits, for example.[33] The strange pain in his ears continued. He was the genius of the family, certainly, but also the most vulnerable. A photograph taken a few years later on the balcony with his brother shows him looking foppish and insolent in a suit, his hair mussed.

His trips to the sanatorium continued, much to Albert's growing despair and disgust. It was wrong, he thought, that Tete had been deprived of his youth. He once wrote begging Tete to join him at the seacoast, if he didn't have to go back to school just yet. "Do not ask any doctor for advice. They are the successors of the priests (derogatory term) who hold dominion over us in the name of our corpses."[34]

The publicity searchlight that was now trained on Einstein did not alight on Mileva and the boys. Mileva never talked about the divorce or the years with Albert, and nobody, it seems, ever asked her. Once a German newspaper printed a photograph of Albert with Ilse and Margot, labeling them his daughters.[35] Mileva was furious, and Hans Albert had what was described as "an unpleasant and touchy scene" with his father. As usual, Albert blamed Mileva for putting Albertli in the middle. He himself, he said, had no influence on the newspapers. "I cannot waste my time on such petty trifles without appearing absurd—even to myself." As a youth he had often brought his mother to her knees in tears with his aloofness from bourgeois conventions; as an adult, a giant among mankind, Albert would not be any more bound by the rules that society tried to fashion like lassos around his soul. He was his own man now.

In the years to come he would fight more often with Hans Albert, at whose intention to be an engineer he still chafed. But as far as he was concerned, the struggle was over between him and Mileva; all was forgiven, they could be friends again. In a typical passage announcing a forthcoming visit, he good-naturedly chides Mileva: "We will let the past be forgotten, at least the bad things. You must not grumble so continually, but take joy in the good things that life has brought you. For example, the children, the house, and the fact that you are no longer married to me."[36]

IN DECEMBER OF 1919 Michele Besso moved back to Bern and rejoined the patent office, where he and Albert had had such a wonderful time, and Albert had thought his most beautiful thoughts. The years had not changed him; he was still razor-sharp and completely indecisive.

That same month, an overwhelming majority of the members of the Royal Astronomical Society voted to award Albert the society's annual Gold Medal. Albert duly scheduled a trip to England in the spring to accept the prize and have a closer look at the source of all this tomfoolery that had changed his life. The glow of international goodwill was short-lived, however, for a conservative faction on the Society's board of directors managed to block the award. Eddington had to write back to Albert and explain that for the first time in thirty years the Gold Medal would not be given this year. Eddington hoped that Albert would not take the rejection personally, nor despair; the "better spirit," he believed, was progressing.[37]

In Germany it had become a source of shame among the astrophysically inclined that the Germans themselves had had so little to do with the confirmation of their colleague's theory. But there was still the elusive and crucial gravitational redshift to be pinned down. Freundlich accordingly made his greatest career move and drafted an "Appeal for the Einstein Donation Fund," which he sent to the usual suspects in business and industry. "It is an obligation of honor to those who are concerned about Germany's cultural standing to come up with whatever funds they can afford in order to enable at least *one* German observatory to work directly with its creator in testing the theory," urged the letter.[38] The goal was to raise 300,000 marks to build a telescope at the Postsdam Observatory designed to examine the solar spectrum and obtain evidence of the gravitational redshift. An indication of how quickly the mark was inflating can be seen from the fact that, by the time this letter was printed in its final form, the sum had risen to 500,000 and then 1,500,000 marks.

The so-called Einstein Tower was eventually built, and still stands. As designed by the architect Erich Mendelsohn, it looks a little like a lighthouse that has melted and then refrozen, all curves with horizontal bands of windows pushed out on one side like balconies—a tribute, its creator said, to the curvilinear quality of Einstein's universe. Albert called it "organic." It was, of course, a monument—most of all, to him—and it duly took its place on the cover of the *Illustrierte Zeitung.* The Great Work had to go on.

NOT EVERYBODY was enraptured by this general trend of celebrity and idolatry. If you were a conservative, or a German physicist who had won the Nobel Prize (as Einstein had not yet done) without having your face decorate magazine covers and being anointed a new Copernicus, there was something vaguely ominous about the brown-eyed face staring out from the newspapers and magazine covers. It was, after all, a *Jewish* face. And the word "relatively" was being heard entirely too often these days in contexts that had nothing to do with moving trains and the speed of light. It was a joke, it was a code, a shorthand for a certain kind of corruption, a moral rot, "the purest subjective idealism," in the words of the London *Times,* substituting for the old pillars of culture and knowledge.[39]

Berlin, Albert had told Ehrenfest late in 1919, was rife with anti-Semitism, adding that "political reaction is violent, at least among the intelligentsia." Soon he began to see it everywhere.[40]

Albert's relativity lectures at the University of Berlin had become a tourist attraction. In February his students began to complain to university officials about the presence of these unregistered freeloading visitors and the ruckus they caused. Albert felt his classes should be open to anybody, but the university insisted on its rules. Representatives of the students' committee disrupted the class on February 12, and the issue was hashed out, with Albert agreeing to waive all fees and reconstitute the course as a set of public lectures. In the accounts of this incident nasty things were said, including an anti-Semitic jab at Einstein in a Nazi newspaper. Albert supposedly referred to the protestors as "outcasts of mankind," and a left-wing student newspaper, *Vorwärts,* made allusions to an "anti-Semitic mob."

The Ministry of Education put out a statement assuring people that the protests had been neither political nor anti-Semitic in nature. It was all a misunderstanding. Albert went along with this, more or less, in his own statement quoted in *Vorwärts:* "There can be no question of a scandal said to have taken place yesterday; nevertheless, a few remarks that were made testified to a certain animosity toward me. There were no anti-Semitic utterances as such, but their undertone could be so interpreted."[41] Albert quietly canceled plans to teach the course again in the summer.

He was not the only one affected by the changing political climate. Things were getting out of hand throughout Germany. In early March of 1920 a *Vorwärts* article summed up the last two weeks' worth of atrocities: A republican teacher horsewhipped by army officers in Königsberg; a Jewish student named Kahn shot and killed in Baden-Baden; Einstein the target of anti-Semitic riots; troops busting up an amateur labor union theater in the Ruhr; a pacifist rally in Osnabrück invaded by thugs from the notorious Frei Korps Lichtschlag shouting insults at Jews and "vermin," and killing a young man who tried to stop them; and marines beating the speaker at a pacifist rally in Berlin almost to death.

Two days after this article a group of right-wing army officers under the leadership of a reactionary politician named Wolfgang Kapp marched through the Brandenburg Gate and seized the government. The so-called Kapp Putsch foundered four days later after a general strike was called, but not before the rector of the University of Berlin himself had pledged allegiance to "Chancellor Kapp." The putsch leaders and backers withdrew into the shadows to await a more propitious moment.[42]

IN THE MIDST of this turmoil a real revolutionary came to town, namely Niels Bohr. Albert had never met the man who, in reinventing the atom, had taken the step that he, Albert, had abjured as portending the destruction of physics.

With that step in 1913 the young Bohr had assumed the mantle that Albert had worn so fitfully for the previous decade as the champion of the quantum approach to matter and radiation, and become a sort of Einstein of the microworld, a Danish wunderkind. The Carlsberg Brewery organization had just given him funding to establish an Institute of Theoretical Physics in Copenhagen. Planck had invited him to speak to the Berlin Physics Society in April, but the major event of his visit was the meeting with Einstein. Albert knew what Bohr was about almost as well as Bohr did; Bohr represented, after all, the other side of himself, the road not taken. Lately, Bohr had been championing a "complementary" approach to physics, in which light could be a wave or a particle, depending on how it was viewed—a notion vaguely reminiscent of Einstein's prophetic talk back in 1909 at Salzburg. Nevertheless, Albert didn't like it; the quantum revolution he had midwived never failed to bother him. Only the month before he had written to his friend Max Born: "The question of causality worries me a lot. Will the quantum absorption and emission of light ever be grasped in the sense of complete causality, or will there remain a statistical residue? I should be very, very loath to abandon complete causality."[43]

Personally, the two scientists hit it off easily. "I am just as keen on him as you are," Albert later told Ehrenfest. "He is an extremely sensitive lad and goes about the world as if hypnotized."[44]

Their conversation quickly zeroed in on the quantum and causality issue. What do you want? Bohr asked him. Einstein was the one who had invented the notion of light as particles. *Now* he presumed to be concerned with the situation in physics that allowed light to have a dual nature. Bohr invited him ask the German government to ban the use of photoelectric cells, if he thought that light should be waves, or to ban diffraction gratings if he believed that light was exclusively corpuscular.

Albert responded with frustration that two like-minded persons such as themselves could not agree on a common language. Perhaps they should begin by agreeing on certain fundamentals or general propositions. That was Albert, always looking for a principle, *constructive principles,* he called them once—the speed of light always being the same, or the equivalence of gravity and acceleration—some level with which he could grasp and overturn the world and yet remain sane.

Bohr immediately rejected the idea. No, never. It would be the greatest treachery, he said, to enter a new domain of knowledge with any foregone conclusions of the truth.[45] Bohr, recall, was the man who later defined a great truth as a statement whose opposite was also a great truth. His polestar was not principle, but paradox. Albert, listening to the whispers of God, above all wanted to preserve order. Niels was serving notice that he was ready to follow his nose and that nothing was sacrosanct. Nature would have to choose between these two visions and these two young men. For Albert and Niels it was a certain kind of love at first sight. All his life long Albert had needed somebody to argue and fight with in order to clarify his ideas. Mileva had been the first, then Besso, Solovine, Ehrenfest, Abraham, and de Sitter, but in Bohr he found his greatest match. The argument they started that day would go on for the rest of their lives.

After Bohr returned to Copenhagen Albert retreated to his little hideaway attic study. He pored over Bohr's papers, trying to make sense of the great duality and the paradox that physics had threatened to become. When he got stuck, as he later told Bohr, he imagined Bohr's friendly young face across the table from him, smiling and explaining, ever explaining.

It was unnaturally quiet in that upstairs redoubt where Newton still gazed serenely down underneath the slanted light of spring trying to break through. The end had come for Pauline in March, about the time of Albert's birthday, after what seemed like endless, pointless torture. She had fought and clung to life as fiercely and jealously and loudly as she had clutched and dominated her children. It was a wracking ordeal. "We are all exhausted from naked, firsthand experience," Albert reported to Zangger. "One feels in the bones what the bond of blood means."[46] She had been his first love, his first gravity, the sun that would have blotted out all stars as unworthy. In the silent and suddenly weightless space that ensued Albert felt as if he were standing in front of a blank screen. No one could make a picture of the future.

EPILOGUE

ALBERT EINSTEIN WAS AWARDED THE 1921 NOBEL PRIZE FOR PHYSICS in 1922. By the terms of his divorce agreement, he gave the 121,572 Swedish kronor award money to Mileva. Despite various offers to work elsewhere, Albert stuck it out in Berlin until 1933. Then, upon learning that Hitler had taken power, he and his entourage (who happened to be returning from a trip to the United States) went into exile. He eventually made his way to Princeton, where he accepted an offer to join the newly established Institute for Advanced Study. There, padding about the streets in a sweatshirt, sockless, his head wrapped in an increasingly wild corona of white hair, he retreated further into the studied indifference that had enabled him to survive in Berlin the First World War, and gradually became a kind of living shrine, a walking embodiment of cosmic mystery, humility, and atomic guilt. Scientifically, his years in New Jersey were spent grumbling about the element of chance in the newly formulated theory of quantum mechanics, while he groped, ultimately unsuccessfully, for a so-called unified field theory that would combine electromagnetism and general relativity.

Albert remained married to Elsa until her death in 1936, but the union was marked by numerous infidelities and affairs on his part, especially during the 1920s, when he was the toast of Berlin, and indeed Europe. The most serious of these relationships, and the best documented, was with Betty Neumann, the niece of his good friend Hans Mühsam. Albert fell in love with her in 1923, when Betty was twenty-three and visiting from Austria. He arranged to hire her as his secretary at the physics institute in order to keep her in Berlin, and for a period of about a year he saw her twice a week with Elsa's permission, in exchange for not sneaking around. Albert finally broke off the relationship in 1924, saying that he had to seek in the stars what was denied him on Earth. Betty returned to her native Austria and eventually moved to New York, where she died in 1972.

Among his later consorts in Berlin were Margarete Lenbach, referred to by Albert's contemporaries as the "Austrian woman," and Toni Mendel, who on occasion sent her chauffeur to fetch Albert for nights at the theater. It may have been Mendel whom he addressed in a letter that surfaced at an auction held by the Las Vegas Historical Documents Museum in 1982: "Liebchen . . . I work hard, and in the meantime I think happily of you. . . . This letter is written under great difficulty, as Elsa may come at any moment, and so I really have to watch. . . . Yesterday was so beautiful that I am still filled with delight. . . . I will come again at 5 o'clock at the same place, or better still, ten minutes before five if you can arrange it. . . . Be kissed, my dear, from your AE."

After Elsa's death Einstein never remarried. In what was in some sense a return to the circumstances of his youth, when he had always been surrounded by the women of his family, Albert lived on in the house he and Elsa had bought at 112 Mercer Street in Princeton with Margot, Maja, and his secretary, Helen Dukas. He did, however, continue to have affairs. According to letters that were put up for auction at Sotheby's in 1998, Albert maintained a relationship for several years in Princeton with Margarita Konenkova, a Russian emigrée who lived in Greenwich Village. The letters achieved extra notoriety because in his memoirs a Soviet spymaster by the name of Pavel Sudoplatov had listed Konenkova among his operatives, along with scientists like Niels Bohr. There is no evidence that any of them was a spy.

In 1939, at the urging of the Hungarian physicist Leo Szilard, Albert wrote a letter to President Franklin Roosevelt calling attention to the possibility of the development of atomic bombs and suggesting that the United States try to build one before the Germans did so first. Another letter, introducing Szilard, who wanted to express his concern about the wisdom of actually using the bomb, was sitting on Roosevelt's desk when he died. As the Cold War got under way, Albert's pacifism and socialistic tendencies attracted vitriol and suspicion from the right wing. His FBI file was later revealed to amount to more than a thousand pages.

Albert died on April 18, 1955, when an aneurysm in his abdominal aorta—the source of his lifelong stomach problems—burst. In 1948 a team of doctors at Brooklyn Jewish Hospital, led by Rudolph Nissen, had operated on Albert and wrapped his aneurysm with cellophane in an attempt to toughen it. When the aneurysm threatened again, Einstein rejected surgery, saying he preferred to go when it was his time. He was cremated, and his ashes were scattered in a nearby river. Einstein's brain was removed during an autopsy at Princeton Hospital by the pathologist Thomas Harvey, who subsequently received permission from the Einstein family to conduct scientific research on it. Harvey moved to Kansas, where the brain spent several decades in a jar in his office. In 1999, researchers from McMaster University in Ontario announced that Einstein's parietal lobe, a region associated with mathematics and spatial relationships, was 15 percent bigger than a normal person's.

According to a 1994 article in the *Guardian Weekend,* Einstein's eyes were removed during the autopsy by his ophthalmologist, Henry Abrams. They presently reside in a jelly jar in a New Jersey bank vault.

EDUARD "TETE" EINSTEIN grew into a sensitive and intellectual young man and a talented pianist, who exchanged thoughtful philosophical letters with his father. At the gymnasium in Zurich, he impressed his classmates with his literary and musical skills. In 1929 he began studying medicine at the University of Zurich, but his ambitions were shattered by a series of schizophrenic breakdowns beginning in 1930, after the failure of an affair with an older woman. In 1932 he was committed to the Burghölzi psychiatric hospital, where Carl Jung had once worked. He would be in and out of that institution for the rest of his life, subjected to insulin therapy and shock treatment. He saw his father for the last time in 1934 and gradually came to feel that he had been abandoned. "The worst destiny is to have no destiny," he wrote in one youthful poem, "and also to be the destiny of no one else." He died in 1965.

HANS ALBERT EINSTEIN graduated from the Polytechnic, his father's alma mater, with a civil engineering degree in 1927 and a Ph.D. in 1936. He went on to become one of the world's leading experts on silt and sediment transport. In 1927 he married Frieda Knecht, a neighbor on Gloriastrasse in Zurich (where his mother lived), against the will of his father, who thought that Frieda, nine years Albertli's senior, was too old and that her short stature (she was 4 feet 11 inches) was a sign that dwarfism ran in her family. Albert recruited Zangger to convince Hans Albert never to have children with Frieda, but a son, Bernhard Caesar, was born in 1930. A second son, Klaus, born in 1932, died at the age of six of diphtheria. Three years later, in 1942, Hans Albert and Frieda adopted a daughter, Evelyn, in Chicago.

Upon Mileva's death in 1948, some 430 letters between Albert and her and the children passed into Hans Albert's hands. It was only upon reading them, according to Robert Schulmann of the Einstein Papers Project, that Albertli finally learned of the existence of his lost sister, Lieserl, though he made no attempt to find her. An effort by Frieda in 1958 to publish a book about Albert and Mileva, based on the letters, was quashed by Einstein's executor, Otto Nathan, and by Helen Dukas and Margo Einstein.

After Frieda died in 1958, Hans Albert married Elizabeth Roboz, a neuroscientist. Hans Albert died on Martha's Vineyard in 1973.

Bernhard studied at the Federal Polytechnic, like his grandfather and father, and subsequently made his career doing physics for defense companies
in the United States and in Switzerland. He married a medical student,
Aude Ascher, with whom he had five children—Thomas, Eduard, and
Paul, who now live in California; Mira, who lives in Israel; and Charlie,
who lives in Bern. Bernhard and Aude are now divorced. Evelyn attended
the University of California, Berkeley, was a Berkeley policewoman, and
became widely known as a cult deprogrammer until illness forced her to re-
tire. Now confined to a wheelchair, she lives in Albany, California.

In 1984 Elizabeth Roboz Einstein placed the so-called family correspondence and Einstein documents in a trust for the benefit of Bernhard and
Evelyn, to be administered by her grandson Thomas, now an anesthesiolo-
gist in Santa Monica. In 1995 Evelyn sued Thomas and the lawyer who had
drawn up the trust, charging that its existence had been concealed from her,
and demanding that the letters be sold so that her share of the income could
be used for medical expenses. As part of a settlement, the Einstein family
correspondence went on the block at Christie's in November 1996. The
430 love letters exchanged between Albert and Mileva during their
courtship days fetched $400,000. Many of the rest remained unsold and
were split up among the family members.

ILSE EINSTEIN died of cancer in Paris, where she and her husband,
Rudolf Kayser, had fled, in 1934.

MAJA WINTELER EINSTEIN moved into her brother's house on Mercer
Street in Princeton in 1939 after being forced out of Italy. Her husband,
Paul Winteler, was denied entrance into the United States and remained in
Geneva. A stroke made Maja's exile permanent. As her health failed, Albert
read to her every day until her death in 1951.

MILEVA EINSTEIN MARIC used the money from Albert's Nobel Prize
to buy three apartment buildings in Zurich, in one of which, Huttenstrasse
62, despite intermittent pleadings by Albert that she and the children go to
Germany or some other place, she lived a quiet, stoic life, giving private
mathematics lessons, caring for Eduard, and rarely speaking of her life with
Albert, whom she last saw in 1934, when he made a special trip to see Tete.
Albert and Mileva corresponded until the end. In her later years Mileva
suffered from mild strokes, sclerosis, and a growing paranoia. In May 1948
she was hospitalized after a stroke left her paralyzed on one side. She died
that August. Afterward, 80,000 Swiss francs were found in her mattress. She
is now buried in an unmarked grave in Zurich.

ANNELI MEYER-SCHMID'S only child—a daughter named Erika—was
born on March 10, 1910, roughly nine months after Anneli's attempted rendezvous with Einstein. The girl seems to have grown up believing that she

was Einstein's daughter, according to Charles Schaerer, Erika's widower husband. Family lore has it that Albert and Anneli did meet in the spring of 1909 in Basel after Mileva's interception of a letter from Anneli to him caused a major outburst. In 1985, before Erika died, Max Flückiger, a schoolteacher who had founded the Albert Einstein Society in Bern, interviewed her for four hours and left with a pile of documents, letters, and photographs. Subsequently, photographs of Anneli and a chubby baby Erika, as well as Albert's 1899 inscription in her daybook, came to be prominently displayed in the archives of the society, which are now housed in the Swiss National Library in Bern.

According to letters in the Einstein Archives, Anneli's discretion regarding her relationship with Einstein was sorely tested at least once. In 1926 an unnamed American publisher offered her money to tell her story, at which Albert was incredulous.[1] "How did guys in America find out we know each other?" he demanded.[2] "How can they interfere in such a private affair? *Besser weiss es nicht* [Better it's not known]."

Anneli assured him that she had told them nothing. "The fat check from over there had absolutely no appeal, and now I most certainly would not exchange it for your little letter," she wrote. The only problem was that Albert had used the intimate form of address, *Du,* and she didn't dare show it to her husband.

Albert lost touch with Anneli after he moved to the United States in 1933 and had to ask his sister for her address, referring to "das ganz lustige Anneli und die schöne Zeiten [the utterly joyful Anneli and the beautiful times]." "When I think of you, the years fall away," he wrote on his sixtieth birthday in 1939. "How happy I was to be in Mettmenstetten in those times."

MICHELE BESSO worked in the patent office in Bern until his retirement in 1938, when he moved to Geneva. He died in 1955, only a month before Albert. Writing to Michele's son Vero and his wife, Albert said, "What I admired most about Michele was the fact that he was able to live so many years with one woman, not only in peace but also in constant unity, something I have lamentably failed at twice." The letter ended, "So in quitting this strange world he has once again preceded me by a little. That doesn't mean anything. For those of us who believe in physics, this separation between past, present, and future is only an illusion, however tenacious."

MAX ABRAHAM, who dueled with Albert about gravity and relativity, spent the war studying the theory of radio transmission for the Tele-funkengesellschaft. After the war he became a professor at the *technische Hochschule* in Stuttgart and then at the University of Aachen. In 1921 he was stricken with a brain tumor and died after six months of agony in a Munich hospital.

FRITZ ADLER was released from prison after the war ended and quickly elected to the Austrian National Assembly. The thoughts about relativity that his father had hoped would convince a jury he was insane, and which Albert simply thought mistaken, were subsequently published in the form of a long, mostly unread book.

NIELS BOHR continued his study of the quantum. In the 1920s, in what has been called the Second Scientific Revolution, Bohr, along with Werner Heisenberg, Erwin Schrödinger, Wolfgang Pauli, Max Born, and others, formulated a mathematically consistent theory of subatomic behavior known as quantum mechanics, at the expense of such commonsense notions as cause and effect, and predictability. Among its guiding precepts is Heisenberg's uncertainty principle, which says that the position and velocity of a particle cannot both be simultaneously known. Einstein's response was summarized famously by a phrase in a letter to Cornelius Lanczos: "It is hard to sneak a look at God's cards," Albert wrote in 1942, "but that he would choose to play dice with the world . . . is something that I cannot believe for a single moment." To which Bohr is said to have remarked, "Stop telling God what to do." For the rest of their lives Bohr and Einstein, though good friends, argued about quantum mechanics, which is now the foundation for every modern theory in physics, except general relativity.

In 1939, during a visit to Princeton, Bohr and the young physicist John Wheeler wrote a paper that laid out the theoretical basis for nuclear fission. Four years later Bohr, who was Jewish, was smuggled out of occupied Denmark on a fishing boat, just as the Nazis were preparing to deport the Danish Jews to concentration camps, and brought to the United States, where he consulted on the Manhattan Project, which built the first atomic bombs. He died in 1962 in Copenhagen.

PAUL EHRENFEST committed suicide in 1933.

ERWIN FREUNDLICH ran the Einstein Tower, unsuccessfully seeking further experimental proof of general relativity, from 1922 to 1933, when the Nazi takeover forced him into exile at the University of Istanbul. In 1939 he joined the University of St. Andrews in Scotland, where he became the Napier Professor of Astronomy and worked until 1959. He was still hoping to measure the deflection of starlight as late as 1954 when he traveled to an eclipse in Sweden, but was clouded out. He died in Mainz, Germany, in 1964.

FRITZ HABER, despite his continued leadership in secret chemical warfare projects, was driven into exile in 1933. He died in Basel, broken and bitter, but the worst was yet to come. During the Second World War, Zyklon

B, which Haber's institute had pioneered the study of, was used to kill members of Haber's own family, along with thousands of others, at Auschwitz.

PAUL HABICHT, who collaborated with Albert on the *Maschinchen,* became an electrical designer in Schaffhausen. Although he obtained many patents on electrical devices, commercial success eluded him, as it had Einstein's father. He died in 1948.

MAX VON LAUE, who as Planck's graduate student was one of the first German physicists to befriend Einstein and visit him in Bern, won the Nobel Prize in 1914 for his studies of X-ray diffraction in crystals. He continued to champion relativity even after it became unpopular, and when Einstein resigned from the Prussian Academy and the vice president of the academy remarked that it was no loss, von Laue was the only member to protest. He died in 1960 as a result of injuries suffered when a motorcyclist ran into his car.

HENDRIK LORENTZ continued to cling to his beloved aether after the advent of general relativity, and Albert humored him to the extent of titling a talk he gave in Leiden in 1920 "Aether and Relativity Theory." In the postwar years Lorentz campaigned to get German scientists readmitted to international scientific organizations. When he died, in 1928, the Dutch telegraph and telephone services were suspended for three minutes in his honor. His funeral was attended by government and scientific dignitaries from around the world, including Einstein, who called Lorentz "the greatest and noblest man of our times."

GEORG NICOLAI was turned out of his university lectureship by right-wing student protests in 1920. After losing a lawsuit to regain his post, he accepted an appointment as head of the department of physiology at the University of Córdoba and emigrated to Argentina in 1922. In 1936 he moved to the University of Chile and died in Santiago in 1964.

MAX PLANCK retired from the University of Berlin in 1928. Like Einstein, he never reconciled himself to the new statistical view of nature embodied in quantum mechanics. Although he opposed Hitler's policies, Planck felt that it was his duty to remain in Germany when the Nazis came to power in order to preserve what he could of German science. He continued as secretary of the mathematics and physics section of the Prussian Academy of Sciences and president of the Kaiser Wilhelm Society until the late 1930s. Einstein never forgave him for not speaking out in 1933 when Jews were "cleansed" from the university faculties.

Planck's house and belongings, including his papers, were destroyed in the Allied bombing in 1944. That same year, his younger son, Erwin, was implicated in an attempt to assassinate Hitler and sentenced to death. The authorities allegedly offered to spare Erwin's life if Planck would join the

Nazi party. Planck refused, and his son was executed a year later. Planck spent the remainder of the war sleeping in a haystack doubled over with pain from fused vertebrae. He died of a stroke in 1947 in Göttingen.

HENRI POINCARÉ died unexpectedly in 1912, after a supposedly successful operation, never having accepted Einstein's version of relativity. "What shall be our position in view of these new conceptions? Shall we be obliged to modify our conclusions?" he asked rhetorically in a lecture near the end of his life. "Certainly not . . . ," he concluded. "Today some physicists want to adopt a new convention. It is not that they are constrained to do so; they consider this new convention more convenient; that is all. And those who are not of this opinion can legitimately retain the old one in order not to disturb their old habits. I believe, just between us, that this is what they shall do for a long time to come."

WILLEM DE SITTER was awarded the Gold Medal of the Royal Astronomical Society in London in 1931. That same year he collaborated with Einstein on a cosmological solution to general relativity that described an expanding universe with no cosmological constant, which has served as the basis for cosmic theorizing ever since. Despite repeated episodes of illness, de Sitter served as director of the Leiden Observatory, reorganizing and modernizing it, and arranging for a Southern Hemisphere observatory in Johannesburg, South Africa. He died in 1934 in Leiden.

MAURICE SOLOVINE, Albert's old Olympia Academy friend, became an editor and publisher in Paris, where he published French translations of many of Einstein's works, as well as other classic philosophical and scientific works. In 1952 he conceived and published a series of books called *Science and Civilization* and *Masters of Scientific Thought*. After the war he visited Einstein in Princeton. He died in Paris in 1958.

OTTO STERN, Albert's graduate student, with whom he might have caroused in Prague and Zurich, was awarded the Nobel Prize in 1943 for studies of molecules and the magnetic properties of protons. He died in Berkeley, California, in 1962.

MARIE WINTELER married Albert Müller, the manager of a watch factory near Bern, in 1911, and raised two children while giving music lessons on the side. In 1927 she divorced Müller and moved to Zurich, where she wrote poetry and continued her music teaching. In 1950 she wrote to Albert, asking his help to emigrate to the United States, but she never left Switzerland. She died in Meiringen in 1957.

HEINRICH ZANGGER published his most important book, *Medizin und Recht* (*Medicine and the Law*), in 1920. During the 1930s and '40s he worked with the Committee of the International Red Cross to combat hunger and disease. He died in 1957 after a long illness.

NOTES

Abbreviation: *CP* = *The Collected Papers of Albert Einstein*

Chapter 1

1. Albert Einstein to Pauline Winteler, June 7, 1897, *CP*, vol. 1.
2. Maja Winteler-Einstein, excerpted in *CP*, vol. 1.
3. Ibid.
4. Lewis Pyenson, *The Young Einstein: The Advent of Relativity*.
5. Aude Einstein, the former wife of Albert's grandson, Bernhard, still has the tablecloth.
6. Albert Einstein to Hans Albert Einstein, Jan. 8, 1917.
7. "Autobiographical Notes."
8. Max Talmey, *New York Post*, Feb. 25, 1931.
9. Jürgen Renn, personal communication.
10. "Autobiographical Notes."
11. Maja Winteler-Einstein.
12. *CP*, vol. 1, footnote, p. 62, and Pyenson, *The Young Einstein*.
13. Anton Reiser (Rudolf Kayser), *Albert Einstein: A Biographical Portrait*.
14. Maja Winteler-Einstein, and *CP*, vol. 1.
15. *CP*, vol. 1.
16. Ibid.
17. Interview with Hans Albert Einstein in *Ladies' Home Journal*, April 1951, related in Ronald Clark, *Einstein: The Life and Times*.
18. *CP*, vol. 1.
19. Ibid.
20. Peter Skiff, personal communication.
21. Carl Seelig, *Albert Einstein: A Documentary Biography*.
22. Albert Einstein to Elsa Löwenthal-Einstein, April 30, 1912, trans. Valerie Tekavec.
23. Albert Einstein to Marie Winteler, *CP*, vol. 1.
24. *CP*, vol. 1, biographical notes.
25. Aude Einstein, personal communication.
26. *CP*, vol. 1.
27. Maja Winteler-Einstein.
28. Seelig, *Albert Einstein: A Documentary Biography*.
29. Ibid.
30. *Bartlett's Familiar Quotations*, 16th ed., Justin Kaplan, gen. ed. (Boston: Little, Brown, 1992), p. 537.
31. C. G. Jung, *Memories, Dreams, Reflections*.

Chapter 2

1. Maja Winteler-Einstein.
2. Carl Seelig, *Albert Einstein: A Documentary Biography,* p. 34.
3. Marie Winteler to Albert Einstein, Nov. 4, 1896, *CP,* vol. 1.
4. Albert Einstein to Mileva Maric, Sept. 28, 1899. In Jürgen Renn and Robert Schulmann, eds., *Albert Einstein, Mileva Maric: The Love Letters.*
5. Marie Winteler to Albert Einstein, Nov. 4, 1896, *CP,* vol. 1.
6. Marie Winteler to Albert Einstein, Nov. 30, , 1896, *CP,* vol. 1.
7. Seelig, p. 28.
8. Hans Christian von Baeyer, *Taming the Atom: The Emergence of the Visible Microworld.*
9. Donald K. Yeomans, *Comets: A Chronological History of Observations, Science, Myth, and Folklore* (New York: John Wiley & Sons, 1991).
10. *Bartlett's Familiar Quotations,* p. 282.
11. Seelig, p. 36.
12. *CP,* vol. 1, p. 56.
13. Heinrich A. Medicus, "The Friendship Among Three Singular Men: Einstein and His Swiss Friends Besso and Zangger."
14. Albert Einstein to Mileva Maric, April 4, 1901, *CP,* vol. 1.
15. *Bartlett's Familiar Quotations,* p. 537.
16. Seelig, p. 38.
17. Ibid., p. 40.
18. Pauline Einstein to Marie Winteler, March 24, 1897, *CP,* vol. 1.
19. Albert Einstein to Pauline Winteler, June 7, 1897, *CP,* vol. 1.

Chapter 3

1. The material in this chapter regarding Mileva's background and childhood is based largely on Desanka Trbuhovic-Gjuric, *Im Schatten Albert Einsteins: Das tragische Leben der Mileva Einstein-Maric,* and an appendix by Dord Krstic to *Hans Albert Einstein: Reminiscences of Our Life Together,* by Elizabeth Roboz Einstein. Krstic was a Yugoslav physicist who researched Mileva's life. When there were contradictions between the two, I have usually leaned in the direction of Krstic, since Trbuhovic-Gjuric's book is rather light on documentation.
2. Krstic, in Roboz Einstein, p. 86.
3. Trbuhovic-Gjuric, p. 9.
4. Ibid., p. 18. Another theory for Mileva's limp, that her joints were damaged in childhood by tuberculosis, is mentioned by Folsing.
5. Ibid.
6. Ibid.
7. Ibid., p. 21.
8. Ibid., p. 53.
9. Ibid., p. 38.
10. Krstic and Trbuhovic-Gjuric both talk about this.
11. Mileva Maric to Albert Einstein, Oct. 20, 1897, *CP,* vol. 1.
12. Albert Einstein to Mileva Maric, Feb. 16, 1898, *CP,* vol. 1.
13. Albert Einstein to Maja Einstein, 1898, *CP,* vol. 1.
14. *CP,* vol. 1, p. 69.
15. Albert Einstein to Mileva Maric, April 16, 1898, *CP,* vol. 1.
16. Albert describes this aspect of his personality in Alexander Moszkowski, *Einstein: The Searcher,* p. 4.
17. Carl Seelig, *Albert Einstein: A Documentary Biography,* p. 38.
18. Albert Einstein to Mileva Maric, Dec. 12, 1901, *CP,* vol. 1.

Chapter 4

1. Charles Zajonc, *Catching the Light* (New York: Bantam, 1993), p. 93.
2. Peter Skiff of Bard College tells me that the exact quote is "Maxwell's theory is Maxwell's equations," from *Electric Waves,* translated by D. E. Jones (London: Macmillan, 1893), p. 28.
3. Albert Einstein to Mileva Maric, March 13, 1899, *CP,* vol. 1.
4. Unfortunately, just what he did call her remains a mystery. The salutation on the letter reads, as well as can be deciphered, "Dear Saud," but *Saud* is not a word in German.

   "We puzzled it as much as we could," says Robert Schulmann of the Einstein Papers Project. "We didn't want to go out on a limb, because there weren't any limbs to go out on." Schulmann admitted to being intrigued by the fact that the first three letters of the mystery noun correspond to the German word for "pig," which in southern Germany can be a term of affection. According to my own dictionary, another definition for *Sau* is "slut." In that book I also found the word *Saudumm,* for which *Saud* could be an abbreviation, which means "painfully stupid" or "thick-headed."

   In his first affectionate greeting it is possible that Albert Einstein was calling his inamorata a slut or a pig-head, among other possibilities lost over the horizons of history, language, and legibility. Such is the nature of Einstein scholarship: a million "Rosebuds" scattered like wind-blown petals through the archives, attics, old trunks, and shoe boxes of Europe and North America.
5. Abraham Pais, *"Subtle Is the Lord . . . ,"* p. 111.
6. Zajonc, *Catching the Light,* p. 121.
7. Alfred E. Moyer, "Michelson in 1887," *Physics Today,* May 1987.
8. Pais, *"Subtle Is the Lord . . . ,"* p. 118.
9. Ibid., p. 127.
10. *CP,* vol. 1, p. 9.
11. Carl Seelig, *Albert Einstein: A Documentary Biography,* p. 46.
12. Ibid., p. 30.
13. Albert Einstein to Julia Niggli, July 28, 1899, *CP,* vol. 1.
14. Albert Einstein to Mileva Maric, early Aug. 1899, *CP,* vol. 1.
15. Albert Einstein to Mileva Maric, Aug. 10, 1899, *CP,* vol. 1.
16. Ibid.
17. *CP,* vol. 1, p. 220.
18. Albert Einstein to Julia Niggli, Aug. 1899, *CP,* vol. 1.
19. Mileva Maric to Albert Einstein, Aug. 10, 1899, vol. 1.
20. Albert Einstein to Mileva Maric, Sept. 10, 1899, *CP,* vol. 1.
21. *CP,* vol. 1, p. 224.
22. Albert Einstein to Mileva Maric, Sept. 28, 1899, *CP,* vol. 1.
23. Ibid.

Chapter 5

1. Albert Einstein to Mileva Maric, undated, 1900, *CP,* vol. 1.
2. Lewis Pyenson, "Einstein's Natural Daughter."
3. Mileva Maric to Helene Kaufler, June 4, 1900, *CP,* vol. 1.
4. Ibid.
5. Milana Bota to her mother, June 7, 1900, *CP,* vol. 1, p. 64.
6. Albert Einstein to Mileva Maric, Oct. 10, 1899, *CP,* vol. 1.
7. *CP,* vol. 1, p. 244.
8. Mileva Maric to Helene Kaufler, March 9, 1900, *CP,* vol. 1.
9. Carl Seelig, *Albert Einstein: A Documentary Biography,* p. 30.
10. Ibid., p. 43.

11. Albert Einstein to Mileva Maric, July 29, 1900, *CP,* vol. 1.
12. Ibid.
13. Albert Einstein to Mileva Maric, Aug. 6, 1900, *CP,* vol. 1.
14. Albert Einstein to Mileva Maric, Aug. 1, 1900, *CP,* vol. 1.
15. Albert Einstein to Mileva Maric, Aug. 9, 1900, *CP,* vol. 1.
16. Albert Einstein to Mileva Maric, Aug. 14, 1900, *CP,* vol. 1.
17. Ibid.
18. Albert Einstein to Mileva Maric, Aug. 20, 1900, *CP,* vol. 1.
19. Albert Einstein to Mileva Maric, Aug. 30, 1900, *CP,* vol. 1.
20. Ibid.
21. Albert Einstein to Mileva Maric, Sept. 13, 1900, *CP,* vol. 1.
22. *CP,* vol. 1, p. 265.
23. Albert Einstein to Mileva Maric, Sept. 19, 1900, *CP,* vol. 1.
24. Albert Einstein to Mileva Maric, Oct. 3, 1900, *CP,* vol. 1.
25. Mileva Maric to Helene Kaufler, Oct. 9, 1900, *CP,* vol. 1.
26. Robert Schulmann, "Einstein at the Patent Office: Exile, Salvation, or Tactical Retreat?," in Mara Beller, Jürgen Renn, and Robert S. Cohen, *Einstein in Context,* pp. 17–24.
27. Ronald Clark, *Einstein: The Life and Times,* p. 40.
28. "Conclusions Drawn from the Phenomena of Capillarity," in *Annalen der Physik* 4 (1901), p. 3.
29. Mileva Maric to Helene Savic, Dec. 20, 1900, *CP,* vol. 1.
30. Mileva Maric to Helene Savic, Dec. 11, 1900, *CP,* vol. 1.
31. Mileva Maric to Helene Savic, Dec. 20, 1900, *CP,* vol. 1.
32. Albert Einstein to Helene Savic, Dec. 20, 1900, *CP,* vol. 1.
33. *CP,* vol. 1, footnote, p. 272.
34. *CP,* vol. 1, p. 277.
35. Mileva Maric to Helene Savic, Jan. 8, 1901, *CP,* vol. 1.
36. Albert Einstein to Marcel Grossman, April 14, 1901, *CP,* vol. 1.

Chapter 6

1. Albert Einstein to Marcel Grossmann, April 10, 1901, in *Albert Einstein, Mileva Maric: The Love Letters.*
2. Albert Einstein to Mileva Maric, March 23, 1901, *The Love Letters.*
3. Albert Einstein to Mileva Maric, April 4, 1901, *The Love Letters.*
4. Ibid.
5. Albert Einstein to Mileva Maric, March 27, 1901, *The Love Letters.*
6. Albert Einstein to Mileva Maric, April 4, 1901, *The Love Letters.*
7. Albert Einstein to Mileva Maric, April 10, 1901, *The Love Letters.*
8. Hermann Einstein to Wilhelm Ostwald, April 13, 1901, *CP,* vol. 1.
9. Albert Einstein to Mileva Maric, June 4, 1901, *CP,* vol. 1.
10. Albert Einstein to Mileva Maric, April 15, 1901, *CP,* vol. 1.
11. Albert Einstein to Marcel Grossman, April 14, 1901, *CP,* vol. 1.
12. Albert Einstein to Mileva Maric, April 30, 1901, *The Love Letters.*
13. Mileva Maric to Albert Einstein, May 2, 1901, *The Love Letters.*
14. Mileva Maric to Albert Einstein, May 3, 1901, *CP,* vol. 1.
15. Mileva Maric to Helene Savic, May 1901, *CP,* vol. 1.
16. Albert Einstein to Mileva Maric, May 9, 1901, *CP,* vol. 1.
17. Albert Einstein to Mileva Maric, late May 1901 (no. 107), *CP,* vol. 1.
18. Albert Einstein to Mileva Maric, May 28, 1901, *CP,* vol. 1.
19. Albert Einstein to Mileva Maric, late May 1901 (no. 110), *CP,* vol. 1.
20. Lewis Pyenson, "Einstein's Natural Daughter."
21. Albert Einstein to Mileva Maric, May 28, 1901, *CP,* vol. 1.
22. Peter Michelmore, *Einstein: Profile of the Man,* p. 42.

23. Albert Einstein to Mileva Maric, July 7, 1901, *CP,* vol. 1.
24. Mileva Maric to Albert Einstein, July 8, 1901, *CP,* vol. 1.
25. Mileva Maric to Albert Einstein, July 31, 1901, *CP,* vol. 1.

Chapter 7

1. Arthur Schopenhauer, "Counsels and Maxims, Our Relations to Ourselves," in *The Complete Essays of Schopenhauer,* trans. T. Bailey Saunders, section 9, Chapter II (New York: John Wiley & Sons, 1942).
2. Abraham Pais, *"Subtle Is the Lord . . . ,"* p. 61.
3. Ibid., p. 83.
4. Albert Einstein to Marcel Grossman, September 6, 1901, *CP,* vol. 1.
5. Carl Seelig, *Albert Einstein: A Documentary Biography,* p. 51.
6. Albert Einstein to Mileva Maric, Dec. 12, 1901, *CP,* vol. 1.
7. *CP,* vol. 2, p. 175.
8. Albert Einstein to Mileva Maric, early Nov. 1901, *CP,* vol. 1.
9. Mileva Maric to Helene Savic, Nov. 23, 1901, *CP,* vol. 1.
10. Ibid.
11. *CP,* vol. 1, p. 317.
12. Mileva Maric to Albert Einstein, Nov. 1901, *CP,* vol. 1.
13. Mileva Maric to Albert Einstein, Nov. 13, 1901, *CP,* vol. 1.
14. Albert Einstein to Mileva Maric, Nov. 28, 1901, *CP,* vol. 1.
15. Mileva Maric to Helene Savic, Nov. 23, 1901, *CP,* vol. 1.
16. Albert Einstein to Mileva Maric, Nov. 28, 1901, *CP,* vol. 1.
17. Albert Einstein to Mileva Maric, Dec. 12, 1901, *CP,* vol. 1.
18. Ibid.
19. Albert Einstein to Mileva Maric, Dec. 17, 1901, *CP,* vol. 1.
20. Ibid.
21. Albert Einstein to Mileva Maric, Dec. 19, 1901, *CP,* vol. 1.
22. Albert Einstein to Mileva Maric, Dec. 28, 1901, *CP,* vol. 1.
23. That is the contention of Gerald Holton, a Harvard physicist and Einstein scholar who was raised in Austria and has written and lectured about Albert and Mileva's relationship. "They were very proud people," Holton explained to me after a lecture in New York.
24. *CP,* vol. 2, footnote, p. 175.
25. Albert Einstein to Mileva Maric, Feb. 4, 1902, *CP,* vol. 1.
26. Pauline Einstein to Pauline Winteler, Feb. 20, 1902, *CP,* vol. 1.

Chapter 8

1. Albert Einstein to Mileva Maric, Feb. 8, 1902, *Albert Einstein, Mileva Maric: The Love Letters.*
2. Max Talmey, *New York Post,* Feb. 25, 1931.
3. Maurice Solovine, *Letters to Solovine,* trans. Wade Baskin (New York: Citadel Press, 1993), p. 8.
4. Ibid., p. 6.
5. Ibid., p. 8.
6. Albert Einstein to Mileva Maric, June 28, 1902, *The Love Letters.*
7. Albert Einstein to Mileva Maric, Feb. 17, 1902, *CP,* vol. 1.
8. Ronald Clark, *Einstein: The Life and Times,* p. 46.
9. Thomas P. Hughes, "Einstein, Inventors, and Invention," in Mara Beller, Jürgen Renn, and Robert S. Cohen, eds., *Einstein in Context.*
10. Albert Einstein to Michele Besso, Jan. 22, 1903, *CP,* vol. 5.
11. Albert Einstein to Hans Wohlwend, Aug. 15, 1902, *CP,* vol. 5.

12. Solovine, *Letters to Solovine,* p. 13.
13. Quoted from *An Enquiry Concerning Human Understanding* (1748), in *Bartlett's Familiar Quotations,* 13th ed., p. 318.
14. Albert Einstein, in Carl Seelig, *Ideas and Opinions,* p. 22.
15. Arthur I. Miller, *Einstein's Special Theory of Relativity,* p. 130.
16. Henri Poincaré, *Science and Hypothesis,* p. 213.
17. Ibid., p. xxiv.
18. Robert Osserman's *Poetry of the Universe: A Mathematical Exploration of the Cosmos* (p. 63 ff.) has a particularly clear explanation of the nature and origins of non-Euclidean geometry.
19. William Kingdon Clifford, "Philosophy of the Pure Sciences," in *Lectures and Essays* (New York: Macmillan and Co., 1879), vol. 1, p. 300.
20. Duncan Sommerville, *Bibliography of Non-Euclidean Geometry,* quoted in Susan Sutliff Brown, "The Geometry of James Joyce's Ulysses: From Pythagoras to Poincaré," doctoral dissertation, University of South Florida, 1986.
21. Ibid.
22. Linda Dalrymple Henderson, *The Fourth Dimension and Non-Euclidean Geometry in Modern Art,* p. 66.
23. Poincaré, p. 71.
24. Arthur I. Miller, *Imagery in Scientific Thought: Creating 20th-Century Physics,* p. 32.
25. Poincaré, p. 90.
26. Solovine, *Letters to Solovine,* p. 9.
27. Maja Winteler-Einstein, "Albert Einstein: Beitrag für sein Lebensbild."
28. *CP,* vol. 1, p. 389.
29. Albert Einstein to Michele Besso, Jan. 22, 1903, *CP,* vol. 5.
30. Albert Einstein to Carl Seelig, May 5, 1952.
31. Albert Einstein to Michele Besso, Jan. 22, 1903, *CP,* vol. 5.
32. Mileva Maric to Helene Savic, March 20, 1903, in John Stachel, "Einstein and Maric: A Failed Collaboration."
33. Schulmann, "Einstein at the Patent Office: Exile, Salvation, or Tactical Retreat?," in Beller, Renn, and Cohen.
34. Albert Einstein to Michele Besso, March 17, 1903, *CP,* vol. 5.
35. Mileva Maric to Helene Savic, March 20, 1903, and Stachel, "Einstein and Maric."
36. Mileva Maric to Albert Einstein, August 27, 1903, *The Love Letters.*
37. Albert Einstein to Mileva Maric, Sept. 19, 1903, *CP,* vol. 5.
38. Ibid.

Chapter 9

1. Desanka Trbuhovic-Gjuric, *Im Schatten Albert Einsteins,* p. 81.
2. Peter Michelmore, *Einstein: Profile of the Man,* p. 43.
3. Trbuhovic-Gjuric, p. 86.
4. Maurice Solovine, *Letters to Solovine,* p. 10, tells the story without saying when, but 1903 is the only birthday of Einstein's that they could have shared.
5. *CP,* vol. 5, p. 8.
6. Ibid., p. 223.
7. Albert Einstein to Conrad Habicht, Feb. 20, 1904, *CP,* vol. 5.
8. "On the General Molecular Theory of Heat," in *Annalen der Physik* 14 (1904), p. 354.
9. Albert Einstein to Marcel Grossmann, April 6, 1904, *CP,* vol. 5.
10. Both Desanka Trbuhovic-Gjuric, p. 94, and Dord Krstic mention this story.
11. Maja Winteler-Einstein.
12. Trbuhovic-Gjuric, p. 83.
13. Susan Quinn, *Marie Curie: A Life,* p. 195.
14. Abraham Pais, *"Subtle Is the Lord . . . ,"* p. 364.

15. Among them had been Heinrich Weber, who later lectured about his results at the Swiss Polytechnic—it was about as close as Weber ever got to the frontier.
16. John L. Heilbron, *The Dilemmas of an Upright Man: Max Planck as Spokesman for German Science,* p. 29.
17. Ibid., p. 14.
18. Ibid., p. 19.
19. Pais, *"Subtle Is the Lord . . . ,"* p. 370.
20. "Autobiographical Notes," in Paul Arthur Schilpp, *Albert Einstein: Philosopher-Scientist,* p. 45.
21. *CP,* vol. 2, p. 149.
22. "Autobiographical Notes," p. 47.
23. *CP,* vol. 2, p. 208.
24. *CP,* vol. 2, p. 223; *Annalen der Physik* 17 (1905), p. 549.
25. Pais, *"Subtle Is the Lord . . . ,"* p. 104.
26. Albert Einstein to Conrad Habicht, May 25, 1905, *CP,* vol. 5, trans. Valerie Tekavec.

Chapter 10

1. Arthur I. Miller, *Albert Einstein's Special Theory of Relativity,* p. 30.
2. Abraham Pais, *"Subtle Is the Lord . . . ,"* p. 122.
3. Banesh Hoffman, *Relativity and Its Roots,* p. 82.
4. Henri Poincaré, *Science and Hypothesis,* p. 172.
5. Miller, *Albert Einstein's Special Theory,* p. 46.
6. Ibid., p. 55.
7. Ibid., p. xxv.
8. Ibid., p. 75.
9. Gerald Holton, *Thematic Origins of Scientific Thought: Kepler to Einstein,* p. 198.
10. Miller, *Albert Einstein's Special Theory,* p. 126.
11. Ibid.
12. See, for example, the discussion in *CP,* vol. 2, p. 260, and Miller, *Albert Einstein's Special Theory,* p. 87.
13. Holton drew attention to Föppl's book and its probable contribution to Einstein's thinking in *Thematic Origins.* Miller also elaborates on it in *Albert Einstein's Special Theory.*
14. Holton, p. 221.
15. Miller, *Albert Einstein's Special Theory,* p. 151.
16. Michele Besso to Albert Einstein, Aug. 3, 1952, *CP,* vol. 2, p. 309, n. 27.
17. Albert Einstein to Erika Oppenheimer, Sept. 13, 1932, *CP,* vol. 2, pp. 261–62.
18. "Autobiographical Notes," in Paul Arthur Schilpp, *Albert Einstein: Philosopher-Scientist,* p. 53.
19. Alexander Moszkowski, *Einstein the Searcher,* p. 4.
20. "Autobiographical Notes," p. 53.
21. See, for example, his memoirs in Schilpp; Albert Einstein to R. S. Shankland, which was to be read at the Michelson centennial, Dec. 19, 1952, in *CP,* vol. 2, p. 262; and Albert Einstein, unpublished manuscript, 1920, in *CP,* vol. 2, p. 262.
22. *CP,* vol. 2, p. 263.
23. Poincaré, p. 90.
24. Albert Einstein, Kyoto address given in 1922, see *CP,* vol. 2, p. 264; also Albert Einstein, "How I created the theory of relativity," trans. Yoshimasa Ono, from notes taken at the time by J. Ishiwara, *Physics Today,* August 1982, p. 46.
25. Peter Michelmore, *Einstein: Profile of the Man,* p. 45, presumably on the authority of Hans Albert.
26. "On the Electrodynamics of Moving Bodies," *Annalen der Physik* 17 (1905), p. 891, *CP,* vol. 2, p. 282.
27. Maja Winteler-Einstein, "Albert Einstein: Beitrag für sein Lebensbild."
28. Albert Einstein to Conrad Habicht, June 22–Sept. 22, 1905, *CP,* vol. 5.

29. *CP,* vol. 2 p. 314; "Does the Inertia of a Body Depend on Its Energy Content?," *Annalen der Physik* 18 (1905), pp. 639–41.
30. Albert Einstein to Conrad Habicht, July 20, 1905, *CP,* vol. 5.
31. Desanka Trbuhovic-Gjuric, *Im Schatten Albert Einsteins,* p. 92.
32. Ibid., p. 93.
33. Ibid.
34. Trbuhovic-Gjuric places this event in 1907, although no record of Albert and Mileva's visiting Serbia in that year exists in the Einstein paper trail.

Chapter 11

1. Albert Einstein to Solovine, April 27, 1906, *CP,* vol. 5.
2. Albert Einstein to Alfred Schnauder, early 1907, *CP,* vol. 5.
3. *CP,* vol. 5, p. 45.
4. Albert Einstein to Jost Winteler, Nov. 3, 1906, *CP,* vol. 5.
5. Albert Einstein to Michele Besso, Dec. 26, 1911, *CP,* vol. 5.
6. Albert Einstein to Mileva, Nov. 28, 1898, *CP,* vol. 1.
7. Aude Einstein, personal communication.
8. Maja Winteler-Einstein, "Albert Einstein: Beitrag für sein Lebensbild."
9. Arthur I. Miller, *Albert Einstein's Special Theory of Relativity,* p. 225.
10. Arnold Sommerfeld to Hendrik Lorentz, Dec. 26, 1907.
11. A. J. Kox, "Einstein and Lorentz," in Mara Beller, Jürgen Renn, and Robert S. Cohen, *Einstein in Context,* p. 51.
12. Abraham Pais, *"Subtle Is the Lord . . . ,"* pp. 169–70, for example. According to a recent monography, "Henri Poincaré and Special Relativity," by Armand Borel, a mathematician at the Institute for Advanced Study in Princeton, Poincaré confessed to a colleague that as of 1910 understanding relativity caused him the "greatest effort." In a lecture in Göttingen that year he referred to the choice between relativity and Lorentzian electrodynamics as a matter of taste. His opinion was unchanged when he died in 1912.
13. Albert Einstein to Solovine, May 3, 1906, *CP,* vol. 5.
14. Winteler-Einstein.
15. Pais, *"Subtle Is the Lord . . . ,"* p. 150.
16. Ronald Clark, *Einstein: The Life and Times,* p. 109.
17. Paul Ehrenfest to Tatiana Ehrenfest, Jan. 12, 1912.
18. Max von Laue to Jakob Laub, 1907, *CP,* vol. 5, p. 74.
19. Mileva to Helene Savic, Dec. 1906, John Stachel, "Einstein and Maric."
20. Max Planck to Albert Einstein, July 6, 1907, *CP,* vol. 5.
21. Albert Einstein to Johannes Stark, Feb. 17, 1908, *CP,* vol. 5.
22. Johannes Stark to Albert Einstein, Feb. 19, 1908, *CP,* vol. 5.
23. Albert Einstein to Johannes Stark, Feb. 22, 1908, *CP,* vol. 5.
24. Miller, *Albert Einstein's Special Theory,* p. 48.
25. Ibid., p. 231.
26. Hendrik Lorentz to Henri Poincaré, March 8, 1906, Miller, *Albert Einstein's Special Theory,* p. 234.
27. Miller, *Albert Einstein's Special Theory,* pp. 232–34.
28. *CP,* vol. 1, p. 271.
29. In 1920 Albert described the origins of what would become general relativity in a handwritten manuscript, never published, for *Nature,* "Grundgedanken und Methoden der Relativitätstheorie in ihrer Entwicklung dargestellt," which now resides in the Pierpont Morgan Library in New York City.
30. *CP,* vol. 2, p. 476.
31. Baron Roland von Eötvös had been carrying out delicate experiments in Hungary since 1889 to measure the difference between the inertial and gravitation masses of various materials and found that they were the same to within a few parts per billion.

32. Albert Einstein to Conrad Habicht, Dec. 24, 1907, *CP,* vol. 5.
33. Jakob Laub to Albert Einstein, March 1, 1908, *CP,* vol. 5.
34. Albert Einstein to Mileva, April 17, 1908, *CP,* vol. 5.
35. Robert Schulmann, "Einstein at the Patent Office," in Beller, Renn, and Cohen, p. 21.
36. *CP,* vol. 5, p. 48.
37. Albert Einstein to Marcel Grossman, Jan. 3, 1908, *CP,* vol. 5.
38. Alfred Kleiner to Albert Einstein, Jan. 28, 1908, *CP,* vol. 5.
39. Alfred Kleiner to Albert Einstein, Feb. 8, 1908, *CP,* vol. 5.
40. Pais, *"Subtle Is the Lord . . . ,"* p. 185.
41. Maja Winteler-Einstein.
42. Albert Einstein to Conrad and Paul Habicht, July 15, 1907, *CP,* vol. 5.
43. Desanka Trbuhovic-Gjuric, *Im Schatten Albert Einsteins,* p. 83.
44. Paul Habicht to Albert Einstein, Oct. 12, 1908, *CP,* vol. 5.
45. Albert Einstein to Jakob Laub, Nov. 1, 1908, *CP,* vol. 5.
46. Albert Einstein to Albert Gockel, Dec. 3, 1908, *CP,* vol. 5.
47. Trbuhovic-Gjuric, p. 83.
48. Alfred Bucherer to Albert Einstein, July 9, 1908, *CP,* vol. 5.
49. Miller, *Albert Einstein's Special Theory,* p. 350.
50. Ibid., p. 238.
51. Clark, p. 123.
52. Pais, *"Subtle Is the Lord . . . ,"* p. 152.
53. Clark, p. 123.
54. Linda Dalrymple Henderson, *The Fourth Dimension and Non-Euclidean Geometry in Modern Art,* p. 53.

Chapter 12

1. Friedrich Adler to Viktor Adler, June 19, 1908.
2. Carl Seelig, *Albert Einstein: A Documentary Biography,* p. 95.
3. Ibid.
4. Friedrich Adler to Viktor Adler, June 19, 1908.
5. Ibid.
6. Albert Einstein to Jakob Laub, May 19, 1909, *CP,* vol. 5.
7. Albert Einstein to Michele Besso, April 29, 1917, *CP,* vol. 8.
8. Albert Einstein to Jakob Laub, May 19, 1909, *CP,* vol. 8.
9. Abraham Pais, *"Subtle Is the Lord . . . ,"* p. 185.
10. Albert Einstein to Anna Meyer-Schmid, May 12, 1909, *CP,* vol. 5.
11. Mileva to George Meyer, May 23, 1909, trans. Valerie Tekavec.
12. Albert Einstein to George Meyer, June 7, 1909, *CP,* vol. 5.
13. Frieda Einstein's manuscript, p. 21. Frieda, who knew Mileva, writes, "Mileva was also very jealous. When the husband gave reason, with his 'Gypsy blood' as he called it, the jealousy in her grew to a cold, hard hatred for the other woman. She was a Slav and capable of very strong emotions. What made it even more impossible to live together in peace, Mileva, once having been offended or hurt, could not forgive or forget."
14. Albert Einstein to Erika Meyer, July 27, 1951. Mileva, he explained in his note to Anneli's daughter, Erika, in 1951, was pathologically jealous, a trait that often went hand in hand with "uncommon ugliness."
15. Mileva to Helene Savic, Sept. 3, 1909, John Stachel, "Einstein and Maric: A Failed Collaboration.
16. Seelig, p. 93.
17. It is not a ridiculous question. In modern particle physics, the strong force that holds nuclear particles together is indeed characterized by three kinds of a so-called color charge, are all attracted to each other and whimsically named "red," "blue," and "green."

18. "On the Theory of Light Production and Light Absorption," *Annalen der Physik* 20 (1906), p. 199, *CP,* vol. 2, p. 349.
19. *CP,* vol. 2, p. 144.
20. Max von Laue to Albert Einstein, June 2, 1906, *CP,* vol. 5.
21. Pais, *"Subtle Is the Lord . . . ,"* p. 389.
22. Albert Einstein to Jakob Laub, May 17, 1909, *CP,* vol. 5.
23. Max Planck to Albert Einstein, July 6, 1907, *CP,* vol. 5.
24. Max von Laue to Albert Einstein, Dec. 27, 1907, *CP,* vol. 5.
25. Albert Einstein to Arnold Sommerfeld, Jan. 14, 1908, *CP,* vol. 5.
26. *CP,* vol. 2, p. 144.
27. Ibid.
28. Albert Einstein to Hendrik Lorentz, May 23, 1909, *CP,* vol. 5.
29. Henrik Lorentz to Albert Einstein, May 6, 1909, *CP,* vol. 5.
30. Albert Einstein to Johannes Stark, July 31, 1909, *CP,* vol. 5.
31. G. I. Taylor, *Proceedings of the Cambridge Philosophical Society* 15 (1909), p. 114. I am indebted to John Wheeler for remembering and pointing out this obscure paper to me, and then helping me fetch it from the bowels of Princeton's Firestone Library.
32. It would take several decades and a few more incarnations of quantum theory to understand that Taylor's experiment was doomed to failure. According to modern quantum mechanics, the path of any particular light-quantum (or photon, as we call it today) through the optical and smoked-glass apparatus is determined by probability statistics on the basis of all the possible paths the photon could take. The different possibilities interfere with each other just like light waves, to make some paths more probable than others. Even if only one photon per hour passed through the smoked glass and hit Taylor's photographic plate, over the time of 2,000 hours, the 2,000 hits would add up to produce the same interference pattern as if all 2,000 photons had come through at once. As if, in fact, time did not even exist. Modern versions of Taylor's experiment, done sometimes with electrons, have confirmed this spooky result time and time again. A particularly devilish variation called the delayed-choice double-slit experiment, invented by Wheeler himself, allows an experimenter to control whether a photon has been interfered with or not after it has passed through the interfering apparatus. "Spooky" is barely a good enough word.
33. "On the Development of Our Views Concerning the Nature and Constitution of Radiation," presented at the Gesellschaft Deutscher Naturforscher und Ärtze in Salzburg on September 21, 1909. The speech and the ensuing discussion were published in *Physikalische Zeitschrift* 10 (1909), pp. 817–25, reprinted in *CP,* vol. 2, p. 563.

Chapter 13

1. Albert Einstein to Michele Besso, Nov. 17, 1909, *CP,* vol. 5.
2. Heinrich A. Medicus, "The Friendship Among Three Singular Men."
3. Albert Einstein to Carl Seelig, August 20, 1952.
4. Medicus.
5. *CP,* vol. 3, p. 321.
6. Carl Seelig, *Albert Einstein: A Documentary Biography,* pp. 100–101.
7. David Reichinstein, *Albert Einstein: A Picture of His Life and His Conception of the World,* p. 48.
8. Seelig, p. 102.
9. Reichinstein, p. 48.
10. Ibid., p. 30.
11. Ibid., p. 41.
12. Seelig, p. 106.
13. Reichinstein, p. 44.
14. Desanka Trbuhovic-Gjuric, *Im Schatten Albert Einsteins,* p. 107.
15. Seelig, p. 96.

16. Mileva to Helene Savic, late 1909, John Stachel, "Einstein and Maric."
17. Abraham Pais, *"Subtle Is the Lord . . . ,"* p. 399.
18. Albert Einstein to Jakob Laub, March 16, 1910, *CP,* vol. 5.
19. Walther Nernst to A. Schuster, March 17, 1910, *CP,* vol. 3, p. xxiii.
20. He was Franz Oppenheim, the founder of what is now known as Agfa. *CP,* vol. 5, p. 260.
21. Ronald Clark, *Einstein: The Life and Times,* p. 136.
22. Peter Michelmore, *Einstein: Profile of the Man,* p. 54.
23. Albert Einstein to Pauline Einstein, April 29, 1910, *CP,* vol. 5.
24. Albert refers to these occasions in his letters to Elsa in 1913, for example, Dec. 21, 1913.
25. *CP,* vol. 5, p. 243.
26. *CP,* vol. 5, p. 244, footnote to above.
27. Trbuhovic-Gjuric, p. 104.
28. Seelig, p. 107.
29. Albert Einstein to Ludwig Hopf, Aug. 19, 1910, *CP,* vol. 5.
30. Trbuhovic-Gjuric, p. 106.
31. My account of this episode is based on Jozsef Illy's account in *Isis.* Illy examined Austrian records and correspondence to correct the accounts in Frank and Clark.
32. Philipp Frank, *Einstein: His Life and Times,* p. 78.
33. Ibid.
34. Frank, p. 104.
35. Reprinted in *Physikalische Zeitschrift* 10 (1909), pp. 62–75; John L. Heilbron, *The Dilemmas of an Upright Man: Max Planck as Spokesman for German Science,* p. 47.
36. Heilbron, p. 51.
37. A speech given in Königsberg in the fall of 1910 to the Naturforscherversammlung, reprinted in *Acht Vorlesen* (1910), p. 126; Heilbron, p. 52.
38. *Physikalische Zeitschrift* 11 (1910), p. 600; Heilbron, p. 54.
39. Albert Einstein to Ernst Mach, Aug. 9, 1909, *CP,* vol. 5.
40. Frank, p. 104: "According to Einstein's judgment Mach did not give enough credit to the creative mind of the scientist who imagines general laws beyond a mere economic description of facts."
41. Albert Einstein to Ernst Mach, Aug. 17, 1909, *CP,* vol. 5.
42. Frank, p. 104.
43. *Scientific American* 193 (1955), p. 73, quoted in Clark, p. 160.
44. Albert Einstein to Ernst Mach, Dec. 1913, *CP,* vol. 5.
45. Einstein and Hopf, "Statistical Investigation of a Resonator's Motion in a Radiation Field," *Annalen der Physik* 33 (1910), p. 1105; *CP,* vol. 3, p. 269.
46. Albert Einstein to Jakob Laub, Nov. 4, 1910, *CP,* vol. 5.
47. Albert Einstein to Jakob Laub, Nov. 11, 1910, *CP,* vol. 5.
48. Lisbeth Hurwitz diary, Jan. 21, 1911, Trbuhovic-Gjuric, p. 107.
49. Albert Einstein to Fritz Adler, Feb. 9, 1911.
50. Albert Einstein to Hendrik Lorentz, Jan. 27, 1911, *CP,* vol. 5.
51. Albert Einstein to Hendrik Lorentz, Feb. 15, 1911, *CP,* vol. 5.
52. Albert Einstein to Michele Besso, May 13, 1911, *CP,* vol. 5.
53. Mileva to Helene Savic, early 1911, Stachel.
54. Trbuhovic-Gjuric, p. 111.

Chapter 14

1. Peter Michelmore, *Einstein: Profile of the Man,* p. 54.
2. Numerous stories attest to this. In his book *The Private Albert Einstein,* Peter Bucky tells of the time Albert's sailboat was swamped and he fell into the water. Upon being rescued, he shouted that this should count as a bath.
3. Albert Einstein to Michele Besso, May 11, 1911, *CP,* vol. 5.

4. Einstein protested vehemently against Marianoff and Wayne's biography (Clark, for example), but Marianoff was part of Albert's household for years in Berlin and might thus have had some basis for what he said.
5. Jozsef Illy, "Albert Einstein in Prague."
6. Albert Einstein to Marcel Grossmann, April 27, 1911, *CP,* vol. 5.
7. Albert Einstein to Alfred and Clara Stern, March 17, 1912, *CP,* vol. 5.
8. Albert Einstein to Marcel Grossmann, April 27, 1911, *CP,* vol. 5.
9. Philipp Frank, *Einstein: His Life and Times,* p. 118.
10. "On the Influence of Gravitation on the Propagation of Light," *Annalen der Physik* 35 (1911), *CP,* vol. 3, p. 485.
11. Albert Einstein to Jakob Laub, August 1911, *CP,* vol. 5.
12. Albert Einstein, *Relativity: The Special and General Theory.*
13. There are 60 seconds of arc in a minute of arc, and 60 minutes in a degree. The full moon subtends an angle of about half a degree, or 30 minutes of arc.
14. Albert Einstein to Erwin Freundlich, Sept. 1, 1911, *CP,* vol. 5.
15. Albert Einstein to Heinrich Zangger, Aug. 24, 1911, *CP,* vol. 5.
16. *CP,* vol. 5, p. 314, footnote 6.
17. Albert Einstein to Mileva Maric, Dec. 12, 1901, *CP,* vol. 5.
18. Albert Einstein to Michele Besso, Aug. 1911, *CP,* vol. 5.
19. Frank confided this impression in a conversation with his student Gerald Holton, who in turn passed it on to me.
20. Albert Einstein to Carl Seelig, May 5, 1952.
21. Ibid.
22. Mileva to Albert Einstein, Oct. 4, 1911, *CP,* vol. 5.
23. Albert Einstein to Heinrich Zangger, Oct. 22, 1911, *CP,* vol. 5.
24. Elizabeth Roboz Einstein, *Hans Albert Einstein: Reminiscences of Our Life Together,* p. 39.
25. Dimitri Marianoff and Palma Wayne, *Einstein: An Intimate Study of a Great Man.*
26. Jozsef Illy, "Albert Einstein in Prague"; Frank, p. 83.
27. Abraham Pais, *"Subtle Is the Lord . . . ,"* p. 485.
28. The Kafka Museum in Prague displays information about Kafka and Brod's social circle in Prague.
29. Illy.
30. Carl Seelig, *Albert Einstein: A Documentary Biography,* p. 126.
31. Frank, p. 85.
32. Quoted in Frank, p. 86.
33. Frank, p. 85.
34. Albert Einstein to Heinrich Zangger, Sept. 20, 1911, *CP,* vol. 5.
35. Heinrich A. Medicus, "The Friendship Among Three Singular Men."
36. Heinrich Zangger to Ludwig Forrer, Oct. 9, 1911, *CP,* vol. 5.
37. Albert Einstein to Heinrich Zangger, Oct. 22, 1911, *CP,* vol. 5.
38. Albert Einstein to Michele Besso, Oct. 21, 1911, *CP,* vol. 5.

Chapter 15

1. Abraham Pais, *"Subtle Is the Lord . . . ,"* p. 103.
2. Albert Einstein to Heinrich Zangger, Nov. 15, 1911, *CP,* vol. 5.
3. "On the Present State of the Problem of Specific Heats," presented at the Solvay Congress, Nov. 3, 1911, and reprinted in *CP,* vol. 3, p. 520.
4. Albert Einstein to Michele Besso, Dec. 26, 1911, *CP,* vol. 5.
5. Arthur I. Miller, *Albert Einstein's Special Theory,* p. 255; Abraham Pais, *"Subtle Is the Lord . . . ,"* p. 167.
6. Albert Einstein to Heinrich Zangger, Nov. 15, 1911, *CP,* vol. 5.

7. Albert Einstein to Heinrich Zangger, Nov. 7, 1911, *CP*, vol. 5.

8. In a 1947 obituary, quoted in Miller, *Albert Einstein's Special Theory*, p. 388.

9. Susan Quinn, *Marie Curie: A Life*, p. 302.

10. Albert Einstein to Heinrich Zangger, Nov. 7, 1911, *CP*, vol. 5.

11. Albert Einstein to Marie Curie, Nov. 23, 1911, quoted in Quinn, p. 310.

12. Albert Einstein to Hans Albert, Nov. 1925, quoted in "The Einstein Family Correspondence," catalog for Christie's auction, Nov. 25, 1996.

13. Albert Einstein to Michele Besso, Dec. 26, 1911, *CP*, vol. 5. The boy, Paoul, turned out to resemble Albert. According to Aude Einstein, this led to rumors that still circulate among the Einstein-Winteler-Besso women that Albert was the father. However, the baby would have been conceived in November, the same time as the wedding. Albert was in Switzerland only once that month, in the company of Mileva, according to Lisbeth Hurwitz's diary.

14. Ronald Clark, *Einstein: The Life and Times*, p. 158. Einstein heard about the work from George de Hevesy at a conference in Vienna, who wrote to Rutherford about his reaction.

15. "Autobiographical Notes," p. 47.

16. Otto Stern to Thomas Kuhn, interview May 29, 1962. Niels Bohr Library, American Institute of Physics.

17. Clark, p. 158, recounting Bohr's account of what Einstein later told him about his reaction.

18. *CP*, vol. 5, p. 348, n. 5.

19. Albert Einstein to Heinrich Zangger, Nov. 15, 1911, *CP*, vol. 5.

20. Henri Poincaré and Special Relativity," text of a talk given by Armand Borel of the Institute for Adanced Study at the New York Academy of Sciences, Dec. 3, 1998; also Clark, p. 149.

21. Heinrich A. Medicus, "The Friendship Among Three Singular Men."

22. Albert Einstein to Hendrik Lorentz, Nov. 23, 1911, *CP*, vol. 5.

23. Hendrik Lorentz to Albert Einstein, Dec. 6, 1911, *CP*, vol. 5.

24. Albert Einstein to Hendrik Lorentz, Dec. 12, 1911, *CP*, vol. 5.

25. Paul Habicht to Albert Einstein, Dec. 27, 1911, *CP*, vol. 5.

26. Heinrich Zangger to Albert Einstein, Jan. 30, 1912, *CP*, vol. 5.

27. Hendrik Lorentz to Albert Einstein, Feb. 13, 1912, *CP*, vol. 5.

28. Albert Einstein to Hendrik Lorentz, Feb. 18, 1912, *CP*, vol. 5.

29. Martin J. Klein, *Paul Ehrenfest: The Making of a Theoretical Physicist*.

Chapter 16

1. *Neue Freie Presse*, no. 17223, Aug. 5, 1912, *CP*, vol. 5.

2. Gerald Holton, "Ernst Mach and the Fortunes of Positivism in America," *Isis* 83 (1992), p. 27. Lenin's attack was published in the book *Materialism and Empirical Criticism*. It came in handy years later during the McCarthy era, when Frank, by then a Harvard professor, was investigated by the FBI. When agents visited his house, Frank pulled out the passage. The agents practically saluted and left happily.

3. Frank told of the exchange in *Modern Science and Philosophy* (Cambridge, Eng.: Cambridge University Press, 1949).

4. A corollary to the death of metaphysics in this circle was an emphasis on breaking down the barriers between different branches of science in order to stress the unity of the scientific method and its application to improving the lot of mankind. "The scientific conception serves life, and in turn is taken up by life," concluded a manifesto published in 1929 and quoted in Holton's *Isis* paper. There, some might say, was the entire modernist project in a nutshell, in all its glory and despair.

   When the Nazi terrors disrupted Europe, about a third of the circle shifted to the United States and busied themselves with organizing congresses and writing encyclopedias around the theme of the unity of science. Frank found shelter at Harvard, where he was one of the ringleaders, founding and serving as president of the Institute for the Unity of Science, and wrote a biography of Albert Einstein in 1947 that emphasized the positivistic aspects of Albert's thought and science.

With no ingrained metaphysical tradition to oppose it, positivism in America flourished and died of its own success, having become so universal as to be invisible. "To its critics," wrote Gerald Holton, who was Frank's student, "one might have applied Einstein's dictum that they were unaware how much they had imbibed of the belief system that they were now berating."

5. Philipp Frank, *Einstein: His Life and Times*, p. 118.
6. Albert Einstein to Heinrich Zangger, Nov. 15, 1911, *CP*, vol. 5.
7. Max von Laue to Albert Einstein, Dec. 27, 1911, *CP*, vol. 5.
8. Albert Einstein to Ludwig Hopf, Feb. 20, 1912, *CP*, vol. 5.
9. Albert Einstein to Paul Ehrenfest, March 10, 1912, *CP*, vol. 5.
10. Albert Einstein to Michele Besso, March 26, 1912, *CP*, vol. 5.
11. Albert Einstein to Ludwig Hopf, Feb. 20, 1912, *CP*, vol. 5.
12. "The Speed of Light and the Statics of the Gravitational Field," *Annalen der Physik* 38 (1912), reprinted in *CP*, vol. 4, p. 129.
13. Albert Einstein to Paul Ehrenfest, June 20, 1912, *CP*, vol. 5.
14. Max Abraham, "Relativität und Gravitation. Erwiderung auf eine Bemerkung des Hrn. A. Einstein," *Annalen der Physik* 38 (1912), pp. 1056–58, quoted in *CP*, vol. 4, p. 126.
15. "Relativity and Gravitation, Reply to a Comment by M. Abraham," *Annalen der Physik* 38 (1912) p. 1059–1064, *CP*, vol. 4, p. 180.
16. "Is There a Gravitational Effect Which Is Analogous to Electrodynamic Induction?" *Vierteljahresschrift für Gerichtliche Medizin und Öffentliche Sanitätswesen* 44 (1912), pp. 37–40, *CP*, vol. 4, p. 174.
17. *CP*, vol. 4, p. 189.
18. Albert Einstein to Alfred Kleiner, April 3, 1912, *CP*, vol. 5.
19. Albert Einstein to Heinrich Zangger, May 20, 1912, *CP*, vol. 5.
20. While Albert was in Berlin another implication of the light-bending theory apparently hit him. If two stars were lined up along the line of sight, the farther one nearly directly behind the nearer, the gravitational field of the closer star might act as a lens to focus and distort the light from the star behind it. Depending on the exact geometry, the more distant star might appear as multiple images, a single brightened image, or even a ring. As reported by Jürgen Renn, Tilman Sauer, and John Stachel, *Science* 275 (Jan. 10, 1975), the notebook Albert carried with him those years contained, interspersed with notations on his Berlin appointments, eight pages of calculations and formulas describing the intensification and distortion of star images by such gravitational lensing. Apparently Albert concluded that the chances of observing this effect were even more hopeless than eclipse light-bending because he never mentioned the subject until 1936, when he published a paper in response to the urgings of an amateur astronomer, Rudi Mandl. It contained virtually the same formulas as he had derived in 1912. Astronomers discovered their first gravitational lens, a quasar doubly focused by a cluster of galaxies, in 1979. In 1994 the refurbished Hubble Space Telescope recorded a spectacular image of the galaxy cluster Abell 2218 that contained dozens of arcs of light representing lensed images of background galaxies. Today the observation and analysis of such images is a growing branch of astronomy.
21. Schöneberg is famous today for its town hall, or *Rathaus*, where John Kennedy gave his famous "Ich bin ein Berliner" speech in 1963.
22. That gauzy past included, apparently, some kind of romantic interlude between Albert and Paula—shy fumblings behind the bushes in the old walled garden on holiday evenings?—memories that appalled Albert as he stood before Elsa eighteen years later. What could he have seen in Paula? "She was young, a girl, and obliging," he explained. "That was enough. The rest belies a sweet fantasy. Whomever she has not lied to, he does not know the meaning of happiness." Albert Einstein to Elsa, April 30, 1912, Valerie Tekavec translation.

23. Dimitri Marianoff and Palma Wayne, *Einstein: An Intimate Study of a Great Man,* p. 5.
24. Evelyn Einstein, personal communication.
25. Albert Einstein to Elsa, May 7, 1912, *CP,* vol. 5.
26. Marianoff and Wayne, p. 5.
27. Albert Einstein to Elsa, April 30, 1912, *CP,* vol. 5.
28. Elizabeth Roboz Einstein, p. 18.
29. Albert Einstein to Elsa, May 7, 1912, *CP,* vol. 5.
30. Albert Einstein to Elsa, May 21, 1912, *CP,* vol. 5.

Chapter 17

1. Albert Einstein to Helene Savic, Dec. 17, 1912, *CP,* vol. 5.
2. Desanka Trbuhovic-Gjuric, *Im Schatten Albert Einsteins,* p. 121.
3. Lisbeth Hurwitz diary, Oct. 19, 1912, Trbuhovic-Gjuric, p. 120.
4. Albert Einstein to Paul Ehrenfest, Dec. 20, 1912, *CP,* vol. 5.
5. David Reichinstein, *Albert Einstein: A Picture of His Life and His Conception of the World,* p. 47.
6. Abraham Pais, *"Subtle Is the Lord . . . ,"* p. 486. In an oral history interview conducted by Thomas S. Kuhn, May 29, 1962 for the Niels Bohr Library, American Institute of Physics, Stern recalled Einstein's boss, George Pick, laughing at the two of them because they were talking about atoms. A Machian attitude persisted, surprisingly, even in Zurich, where a professor in the physical chemistry department told Stern that his belief in atoms would have disqualified him for a Ph.D.
7. Roger Highfield and Paul Carter, *The Private Lives of Albert Einstein,* p. 284, report that an interview with Stern by the Swiss physicist Res Jost contained references to a brothel visit but was censored.
8. John Stachel, "Einstein and the Rigidly Rotating Disk," in *Einstein and the History of General Relativity,* ed. D. Howard and J. Stachel (Boston: Birkhäuser, 1989).
9. Another was the famous "twin paradox" first enunciated by Lorentz in 1909 and which generated so much argument that *Nature* magazine finally decreed a moratorium on papers about it in 1939.
10. Einstein made the argument in a letter to Hyman Levy in the winter of 1939–40, which is summarized in Stachel, "The Rigidly Rotating Disk."
11. Abraham Pais, *"Subtle Is the Lord . . . ,"* p. 212.
12. Charles W. Misner, Kip S. Thorne, and John Archibald Wheeler, *Gravitation,* p. 220.
13. If Riemann had been willing to consider time as his fourth dimension, he might have invented a version of the big bang theory of the universe that is often used in theoretical calculations by Stephen Hawking, among others, with the antipodal point representing the birth of the universe, the big bang itself. In his book *Poetry of the Universe,* the Stanford mathematician Robert Osserman points out that this interpretation of the Riemannian universe bears at least a poetic resemblance to Virgil's journey in the "Paradiso" section of Dante's *Divine Comedy.* Under Beatrice's guidance the poet climbs upward through the Primum Mobile, the realm of the planets, until he suddenly finds himself looking down into the Empyrean realm of the angels, at the center of which lies the dazzling Empyrean point. "I saw a point which radiated a light so keen that the eye on which it blazes must close . . ." ("Paradiso," Canto 28, lines 1–129).
14. Albert Einstein, "Errinerungen—Souvenirs," *Schweizersiche Hochschulzeitung* 28 (Sonderheft) (1955), quoted in John Stachel, "Einstein's Search for General Covariance 1912-1915," in Don Howard and John Stachel, eds., *Einstein and the History of General Relativity,* pp. 63-100.
15. Albert Einstein to Ludwig Hopf, Aug. 16, 1912, *CP,* vol. 5.
16. Albert Einstein to Arnold Sommerfeld, Oct. 29, 1912, *CP,* vol. 5.
17. Albert Einstein and Marcel Grossmann, *Outline of a Generalized Theory of Relativity and of a Theory of Gravitation* (Leipzig: Teubner, 1913); reprinted in *CP,* vol. 4, p. 302.

18. Albert Einstein to Hendrik Lorentz, Aug. 14, 1913, *CP,* vol. 5.
19. Besides Stachel, the members of the consortium include John Norton, from the University of Pittsburgh; Jürgen Renn, from the Max Planck Institute of the History of Science, in Berlin; Tilman Sauer, from Göttingen University; and Michel Janssen, now at the University of Minneapolis. A book based on their work is to be published by Birkhäuser. I am indebted to Michel Janssen, in particular, for many discussions of the evolving view of Einstein's adventures into general relativity.
20. John Norton, "How Einstein Found His Field Equations," pp. 101–59, in Stachel and Howard. See also Stachel, "Einstein's Search for General Covariance, 1912–1915," and, more recently, Jürgen Renn, "The Third Way to General Relativity: Einstein and Mach in Context," to appear in *Mach's Principle from Newton's Bucket to Quantum Gravity,* ed. Julian Barbour and Herbert Pfister, Einstein Studies (Boston: Birkhäuser, forthcoming).
21. Albert Einstein to Elsa, March 21, 1913, *CP,* vol. 5, trans. Valerie Tekavec.
22. Trbuhovic-Gjuric, p. 122.
23. Ibid., p. 120–22.
24. Albert Einstein to Elsa, March 14, 1913, *CP,* vol. 5, trans. Valerie Tekavec.
25. Albert Einstein to Clara Stern, March 14, 1913, *CP,* vol. 5.

  Was it in fact a toothache? Or was Lisbeth's pointed mention of Mileva's swollen jaw her typically Swiss discreet way of suggesting that Albert had hit her? This incident has become Exhibit A in the attempts to convict Einstein of wife-beating, with the argument running as follows: Already stressed out from his gravitational work, Albert is confronted by Mileva after she finds the letter from Elsa. Mileva throws another jealous tantrum, as she had in the case of Anneli, and perhaps even slaps him. Albert loses his temper and hits her back, exasperated by her pathologically insane bitterness.

  As part of his divorce testimony six years later, Albert admitted in a telegram to his lawyer, Emil Zürcher, that such fracases had occurred in the household. But if he had hit her it was only because she had hit him first: "In the marriage there were many scenes as a result of differences of opinion that came about whereby there was also cursing and physical violence on the part of the plaintiff which I, in a situation of irritation, also returned" (November 18, 1918, Protocol 1386/1918 of the Zurich District Court, Department 2).

26. Albert Einstein to Elsa, March 21, 1913, *CP,* vol. 5, trans. Valerie Tekavec.
27. Albert Einstein to Elsa, April 3, 1913, *CP,* vol. 5, trans. Valerie Tekavec.
28. Albert Einstein to Marie Curie, April 3, 1913, in Susan Quinn, *Marie Curie: A Life.*
29. Albert Einstein to Paul Ehrenfest, May 28, 1913, *CP,* vol. 5.
30. Albert Einstein to Ernst Mach, June 25, 1913, *CP,* vol. 5.
31. Quoted in Ronald Clark, *Einstein: The Life and Times,* pp. 160–61.
32. Max Planck to Wilhelm Wien, June 29, 1913, quoted in *CP,* vol. 5 footnote, p. 584.

Chapter 18

1. Fritz Haber to Hugo Kreuss, Jan. 4, 1913, *CP,* vol. 5.
2. Albert refers to this meeting in two lettes to Elsa, before and after Aug. 11, 1913.
3. M. Planck, W. Nernst, H. Rubens, E. Warburg, "Proposal for Einstein's Membership in the Prussian Academy of Sciences," June 12, 1913, *CP,* vol. 5, p. 526.
4. Ronald Clark, *Einstein: The Life and Times,* for example, p. 168.
5. Albert Einstein to Elsa, July 14, 1913, *CP,* vol. 5, trans. Valerie Tekavec.
6. Albert Einstein to Elsa, Aug. 11, 1913, *CP,* vol. 5, trans. Valerie Tekavec. Albert discusses Mileva's misgivings.
7. Susan Quinn, *Marie Curie: A Life,* p. 348.
8. Albert Einstein to Elsa, Aug. 11, 1913, *CP,* vol. 5, trans. Valerie Tekavec.
9. Albert Einstein to Elsa, after Aug. 11, 1913, *CP,* vol. 5. trans. Valerie Tekavec.
10. Albert Einstein to Hendrik Lorentz, Aug. 16, 1913, *CP,* vol. 5.

11. Albert Einstein to Paul Ehrenfest, Nov. 7, 1913, *CP,* vol. 5.

12. "Einstein and Besso: Manuscript on the Motion of the Perihelion of Mercury," June 1913, *CP,* vol. 4, p. 360. In December 1996 the manuscript was sold at auction at Christie's in New York for $360,000 to a collector from the West Coast.

13. Derived by Simon Newcomb in 1895. The modern value is 43.

14. Michel Janssen, "The Einstein-Besso Working Manuscript," Christie's auction catalog for November 25, 1996.

15. Clark, p. 162.

16. George Ellery Hale to Albert Einstein, Nov. 8, 1913, *CP,* vol. 5.

17. Albert Einstein to Heinrich Zangger, Sept. 30, 1913, *CP,* vol. 5.

18. Dord Krstic, in Elizabeth Roboz Einstein, *Hans Albert Einstein: Reminiscences of Our Life Together,* p. 97; Desanka Trbuhovic-Gjuric, *Im Schatten Albert Einsteins,* pp. 124–25.

19. "On the Present State of the Problem of Gravitation," reprinted in *Physikalische Zeitschrift* 14 (1913), pp. 1249–62, *CP,* vol. 4, p. 486.

20. *CP,* vol. 4, p. 504.

21. Felix Ehrenhaft, "My Experiences with Einstein," quoted in Clark, p. 150.

22. Linda Dalrymple Henderson, *The Fourth Dimension and Non-Euclidean Geometry in Modern Art,* p. 362.

23. Albert Einstein to Elsa, Oct. 10, 1913, *CP,* vol. 5.

24. Albert Einstein to Elsa, Nov. 22, 1913, *CP,* vol. 5.

25. Albert Einstein to Elsa, Oct. 16, 1913, *CP,* vol. 5.

26. Albert Einstein to Elsa, before Dec. 2, 1913, *CP,* vol. 5.

27. Albert Einstein to Elsa, after Dec. 2, 1913, *CP,* vol. 5.

28. Albert Einstein to Elsa, Nov. 17, 1913, *CP,* vol. 5.

29. *CP,* vol. 5, p. 582.

30. Albert Einstein to Elsa, after Dec. 2, 1913, *CP,* vol. 5.

31. Albert Einstein to Elsa, Dec. 21, 1913, *CP,* vol. 5.

32. Albert Einstein to Elsa, before Dec. 2, 1913, *CP,* vol. 5.

33. Ibid.

34. Albert Einstein to Elsa, Dec. 21, 1913, *CP,* vol. 5.

35. Ibid.

36. Albert Einstein to Elsa, Dec. 27, 1913, *CP,* vol. 5.

37. Quoted by Carlo Cattani and Michelangelo De Maria, "Abraham and Relativity in Italy," in Don Howard and John Stachel, eds., *Einstein and the History of General Relativity,* pp. 160–74.

38. Albert Einstein to Michele Besso, Jan. 1, 1914, *CP,* vol. 5.

39. Albert Einstein to Heinrich Zangger, Jan. 20, 1914, *CP,* vol. 5.

40. Albert Einstein to Erwin Freundlich, Dec. 7, 1913, *CP,* vol. 5.

41. Albert Einstein to Heinrich Zangger, Jan. 20, 1913, *CP,* vol. 5.

42. Albert Einstein to Heinrich Zangger, March 10, 1914, *CP,* vol. 5.

43. Albert Einstein to Elsa, Jan. 28, 1914, *CP,* vol. 5, trans. Valerie Tekavec.

44. Albert Einstein to Elsa, Feb. 1914, *CP,* vol. 5, trans. Valerie Tekavec.

45. Trbuhovic-Gjuric, p. 126.

46. Albert Einstein to Mileva, Hans Albert, and Eduard, March 23, 1914, *CP,* vol. 5.

Chapter 19

1. Albert Einstein to Adolf Hurwitz, May 14, 1914, *CP,* vol. 8.

2. Philipp Frank, *Einstein: His Life and Times,* p. 113.

3. Related in Minna Mühsam's remembrance of Einstein, in the Einstein Archives, Hebrew University of Jerusalem. Trans. Valerie Tekavec.

4. Ronald Clark, *Einstein: The Life and Times,* p. 175.

5. Anna Besso to Heinrich Zangger, Feb. 20, 1918, *CP,* vol. 8, p. 1032, is an account of the breakup based on conversations with Mileva and serves as the foundation for this account, along with Albert's letters to Elsa. Trans. Valerie Tekavec.

6. In an undated memo to Mileva stating his conditions for a return home, Albert repeats remarks he calls "bad jokes" that apparently have been said to him by the family. Trans. Valerie Tekavec.

7. Haber was no stranger to such miseries. His first wife had committed suicide, and Clara, his present wife, would eventually divorce him.

8. Albert's undated memo, trans. Valerie Tekavec.

9. Albert Einstein to Mileva, July 18, 1914, *CP,* vol. 8.

10. Ibid.

11. Anna Besso to Heinrich Zangger Feb. 20, 1918, *CP,* vol. 8, trans. Valerie Tekavec.

12. Albert Einstein to Elsa, July 26, 1914, *CP,* vol. 8.

13. Ibid.

14. Albert Einstein to Elsa, Aug. 3, 1914, *CP,* vol. 8.

15. Albert Einstein to Paul Ehrenfest Aug. 19, 1914, *CP,* vol. 8.

16. John Earman and Clark Glymour, "Relativity and Eclipses: The British Eclipse Expeditions of 1919 and Their Predecessors."

17. Dord Krstic in Elizabeth Roboz Einstein, *Hans Albert Einstein,* p. 27, and Peter Michelmore, *Einstein: Profile of the Man,* p. 67, mention Hans Albert's reactions.

18. Albert Einstein to Mileva, Sept. 15, 1914, *CP,* vol. 8.

19. Plesch quoted in Clark, p. 191.

20. Abraham Pais, *"Subtle Is the Lord . . . ,"* p. 489.

21. Minna Mühsam.

22. This would not turn out to be a propitious line of research either for Germany or for Haber himself, since one of the long-range results would be the Zyklon gas with which Haber's relatives and millions of others would be exterminated in Nazi concentration camps during World War II.

23. John L. Heilbron, *The Dilemmas of an Upright Man,* p. 72.

24. Hubert Gönner and Giuseppe Castegnetti, "Albert Einstein as a Pacifist and Democrat During the First World War."

25. Heilbron, p. 74.

26. Nicolai's story is engagingly told in Wolf Zuelzer, *The Nicolai Case* (Detroit: Wayne State University Press, 1982).

27. Ibid., p. 21.

28. Ibid., p. 25.

29. Ibid., p. 28.

30. The story was told by Margot Einstein to Wolf Zuelzer, Nicolai's biographer, in an interview in Princeton in 1972. Zuelzer's notes are in his archive at the University of Wyoming.

31. Zuelzer, p. 228.

32. Margot Einstein admitted this to Zuelzer in his 1972 interview.

33. Gönner and Castagnetti.

34. Quoted in Gönner and Castagnetti.

35. Albert Einstein to Paul Ehrenfest, Aug. 23, 1915, quoted in Gönner and Castagnetti.

36. Thomas P. Hughes, "Einstein, Inventors, and Invention," in Mara Beller, Jürgen Renn, and Robert S. Cohen, *Einstein in Context.*

37. Albert Einstein to Michele Besso, Feb. 12, 1915, *CP,* vol. 8.

38. Albert Einstein to Mileva, Dec. 12, 1914, *CP,* vol. 8.

39. Evelyn Einstein, personal communication.

40. Albert Einstein to Mileva, Dec. 12, 1914, *CP,* vol. 8.

41. Albert Einstein to Mileva, Jan. 12, 1915, *CP,* vol. 8.

42. Desanka Trbuhovic-Gjuric, *Im Schatten Albert Einsteins,* p. 134.

43. Trbuhovic-Gjuric, p. 132.
44. Albert Einstein to Heinrich Zangger, May 17, 1915, *CP*, vol. 8.
45. Albert Einstein to Heinrich Zangger, April 15, 1915, *CP*, vol. 8.
46. Gönner and Castagnetti.
47. Albert Einstein to Heinrich Zangger, May 17, 1915, *CP*, vol. 8.
48. Clark, p. 184.
49. Gönner and Castagnetti.

Chapter 20

1. Albert Einstein to Hendrik Lorentz, Feb. 2, 1915, *CP*, vol. 8.
2. Don Howard and John Stachel, eds., *Einstein and the History of General Relativity*, p. 71.
3. Albert Einstein to Paul Ehrenfest, late 1913, *CP*, vol. 5.
4. "The Formal Foundation of the General Theory of Relativity," presented to the Prussian Academy of Sciences, October 29, 1914, and reprinted in *CP*, vol. 6, p. 72, discussed in Howard and Stachel, p. 73.
5. Albert Einstein to Michele Besso, March 10, 1914, *CP*, vol. 8.
6. Hendrik Lorentz to Albert Einstein, Jan. 1915, quoted in A. J. Kox, "Hendrik Antoon Lorentz, the Ether, and the General Theory of Relativity," in Howard and Stachel, p. 202.
7. See Carlo Cattani and Michelangelo De Maria, "Max Abraham and the Reception of Relativity in Italy: His 1912 and 1914 Controversies with Einstein," and "The 1915 Epistolary Controversy Between Einstein and Tullion Levi-Civita," in Howard and Stachel, pp. 160–74 and 175–200.
8. Albert Einstein to Tullio Levi-Civita, March 5, 1915, quoted by Cattani and De Maria, "Epistolary Controversy," p. 186.
9. Albert Einstein to Tullio Levi-Civita, April 2, 1915, Cattani and De Maria, "Epistolary Controversy, p. 190.
10. Ibid., p. 192.
11. Albert Einstein to Heinrich Zangger, April 10, 1915, *CP*, vol. 8.
12. Albert Einstein to Tullio Levi-Civita, April 14, 1915, Cattani and De Maria, "Epistolary Controversy, p. 193.
13. Ibid., April 20, 1915, Cattani and De Maria, "Epistolary Controversy p. 193.
14. Albert Einstein to Heinrich Zangger, July 1915, *CP*, vol. 8.
15. Don Howard and John Norton "Out of the Labyrinth? Einstein, Hertz, and the Göttingen Answer to the Hole Argument," in John Earman, Michel Janssen, and J. D. Norton, *The Attraction of Gravitation*, p. 30.
16. Albert Einstein to Hendrik Lorentz, Jan. 1, 1916, Howard and Stachel, p. 138.
17. Michel Janssen, "The Einstein-Besso manuscript," Christie's catalog, November 1996.
18. David Hilbert to Albert Einstein, Nov. 13, 1915, in Abraham Pais, *"Subtle Is the Lord . . . ,"* p. 259.
19. "On the General Theory of Relativity," presented to the Prussian Academy of Sciences, Nov. 4, 1915, reprinted in *CP*, vol. 8, p. 214; also in Pais, *"Subtle Is the Lord . . . ,"* p. 250.
20. Albert Einstein to David Hilbert, Nov. 18, 1915, quoted in Pais, *"Subtle Is the Lord . . . ,"* p. 243.
21. Albert Einstein to Michele Besso, Nov. 17, 1915, quoted in Heinrich Medicus, "A Comment on the Relations Between Einstein and Hilbert," *American Journal of Physics* 52, no. 3 (March 1984), p. 206.
22. Albert Einstein to de Haas, quoted in Pais, *"Subtle Is the Lord . . . ,"* p. 253.
23. Albert Einstein to Paul Ehrenfest, Jan. 17, 1916, quoted in Pais, *"Subtle Is the Lord . . . ,"* p. 253.
24. Albert Einstein to Heinrich Zangger, Oct. 22, 1919, trans. Valerie Tekavec.
25. Pais, *"Subtle Is the Lord . . . ,"* p. 257.
26. David Hilbert to Albert Einstein, Nov. 17, 1915, quoted in Medicus, "A Comment."
27. Albert Einstein to David Hilbert, Nov. 15, 1915, quoted in Pais, *"Subtle Is the Lord . . . ,"* p. 260.

28. Albert Einstein to David Hilbert, Nov. 19, 1916, Pais, *"Subtle Is the Lord . . . ,"* p. 253.

29. "The Field Equations of Gravitation," presented to the Prussian Academy of Sciences, Nov. 25, 1915, *CP,* vol. 8, p. 244.

30. Albert Einstein to Michele Besso, Dec. 10, 1915, *CP,* vol. 8.

31. Medicus, "A Comment."

32. Albert Einstein to Heinrich Zangger, Nov. 26, 1915, quoted in Medicus, "A Comment."

33. Albert Einstein to Michele Besso, Nov. 30, 1915, *CP,* vol. 8.

34. Pais was told this by E. G. Strauss, who was Einstein's assistant in Princeton in the 1940s.

35. Albert Einstein to David Hilbert, Dec. 20, 1915, quoted in Pais, *"Subtle Is the Lord . . . ,"* p. 260.

36. Medicus, "A Comment."

37. Albert Einstein to Paul Ehrenfest, May 24, 1915 in J. D. Norton, Howard and Stachel.

38. Hertz's suggestion was reconstructed from Einstein's answers to him by Don Howard and John Norton in a paper titled "Out of the Labyrinth? Einstein, Hertz, and the Göttingen Answer to the Hole Argument," in Earmann, Janssen, and Norton, *The Attraction of Gravitation.* I am grateful to John Norton for pointing it out.

39. Howard and Norton, "Out of the Labyrinth?"

40. Albert Einstein to Paul Ehrenfest, Dec. 26, 1915, *CP,* vol. 8.

41. Albert Einstein to Michele Besso, Jan. 3, 1916, John Stachel, "Einstein and the Rigidly Rotating Disk," in Howard and Stachel.

42. "The Foundation of the General Theory of Relativity," *Annalen der Physik* 49 (1916), reprinted in *CP,* vol. 6, p. 283.

43. It was a subtle point. Even the great Hilbert fell for the hole argument, according to John Norton, who has studied the recently discovered page proofs (the ones into which he inserted Einstein's field equations at the last minute) for Hilbert's gravitation paper.

44. Howard and Norton, "Out of the Labyrinth."

45. Albert Einstein to Hans Albert, Nov. 23, 1915, *CP,* vol. 8.

46. Albert Einstein to Hans Albert, Nov. 30, 1915, *CP,* vol. 8.

47. Albert Einstein to Mileva, Dec. 10, 1915, *CP,* vol. 8.

Chapter 21

1. Heinrich Zangger to Michele Besso, late 1915, Heinrich A. Medicus, "The Friendship Among Three Singular Men."

2. Michele Besso to Heinrich Zangger, late 1915, quoted in Medicus.

3. Albert Einstein to Heinrich Zangger, Nov. 26, 1915, *CP,* vol. 8.

4. Albert Einstein to Mileva, Feb. 6, 1916, *CP,* vol. 8.

5. Albert Einstein to Mileva, April 1, 1916, *CP,* vol. 8.

6. Ibid.

7. Albert Einstein to Mileva, April 8, 1916, *CP,* vol. 8.

8. Desanka Trbuhovic-Gjuric, *Im Schatten Albert Einsteins,* p. 138.

9. Pauline Einstein to Elsa, Aug. 6, 1916, Einstein Archives, Hebrew University of Jerusalem.

10. Albert Einstein to Michele Besso, July 14, 1916, *CP,* vol. 8.

11. Albert Einstein to Michele Besso, July 21, 1916, *CP,* vol. 8.

12. Albert Einstein to Michele Besso, later on July 21, 1916, *CP,* vol. 8.

13. Albert Einstein to Heinrich Zangger, July 26, 1916, *CP,* vol. 8.

14. Albert Einstein to Michele Besso, Sept. 6, 1916, *CP,* vol. 8.

15. Albert Einstein to Helene Savic, Sept. 8, 1916, John Stachel, "Einstein and Maric."

16. Abraham Pais, *"Subtle Is the Lord . . . ,"* p. 357.

17. Tetrode (1895–1931) lived in Amsterdam and was both prodigally brilliant and a painfully shy recluse. Ehrenfest tried to recruit him to Leiden once, but failed. Einstein was a great admirer of Tetrode's and once went to his house with Ehrenfest. Tetrode wouldn't come to the door. Heinrich Casimir told this story at a symposium at Yale in honor of Martin Klein in June 1994.

18. Albert Einstein to Michele Besso, Aug. 11, 1916, *CP,* vol. 8.
19. Ibid.
20. Albert Einstein to Michele Besso, Sept. 6, 1916, *CP,* vol. 8.
21. "On the Quantum Theory of Radiation," *Physikalische Zeitschrift* 18 (1917), pp. 121–28, *CP,* vol. 6, p. 381.
22. Albert Einstein to Michele Besso, Sept. 11, 1911, *CP,* vol. 8.
23. Albert Einstein to Michele Besso, March 9, 1917, *CP,* vol. 8.
24. See Albert Einstein to Walter Dallenbach, Nov. 1916, *CP,* vol. 8, also quoted in John Stachel, "The Other Einstein: Einstein Contra Field Theory," in Mara Beller, Jürgen Renn, and Robert S. Cohen, *Einstein in Context.*
25. V. Ya. Frenkel and B. E. Yavelov, "What May Happen to the Person Who Thinks a Great Deal but Reads Very Little," a draft manuscript transmitted by Jozsef Illy.
26. "Elementary Theory of Water Waves and of Flight," *Die Naturwissenschaften* 4 (1916), *CP,* vol. 6, p. 399.
27. Carl Seelig, *Albert Einstein: Leben und Werk eines Genies unserer Zeit* (Zurich: Europa Verlag, 1960), quoted in Frenkel and Yavelov.
28. Albert Einstein to Paul Georg Erhardt, Sept. 7, 1954, quoted by Frenkel and Yavelov.
29. Trbuhovic-Gjuric, p. 155.
30. Michele Besso to Albert Einstein, Dec. 5, 1916, *CP,* vol. 8.
31. Albert Einstein to Michele Besso, March 9, 1917, *CP,* vol. 8.

Chapter 22

1. Paul Ehrenfest to Hendrik Lorentz, Jan. 1916, quoted in Abraham Pais, *"Subtle Is the Lord . . . ,"* p. 271.
2. Quoted in Gerald Holton, *Thematic Origins of Scientific Thought,* p. 248. Mach's papers were destroyed by his son Ludwig. In 1987 the historian Gereon Wolters argued that Ludwig had forged the preface in question, perhaps under the influence of a friend, the philosopher Hugo Dingler, who hated relativity. Modern scholars have not followed his view. See Holton, *Science and Anti-Science* (Cambridge, Mass.: Harvard University Press, 1993).
3. *The Quotable Einstein,* p. 72.
4. "Regarding the Gravitational Field of a Point Mass According to Einsteinian Theory," presented to the Prussian Academy of Sciences and published in their *Proceedings,* 1916, p. 189–96, quoted in *Dictionary of Scientific Biography,* p. 251.
5. Quoted in Jean Eisenstaedt, "The Early Interpretation of the Schwarzschild Solution," in Don Howard and John Stachel, eds., *Einstein and the History of General Relativity,* p. 216.
6. Ibid., p. 217.
7. Half a century later, when astrononmical observations of mysterious energetic objects called quasars convinced astronomers that such dense, weirdly sarcophagally contracted objects did exist in the far reaches of the sky, this would be recognized as a somewhat hyperbolic description of a black hole. The name was dreamed up by John Wheeler in 1968 to dramatize what he saw as the greatest crisis and greatest opportunity in physics. Theoretical calculations had shown by then that a star of sufficient mass, once it had exhausted its thermonuclear fuel, could collapse inward past its Schwarzschild radius and, in fact, keep collapsing and shrinking forever. Past that point, the star would bend space so severely that light would only orbit around it and never escape, hence the name "black hole." In this sense, the Schwarzschild radius really was a boundary of space-time because anything that crossed it could never come back out: no light, no spaceships, no information. Wheeler was fond of pointing out that physics as presently known could not say what happened at the center of the black hole, where the density and pressure would be infinite and Einstein's equations would break down. With them would break down space and time, matter, and the rest of physics. "Smoke pours from the computer," he liked to say. At issue was nothing less than the annihilation and the creation of the universe. Whatever theory could explain the death of space-time at the cen-

ter of the black hole might also be able to explain the birth of space-time—the ancient burning riddle of why there is something rather than nothing at all. Perhaps, as Wheeler often said, it would take the "fiery marriage" of general relativity and quantum theory. As of 1999, the black hole remained the dark portal of physicists' dreams.

8. Albert Einstein to Karl Schwarzschild, Jan. 9, 1916, quoted in Eisenstaedt, "The Early Interpretation."

9. Modern observations put the diameter of the Milky Way at about 100,000 light-years. The sun is about 30,000 light-years from the center, in the relative suburbs.

10. Astronomers have since recognized that this explosion was a supernova, the cataclysmic death in which a star can briefly outshine an entire galaxy.

11. Willem de Sitter to Albert Einstein, Nov. 1, 1916, *CP,* vol. 8.

12. Albert Einstein to Willem de Sitter, Nov. 4, 1916, *CP,* vol. 8.

13. "On Einstein's Theory of Gravitation and Its Astronomical Consequences," *Monthly Notices of the Royal Astronomical Society* 77 (December 1916), pp. 155–83, quoted in Pierre Kerszberg, "The Einstein–de Sitter Controversy of 1916–1917 and the Rise of Relativistic Cosmology," in Howard and Stachel, pp. 325–66. See also the essay in *CP,* vol. 8, by Michel Janssen.

14. Albert Einstein to Paul Ehrenfest, Feb. 4, 1917, quoted in Pais, *"Subtle Is the Lord . . . ,"* p. 285.

15. *CP,* vol. 6, p. 540. See also the treatment in Jeremy Bernstein and Gerald Feinberg, *Cosmological Constants.*

16. Quoted in Bernstein and Feinberg, p. 21.

17. In the late 1990s, astronomers came to the reluctant conclusion that some force was propelling the universe to expand faster and faster with time, and the leading candidate for this force, as of this writing, is Einstein's cosmological constant. See "The Golden Age," in the 2nd ed. of Dennis Overbye, *Lonely Hearts of the Cosmos* (Boston: Little, Brown, 1999).

18. Albert Einstein to Willem de Sitter, March 24, 1917, *CP,* vol. 8.

19. Albert Einstein to Willem de Sitter, March 12, 1917, *CP,* vol. 8.

20. Ibid.

21. See Overbye, *Lonely Hearts,* pp. 45–47. Besides, it took bigger telescopes than Slipher had at his disposal to gather in and dissect the faint light of these nebulae. Such telescopes as the fogweary astronomers of Europe had never dreamed of were being erected on the mountaintops of California. Slipher had tripped over the toe of a giant, and had he persevered he might have been in position to recognize the most elemental and enigmatic fact of astronomy, which remains today unfathomable: The nebulae, island universes, in general are receding from us, flying away at speeds up to the speed of light, as the night billows outward like a sail. The universe is expanding. It would be fifteen years before another astronomer, Edwin Hubble, obsessed with nebulae, using those giant telescopes, put the pieces together. By then Albert was wishing he had never invented the cosmological constant.

22. "On Einstein's Theory of Gravitation and Its Astronomical Consequences. Third Paper," in *MNRAS* 78 (1917), pp. 3–28, quoted in Bernstein and Feinberg, p. 47.

23. A reference, presumably, to Slipher. In a historic 1929 paper, the Mount Wilson astronomer Edwin Hubble would announce that there was a systematic relation between the distances and redshifts of the nebulae, which he had already demonstrated were distant galaxies: The farther away the galaxy, the faster it appeared to be receding. Hubble wondered aloud if this were the "de Sitter effect." Astronomers now agree that it is not. Transformed into proper coordinates, de Sitter's original universe turned out to do nothing. In 1931, however, he and Einstein collaborated on a model of an expanding universe without a cosmological constant, known as the Einstein–de Sitter universe, that was the basis of cosmology for the rest of the century.

24. Albert Einstein to Willem de Sitter, March 24, 1917, *CP,* vol. 8.

25. *CP,* vol. 8, footnote, p. 423.

26. Albert Einstein to Willem de Sitter, April 1, 1917, *CP,* vol. 8.

27. Albert Einstein to Willem de Sitter, Aug. 8, 1917, *CP,* vol. 8, footnote, p. 502.

28. *CP,* vol. 8, p. 354.
29. Pais, *"Subtle Is the Lord . . . ,"* p. 288.

Chapter 23

1. Ronald Clark, *Einstein: The Life and Times,* p. 191.
2. Albert Einstein to Hans Albert, 1917, *CP,* vol. 8
3. Albert Einstein to Michele Besso, March 9, 1917, *CP,* vol. 8.
4. Albert Einstein to Heinrich Zangger, March 10, 1917, *CP,* vol. 8.
5. Albert Einstein to Michele Besso, March 9, 1917, *CP,* vol. 8.
6. Albert Einstein to Michele Besso, May 8, 1917, *CP,* vol. 8.
7. Albert Einstein to Heinrich Zangger, March 10, 1917, *CP,* vol. 8.
8. Albert Einstein to Heinrich Zangger, July 6, 1917, *CP,* vol. 8.
9. Wolf Zuelzer, *The Nicolai Case,* p. 211.
10. Ibid., p. 194.
11. Albert Einstein to Georg Nicolai, Feb. 28, 1916, *CP,* vol. 8.
12. Albert Einstein to Georg Nicolai, March 17, 1917, *CP,* vol. 8.
13. Hubert Gönner and Giuseppe Castagnetti, *Albert Einstein as a Pacifist and Democrat During the First World War,* p. 30.
14. Albert Einstein to Katya Adler, Feb. 1917, *CP,* vol. 8.
15. Albert Einstein to Michele Besso, April 29, 1917, *CP,* vol. 8.
16. Philipp Frank, *Einstein: His Life and Times,* p. 174.
17. Albert Einstein to Michele Besso, April 29, 1917, *CP,* vol. 8.
18. Michele Besso to Albert Einstein, May 5, 1917, *CP,* vol. 8.
19. "Friedrich Adler als Physiker," typewritten manuscript labeled *Vossische Zeitung,* Berlin, May 23, 1917, no. 259, morning edition, in the Adler folder in the Einstein Archives, Hebrew University of Jerusalem.
20. Albert Einstein to Heinrich Zangger, July 6, 1917, *CP,* vol. 8.
21. Albert Einstein to Paul and Maja Winteler, Aug. 1, 1917, *CP,* vol. 8.
22. Albert Einstein to Heinrich Zangger, July 6, 1917, *CP,* vol. 8.
23. Heinrich Zangger to Albert Einstein, Dec. 6, 1917, *CP,* vol. 8.
24. Zorka to Albert Einstein, 1917, *CP,* vol. 8, trans. Valerie Tekavec.
25. Desanka Trbuhovic-Gjuric, *Im Schatten Albert Einsteins,* p. 141.
26. December 16, 1917.
27. All these details are from a paper by Hubert Gönner and Giuseppe Castagnetti, *Directing a Kaiser Wilhelm Institute: Einstein, Organizer of Science?* (Berlin: Max Planck Institute for the History of Science, 1997).
28. Albert's adventures as Freundlich's boss and patron are richly recounted in Klaus Hentschel, *The Einstein Tower.*
29. Albert Einstein to Michele Besso, Jan. 5, 1918, *CP,* vol. 8.
30. Albert Einstein to Hans Albert, Jan. 25, 1918, *CP,* vol. 8.
31. Ilse Einstein to Fräulein von Kampen, Jan. 22, 1918, Nicolai archive, Institute for Contemporary History, Munich, trans. Valerie Tekavec.
32. Heinrich Zangger to Albert Einstein, Jan. 28, 1918, *CP,* vol. 8.

Chapter 24

1. John L. Heilbron, *The Dilemmas of an Upright Man,* p. 129.
2. Reprinted in Carl Seelig, ed., *Ideas and Opinions,* p. 224.
3. Albert Einstein to Mileva Maric, Jan. 31, 1918, *CP,* vol. 8.
4. Heinrich Zangger to Michele Besso, Feb. 6, 1918, quoted in Heinrich A. Medicus, "The Friendship Among Three Singular Men."

5. Anna Besso to Heinrich Zangger, March 4, 1918, *CP,* vol. 8, p. 1032, trans. Valerie Tekavec.
6. Heinrich Zangger to Albert Einstein, March 4, 1918, *CP,* vol. 8.
7. Anna Besso to Heinrich Zangger, March 4, 1918, *CP,* vol. 8.
8. Albert Einstein to Anna Besso, after March 4, 1918, *CP,* vol. 8.
9. Anna Besso to Albert Einstein, after March 4, 1918, *CP,* vol. 8.
10. Albert Einstein to Mileva, June 4, 1918, *CP,* vol. 8.
11. Michele Besso to Heinrich Zangger, March 12, 1918, quoted in Medicus.
12. Albert Einstein to Mileva, March 17, 1918, *CP,* vol. 8.
13. Albert Einstein to Mileva, before April 15, 1918, *CP,* vol. 8.
14. Heinrich Zangger to Albert Einstein, before. Aug. 11, 1918, *CP,* vol. 8.
15. Albert Einstein to Heinrich Zangger, before Aug. 11, 1918, *CP,* vol. 8.
16. Albert Einstein to Mileva, Nov. 9, 1918, *CP,* vol. 8.
17. Filed in Protocol 1386/1918 of the Zurich District Court, Department 2, and in *CP,* vol. 8, p. 794, trans. Valerie Tekavec.
18. The issue became moot in 1922 when Einstein was awarded the previous year's Nobel Prize for his work on the photoelectric effect. In 1923 he duly he endorsed a check from the Swedish Academy of Sciences for 121,572 Swedish kronor (about $32,000) over to Mileva.
19. Albert Einstein to Mileva, before July 9, 1918, *CP,* vol. 8.
20. Albert Einstein to Ilse Einstein, May 12, 1918, *CP,* vol. 8.
21. Ilse to Georg Nicolai, May 22, 1918, *CP,* vol. 8, trans. Valerie Tekavec.
22. Nicolai, however, was an egotist who kept everything, so he saved the letter, which was found by Wolf Zuelzer in Nicolai's *Nachlass* (papers) when he was working on a biography of the rebel pacifist. Although Zuelzer did not publish the letter in his biography, he felt strongly that it should be preserved for posterity, and he hung on to it, eventually returning it to the Nicolai archive at Munich's Institute for Contemporary History, despite pressure from Margot Einstein, Helen Dukas, and lawyers representing the Einstein estate to surrender it or destroy it. The tale, an example of the difficulties scholars have faced in telling the Einstein story, is preserved in Zuelzer's correspondence in the American Heritage archive at the University of Wyoming.
23. Margot Einstein told Wolf Zuelzer this in an interview in June 1972.
24. Zuelzer's notes from his interview with Margot Einstein say that Ilse induced Margot to write the song, but in his book he ascribes authorship to Ilse.
25. Albert Einstein to Tete, April 10, 1936, quoted in Christie's catalog.
26. As he once so memorably put it, "I have little patience for scientists who take a board of wood, look for its thinnest part, and drill a great number of holes when the drilling is easy." *The Quotable Einstein,* p. 185.
27. Albert Einstein to Michele Besso, July 29, 1918, *CP,* vol. 8.
28. Albert Einstein to Heinrich Zangger, before Aug. 11, 1918, *CP,* vol. 8.
29. Albert Einstein to Michele Besso, Aug. 20, 1918, *CP,* vol. 8.
30. Albert Einstein to Hedwig and Max Born, Aug. 2, 1918, *CP,* vol. 8.
31. Hubert Gönner and Giuseppe Castagnetti, *Albert Einstein as a Pacifist and Democrat During the First World War,* p. 35.
32. Albert Einstein to Michele Besso, Dec. 4, 1918, *CP,* vol. 8. Albert also referred to himself as a "supersocialist" in a postcard to his mother, quoted in Gönner and Castagnetti, p. 37.
33. This is based on the account Born gave, quoted in Gönner and Castagnetti, p. 37.
34. Albert Einstein to Hans Albert and Tete, Dec. 10, 1918, *CP,* vol. 8.
35. Margot told this story to Zuelzer.
36. Albert Einstein to Emil Zürcher, Nov. 18, 1918, in Protocol 1386/1918 of the Zurich District Court, Department 2.
37. Abraham Pais, *Einstein Lived Here,* p. 19.
38. Albert Einstein to Michele Besso, Dec. 4, 1918, *CP,* vol. 8.
39. "Deposition in Divorce Proceedings," Feb. 23, 1918, *CP,* vol. 8, p. 974, trans. Valerie Tekavec.
40. Notebook for Judgement of the District Court, Feb. 14, 1919, trans. Valerie Tekavec.

41. Albert Einstein to Mileva, 1926–27, trans. Valerie Tekavec, Einstein Family Correspondence, Einstein Archives, Hebrew University of Jerusalem.
42. Albert Einstein to his mother, May 1919, trans. Valerie Tekavec.

Chapter 25

1. Recounted by John Earman and Clark Glymour, "Relativity and Eclipses."
2. One thing they didn't have to bother dreaming about, should they prove Einstein correct, was the Nobel Prize. According to legend, Nobel's wife had slept with a mathematician. Because there was a long historical connection between astronomy and mathematics, her betrayal had tarred the stargazers as well as the mathematicians in Nobel's estimation.
3. Jeffrey Crelinsten, "Campbell and the Theory of Relativity," *Historical Studies in the Physical Sciences* 14 (1983), p. 1.
4. *New York Times,* June 10, 1918, quoted in Crelinsten, "Campbell."
5. Crelinsten, "Campbell."
6. Eddington, quoted by Earman and Glymour, p. 49.
7. Ronald Clark, *Einstein: The Life and Times,* p. 229.
8. Ibid.
9. Crelinsten, "Campbell."
10. Heinrich Zangger to Albert Einstein, Oct. 22, 1919, trans. Valerie Tekavec.
11. Albert Einstein to Pauline Einstein, Sept. 5, 1919, trans. Valerie Tekavec.
12. Albert Einstein to Heinrich Zangger, Dec. 25, 1919, trans. Valerie Tekavec.
13. Abraham Pais, *"Subtle Is the Lord . . . ,"* p. 303.
14. Clark, p. 230.
15. *Berliner Tageblatt,* October 8, 1919.
16. Earman and Glymour.
17. Albert Einstein to Max Planck, Oct. 23, 1919, quoted in Clark.
18. Heinrich Zangger to Albert Einstein, Oct. 22, 1919, trans. Valerie Tekavec.
19. See, for example, Clark, p. 232.
20. Earman and Glymour.
21. Ibid.
22. Albert Einstein to Mileva, Nov. 6, 1919, trans. Valerie Tekavec.
23. November 15, 1919, quoted in Clark.
24. *New York Times,* November 9, 1919, p. 17.
25. Albert Einstein to Heinrich Zangger, Dec. 25, 1919, trans. Valerie Tekavec.
26. November 28, 1919, quoted in Clark.
27. Albert Einstein to Arthur Stanley Eddington, Dec. 19, 1919.
28. Albert Einstein to Michele Besso, Dec. 12, 1919, trans. Valerie Tekavec.
29. Albert Einstein to Heinrich Zangger, Jan. 1920, trans. Valerie Tekavec.
30. Albert Einstein to Mileva, Dec. 5, 1919, trans. Valerie Tekavec.
31. Desanka Trbuhovic-Gjuric, *Im Schatten Albert Einsteins,* p. 152.
32. Ibid., p. 150.
33. Ibid., p. 149.
34. Albert Einstein to Tete, Aug. 10, 1922, trans. Valerie Tekavec.
35. Albert Einstein to Mileva, Aug. 15, 1922, trans. Valerie Tekavec.
36. Albert Einstein to Mileva, May 12, 1924, quoted in Christie's catalog.
37. Arthur Stanley Eddington to Albert Einstein, 1920, in Clark, p. 246.
38. Klaus Hentschel, *The Einstein Tower,* p. 49.
39. Frank, p. 139.
40. Clark, p. 255.
41. Hubert Gönner, "The Anti-Einstein Campaign in Germany in 1920," in Mara Beller, Jürgen Renn, and Robert S. Cohen, *Einstein in Context,* p. 107.
42. Ibid.

43. Albert Einstein to Max Born, quoted in Clark, p. 334.
44. Albert Einstein to Paul Ehrenfest, quoted in Clark, p. 252.
45. The conversation is recounted in Abraham Pais, *Niels Bohr's Times,* and Clark, p. 253.
46. Albert Einstein to Heinrich Zangger, March 1920, trans. Valerie Tekavec.

Epilogue

1. Anneli Meyer-Schmid to Albert Einstein, Dec. 23, 1926, trans. Valerie Tekavec.
2. Albert Einstein to Anneli Meyer-Schmid, Dec. 26, 1926, trans. Valerie Tekavec.

BIBLIOGRAPHY

Primary Sources

Albert Einstein, Mileva Maric: The Love Letters. Edited by Jürgen Renn and Robert Schulmann. Translated by Shawn Smith. Princeton, N.J.: Princeton University Press, 1992.

"Autobiographical Notes." In *Albert Einstein: Philosopher-Scientist.* Edited and translated by Paul Arthur Schilpp. LaSalle, Ill.: Open Court, 1982.

The Collected Papers of Albert Einstein, volumes 1 and 2. Edited by John Stachel. Princeton, N.J.: Princeton University Press, 1987, 1989.

The Collected Papers of Albert Einstein, volumes 3 and 4. Edited by Martin J. Klein, A. J. Kox, Jürgen Renn, and Robert Schulmann. Princeton, N.J.: Princeton University Press, 1993, 1995.

The Collected Papers of Albert Einstein, volumes 5 and 6. Edited by Martin J. Klein, A. J. Kox, and Robert Schulmann. Princeton, N.J.: Princeton University Press, 1993, 1996.

The Collected Papers of Albert Einstein, volume 8. Edited by Robert Schulmann, A. J. Kox, Michel Janssen, and Jozsef Illy. Princeton, N.J.: Princeton University Press, 1998. [A note on *The Collected Papers:* Anna Beck provided translations for paperback versions of volumes 1 through 5. Volume 6 was translated by Alfred Engel. Volume 8 was translated by Ann M. Hentschel. In the citations for Einstein's letters, these translations are used unless otherwise noted.]

Einstein's Miraculous Year: Five Papers That Changed the Face of Physics. Edited by John Stachel. Princeton, N.J.: Princeton University Press, 1998.

Ideas and Opinions. Edited by Carl Seelig. Translated by Sonja Bargmann. New York: Wings Books, 1954

The Quotable Einstein. Edited by Alice Calaprice. Princeton, N.J.: Princeton University Press, 1996.

Relativity: The Special and General Theory. Albert Einstein. Translated by Robert Lawson. New York: Crown Publishers, 1961.

Unpublished manuscript by Frieda Einstein-Knecht.

Winteler-Einstein, Maja. "Albert Einstein: Beitrag für sein Lebensbild." 1924. Einstein Archives, Hebrew University of Jerusalem.

Archives

The Albert Einstein Archives, The Jewish National and University Library, The Hebrew University of Jerusalem, Israel.

Mugar Library, Special Collections Department, Boston University, Boston, Massachusetts.

Firestone Library, Department of Rare Books and Special Collections, Princeton University, Princeton, New Jersey.

Institut für Zeitgeschichte, Munich, Germany.

Staatsarchiv des Kantons, Zurich, Switzerland.

Schweizerisches Literarchiv, Schweizerisches Landesbibliothek, Bern, Switzerland.

Neils Bohr Library, American Institute of Physics, College Park, Maryland.

Secondary Sources

Beller, Mara, Jürgen Renn, and Robert S. Cohen, eds. *Einstein in Context.* Special issue of *Science in Context* 6, no. 1 (Spring 1993). Cambridge, Eng.: Cambridge University Press.

Bernstein, Jeremy, and Gerald Feinberg. *Cosmological Constants: Papers in Modern Cosmology.* New York: Columbia University Press, 1986.

Bucky, Peter. *The Private Albert Einstein.* Kansas City, Mo.: Andrews and McMeel, 1992.

Clark, Ronald. *Einstein: The Life and Times.* New York: Avon, 1971.

Earman, John, and Clark Glymour, "Relativity and Eclipses: The British Eclipse Expeditions of 1919 and Their Predecessors." *Historical Studies in the Physical Sciences* 11, no. 1 (1980), pp. 49–85.

Earman, John, Michel Janssen, and J. D. Norton, eds. *The Attraction of Gravitation: New Studies in the History of General Relativity.* Einstein Studies, volume 5. Boston: Birkhäuser, 1993.

Einstein, Elizabeth Roboz. *Hans Albert Einstein: Reminiscences of Our Life Together.* Iowa City: Iowa Institute of Hydraulic Research, 1991.

Eisenstaedt, Jean, and A. J. Kox, eds. *Studies in the History of General Relativity.* Einstein Studies, volume 3. Boston: Birkhäuser, 1992.

Folsing, Albrecht. *Albert Einstein: A Biography.* New York: Viking, 1997.

Frank, Philipp. *Einstein: His Life and Times.* New York: Da Capo Press, 1953.

Goenner, Hubert, and Giuseppe Castagnetti. "Albert Einstein as a Pacificist and Democrat During the First World War." Berlin: Max Planck Institute for the History of Science. Preprint 35, 1996.

Grüning, Michael. *Ein Haus für Albert Einstein: Erinnerungen—Briefe—Dokumente.* Berlin: Verlag der Nation, 1990.

Heilbron, John L. *The Dilemmas of an Upright Man: Max Planck as Spokesman for German Science.* Berkeley: University of California Press, 1986.

Henderson, Linda Dalrymple. *The Fourth Dimension and Non-Euclidean Geometry in Modern Art.* Princeton, N.J.: Princeton University Press, 1983.

Hentschel, Klaus. *The Einstein Tower: An Intertexture of Dynamic Construction, Relativity Theory, and Astronomy.* Stanford: Stanford University Press, 1997.

Highfield, Roger, and Paul Carter. *The Private Lives of Albert Einstein.* London: Faber and Faber, 1993.

Hoffman, Banesh. *Albert Einstein: Creator and Rebel.* New York: New American Library, 1972.

———. *Relativity and Its Roots.* New York: W. H. Freeman, 1983.

Holton, Gerald. *Thematic Origins of Scientific Thought: Kepler to Einstein.* Cambridge, Mass.: Harvard University Press, 1973.

———. *Science and Anti-Science.* Cambridge, Mass.: Harvard University Press, 1993.

Howard, Don, and John Stachel, eds. *Einstein and the History of General Relativity.* Einstein Studies, volume 1. Boston: Birkhäuser, 1989.

Illy, Jozsef. "Albert Einstein in Prague." *Isis* 70, no. 251 (1979), pp. 76–84.

Jung, C. G. *Memories, Dreams, Reflections.* New York: Vintage, 1989.

Klein, Martin J., *Paul Ehrenfest: The Making of a Theoretical Physicist.* New York: American-Elsevier, 1970.

Marianoff, Dimitri, and Palma Wayne. *Einstein: An Intimate Study of a Great Man.* New York: Doubleday, Doran, 1944.

McEvoy, J. P., and Oscar Zarate. *Introducing Quantum Theory.* New York: Totem Books, 1996.

Medicus, Heinrich A. "The Friendship Among Three Singular Men: Einstein and His Swiss Friends Besso and Zangger." *Isis* 85, no. 3 (Sept. 1994), pp. 456–78.

Michelmore, Peter. *Einstein: Profile of the Man.* New York: Dodd, Mead, 1962.

Miller, Arthur I. *Albert Einstein's Special Theory of Relativity.* Reading, Mass.: Addison-Wesley, 1981.

———. *Imagery in Scientific Thought: Creating 20th-Century Physics.* Boston: Birkhäuser, 1984.

Misner, Charles W., Kip S. Thorne, and John Archibald Wheeler. *Gravitation.* New York: W. H. Freeman, 1973.

Moszkowski, Alexander. *Einstein the Searcher: His Work Explained from Dialogues with Einstein.* Berlin: Fontane, 1921.

Osserman, Robert. *Poetry of the Universe: A Mathematical Exploration of the Cosmos.* New York: Anchor Books, 1995.

Pais, Abraham. *"Subtle Is the Lord . . .": The Science and the Life of Albert Einstein.* New York: Oxford University Press, 1982.

———. *Niels Bohr's Times.* New York: Oxford University Press, 1991.

———. *Einstein Lived Here.* New York: Oxford University Press, 1994.

Poincaré, Henri. *Science and Hypothesis.* New York: Dover, 1952.

Pyenson, Lewis. *The Young Einstein: The Advent of Relativity.* Bristol and Boston: Adam Hilger, 1985.

———. "Einstein's Natural Daughter." *History of Science* 28, no. 4 (1990).

Quinn, Susan. *Marie Curie: A Life.* New York: Simon and Schuster, 1995.

Reichinstein, David. *Albert Einstein: A Picture of His Life and His Conception of the World.* Prague: Stella Publishing House, 1934.

Reiser, Anton (pseudonym of Rudolf Kayser). *Albert Einstein: A Biographical Portrait.* New York: Albert and Charles Boni, 1930.

Renn, Jürgen. "The Third Way to General Relativity: Einstein and Mach in Context." Berlin: Max Planck Institute for the History of Science. Preprint 8, 1994.

Renn, Jürgen, and Tilman Sauer. "Heuristics and Mathematical Representation in Einstein's Search for a Gravitational Field Equation." Berlin: Max Planck Institute for the History of Science. Preprint 62, 1997.

Schwartz, Joseph, and Michael McGuinness. *Einstein for Beginners.* New York: Pantheon, 1979.

Seelig, Carl. *Albert Einstein: A Documentary Biography.* Originally published as *Albert Einstein und die Schweiz.* Translated by Mervyn Savill. London: Staples Press, 1956.

Speziali, Pierre, ed. *Albert Einstein–Michele Besso Correspondence, 1903–1995.* Paris: Hermann, 1972.

Stachel, John. "Einstein and Maric: A Failed Collaboration." Unpublished manuscript.

Trbuhovic-Gjuric, Desanka. *Im Schatten Albert Einsteins: Das tragische Leben der Mileva Einstein-Maric.* Bern: Verlag Paul Haupt, 1993. Translated for the author by Valerie Tekavec.

Vallentin, Antonia. *The Drama of Albert Einstein.* New York: Doubleday, 1954.

von Baeyer, Hans Christian. *Taming the Atom: The Emergence of the Visible Microworld.* New York: Random House, 1992.

Zuelzer, Wolf. *The Nicolai Case: A Biography.* Detroit: Wayne State University Press, 1982.

INDEX